Greening China's Urban Governance

ARI – SPRINGER ASIA SERIES

Volume 7

The Asia Research Institute (ARI) is a university-level research institute of the National University of Singapore (NUS). Its mission is to provide a world-class focus and resource for research on Asia. The three themes of the ARI-Springer Asia Series – Cities, Religion, and Migration – correspond to three of ARI's research clusters and primary research emphases. ARI's logo depicts rice grains in star-like formation. Rice has been the main staple food for many of Asia's peoples since the 15th century. It forms the basis of communal bonds, an element of ritual in many Asian societies, and a common cultural thread across nations and societies.

More information about this series at http://www.springer.com/series/8425

Jørgen Delman • Yuan Ren
Outi Luova • Mattias Burell
Oscar Almén
Editors

Greening China's Urban Governance

Tackling Environmental and Sustainability Challenges

Springer

Editors
Jørgen Delman
University of Copenhagen
Copenhagen, Denmark

Outi Luova
Centre for East Asian Studies
University of Turku
Turku, Etelä-Suomi, Finland

Oscar Almén
Uppsala University
Uppsala, Sweden

Yuan Ren
School of Social Development and Public Policy
Fudan University
Shanghai, People's Republic of China

Mattias Burell
Hankuk University of Foreign Studies
Seoul, Korea (Republic of)

ISSN 2367-105X ISSN 2367-1068 (electronic)
ARI - Springer Asia Series
ISBN 978-981-13-0739-3 ISBN 978-981-13-0740-9 (eBook)
https://doi.org/10.1007/978-981-13-0740-9

Library of Congress Control Number: 2018952470

This Springer imprint is published by the registered company Springer Nature Singapore Pte Ltd.
The registered company address is: 152 Beach Road, #21-01/04 Gateway East, Singapore 189721, Singapore

Preface

This book examines China's emerging green urban governance. It is a result of work undertaken within the Sino-Nordic Urban Governance for Sustainable Cities Network (UGN). The aim of the Network has been to contribute new insights on the governance of China's green urban transition and to increase Nordic visibility and participation in China's urbanization, especially in relation to Shanghai and the Yangtze River Delta region. The researchers in the Network have primarily come from the universities involved with UGN, i.e. Uppsala University and Lund University in Sweden, the University of Copenhagen in Denmark, Turku University in Finland, and Fudan University and Zhejiang University in China.

The network members have conducted three research workshops so far. The first of these, entitled 'Urban Sustainable Governance and Citizen Participation and Action', was held at Zhejiang University in 2013. The second workshop was hosted at Fudan University in 2014 and combined with a PhD training course: 'Sustainable city development: Theories, methods, challenges, and experiences in China and the Nordic countries'. The final workshop, held in 2015 at the University of Copenhagen, focused on bringing together and discussing the papers for this volume under the theme: 'Governance for urban sustainability in China: Challenges and practices'. We would like to thank all the organizers of these workshops and of the PhD course for their dedicated efforts to make the workshops successful and to ensure the progress of our work.

Through the network's activities, researchers have been able to share their research results and insights, primarily from fresh empirical studies in China. But members of the Network have also found inputs from colleagues working on urban green governance in the Nordic countries helpful and relevant to discuss. Hailing from a variety of disciplines at the Nordic universities within the Network, their participation in our workshops was timely and fruitful. They allowed us to contextualize and theorize our results and to put the Chinese experience in perspective. We wish to express our gratitude to those colleagues for contributing their insights.

The chapters in this book speak to the dynamics of the ongoing development and transformation of the approach of the Chinese party state and of Chinese social actors to China's urban green governance and their respective roles in these processes. The introduction, the theoretical contextualization by Dorthe Hedensted Lund (Chapter 1), and the individual chapters discuss a number of common themes: How can we work with governance theory in a Chinese context? What is the effect of China's rapid urbanization on urban green governance? What are the dynamics and evolutionary processes of China's emerging urban green governance regimes? How does institutional and cultural path dependency shape these emerging regimes? In which ways do economic and administrative reforms alter the picture? What are the theoretical and practical implications of the findings in relation to urban green governance in China and in a wider perspective?

Although some of the papers presented at the workshops have not been included here, they all helped to improve and to deepen our common knowledge and perspectives. We would especially like to thank the authors of the chapters in this volume for their efforts to introduce fresh ideas and insights, not least with regard to how city leaderships and other urban stakeholders in China are dealing with China's severe urban environmental challenges. We would also like to thank the *Nordic Centre at Fudan University* for its crucial role in facilitating network-building activities, as well as for its financial support. Thanks are due as well to *Nordic Institute of Asian Studies* (*NIAS*) and the *Fudan-European Centre for China Studies* at the University of Copenhagen for their financial support to the workshop in Copenhagen in 2015. A generous grant from the *Nordic Council of Ministers* in Copenhagen supported the activities of the Network from 2013 to 2015; this was crucial for making the venture possible and has been much appreciated.

Finally, we would like to thank Peter Mayers for his tireless, thoughtful, and exemplary proofreading of the draft manuscripts and Inger Helen Sørreime for her efficient editorial assistance during the final stages of processing this volume. We would also like to thank the many other student assistants who helped organize the workshops.

Jorgen Delman, Yuan Ren, Outi Luova, Mattias Burell, and Oscar Almén – organizers of UGN and editors of this volume

Contents

About the Authors

Oscar Almèn, PhD is a research fellow at the Department of Government, Uppsala University, where he teaches courses on development studies and methods. He obtained his PhD from the University of Gothenburg in 2005. His research focuses on political participation, political accountability, social movement, and state-civil society relations in China. He is one of the coordinators of the Sino-Nordic Urban Governance for Sustainable Cities Network (UGN). Recent publications include (2016) 'Local Participatory Innovations and Experts as Political Entrepreneurs: The Case of China's Democracy Consultants', *Democratization*, 23:3, and (2013) 'Only the Party Manages Cadres: Limit of Local People's Congress Supervision and Reform in China', *Journal of Contemporary China*, 80:1.

Cunkuan Bao works as a professor (doctoral advisor) in the Department of Environmental Sciences and Engineering in Fudan University. Prof Bao's research focuses on strategic environmental assessment and environmental planning. His recent publications include 'Designing a Management Framework for Strategic Environmental Assessment of Urban Plans in China', *Journal of Environmental Assessment Policy and Management*; 'Theoretical Framework of Environmental Planning Within the Context of Ecological Civilization', *Journal of Fudan University (Natural Science)*; 'Sustainability-Oriented Technical Standard System for Strategic Environmental Assessment: An analysis of the Effectiveness of the "Technical Guidelines for Plan EIA"', *Journal of Urban Planning Forum* (in Chinese); and 'PEIA Should Be Oriented by Value of Eco-civilization: From the Effectiveness Evaluation of PEIA on New-District Development', *Journal of Environmental Economics*. Bao also published quite a lot of policy analysis in newspapers such as *China Environmental News* and *Jiefang Daily* in China.

Stefan Brehm, PhD is a research fellow at the Centre for East and South-East Asian Studies at Lund University in Sweden. He studied economics and Chinese at Eberhard Karls University in Tübingen, Germany, and Fu Jen Catholic University in Taipei, Taiwan. He received a diploma in international economics from Tübingen University in 2002 and was a PhD fellow at the postgraduate research programme

'Global Challenges – Transnational and Transcultural Approaches' from 2002 to 2005. He defended his PhD thesis on 'China's Integration into a Global Financial Order' in 2005 (summa cum laude). The regional focus of his research is China, and his theoretical interest is concerned with new institutional economics and the political economy of regulation. Brehm covers a rather broad range of topics such as financial markets, public administration, and innovation. Currently he is member of a larger research group working on 'Digital China'. In this context, Stefan looks at the role of data and data production for the distribution and redistribution of political power and economic wealth.

Mattias Burell, PhD is a research fellow and teacher at the Department of Government, Uppsala University. He is one of the coordinators of the Sino-Nordic Urban Governance for Sustainable Cities Network (UGN). Since his dissertation (2001), which examined the role of law in China's labour market reforms, he pursued research projects on China's housing market reforms, migrant workers, and citizen trust in government. In 2005, he conducted a major social survey in Wuhan. More recently, his research has focused on state-civil society relations, NGOs, and local environmental governance. His publications include 'China's Housing Provident Fund' in *Housing Finance International* (2006) and the co-edited volume *Making Law Work: Chinese Laws in Context* (2011).

Jørgen Delman, PhD is professor of China Studies, Department of Cross-cultural and Regional Studies (ToRS), at the University of Copenhagen. His research examines state-society relations and political change in contemporary China. Currently, he focuses on China's climate change politics, climate governance at city level, energy and energy security politics, and renewable energy. He is one of the coordinators of the Sino-Nordic Urban Governance for Sustainable Cities Network (UGN). His recent publications include the following: Guan, Ting and Jørgen Delman, 'Energy Policy Design and China's Local Climate Governance: Energy Efficiency and Renewable Energy Policies in Hangzhou', *Journal of Chinese Governance*, 2017(1), pp. 1–23; 'Urban Climate Change Politics in China: Fragmented Authoritarianism and Governance Innovations in Hangzhou' (2016), Brødsgaard, K.E., and A. Mertha (eds.), *Chinese Politics as Fragmented Authoritarianism: Earthquakes, Energy and Environment*, Routledge; 'From "Worn" to "Green" China Model? Energy in the 12th Five-Year Plan in an Environmental and Climate-Change Perspective', Moe, Espen, Paul Midford (eds.) with Ole Odgaard (2014), *The Political Economy of Renewable Energy and Energy Security. Common Challenges and National Responses in Japan, China, and Northern Europe*, Palgrave Macmillan; (2014) 'Climate Change Politics and Hangzhou's Green City Making', Björner, E., P.O. Berg (eds.), *Branding Chinese Mega-Cities: Policies, Practices, and Positioning. Cheltenham*, UK: Edgar Elgar Publishers; and 'Fuels for the Future? The Emerging Architecture in China's Liquid Biofuels Market', Louis Augustin-Jean, Björn Alpermann (eds.), *The Political Economy of Agro-foods Markets in China. The Social Construction of the Markets in the Era of Globalization*, Palgrave Macmillan (with Yu Wang) (2014).

Jia He is a postdoc in the Department of Environmental Sciences and Engineering in Fudan University and has a major in environmental management and planning. She obtained her doctor's degree from the College of Environmental Science and Engineering at Tongji University. The title of her dissertation was 'The Value Preference of Strategic Environmental Impact in China'. Her recent publications in English include 'Experts' Perspective on the Performance of Chinese Technical Guidelines for Plan Environmental Impact Assessment' and 'Framework of Integration of Urban Planning, Strategic Environmental Assessment and Ecological Planning for Urban Sustainability Within the Context of China', *Journal of Environmental Impact Assessment Review*. His current research interests focus on public participation in SEA. The project is funded by the China's National Natural Science Fund Committee.

Feifan Huang is a PhD student at the Department of Political Science of Zhejiang University; she was an associate researcher at the College of Administration in Yuhang District, Hangzhou, from March to August 2015.

Marina Bergen Jensen, PhD works as a professor at the University of Copenhagen. She is heading the Landscape Technology Research Group at the Department of Geosciences and Natural Resource Management. Her research focuses on urban water management and the potential of nature-based solutions for improving the water balance in terms of both quantity and quantity. She explores the conditions for a green transformation of existing infrastructures for water supply and wastewater management towards climate resilience and a reduced hydrological footprint. Her research includes the development of technologies for treatment of contaminated greywater and stormwater runoff and technologies for storage and infiltration of stormwater runoff by redesign of urban surfaces and furniture and by exploitation of geomorphological variation of urban soils. She has been heading numerous research projects and is currently supervising 10+ PhD students. Recent publications include 'Factors Affecting the Hydraulic Performance of Infiltration Based SUDS in clay', in *Urban Water Journal*, online August 2015; 'Green Infrastructure for Flood-Risk Management in Dar es Salaam and Copenhagen: Exploring the Potential for Transitions Towards Sustainable Urban Water Management', in *Water Policy*, 17:126–142, 2015; 'Land Cover in Single-Family Housing Areas and How It Correlates with Urban Form', in *Urban Ecosystems*, 18(4):1103–1123, 2015; and 'Towards Sustainable Urban Water Governance in Denmark: Collective Building of Capabilities in Local Authorities'. *International Journal of Innovation and Sustainable Development*, 10 (2):103–119, 2015.

Youxing Lang is professor and chair of political science at Zhejiang University. His recent publications include *Making Democracy Work: The Crafting and Manipulation of Chinese Village Democracy by Political Elites* (2009); *Political Aspiration and Political Inclusion: The Rich People's Participation in Zhejiang* (2012); and *Reform and Rural China* (2015).

Daisong Liu, PhD is candidate at Fudan University's Social Development and Public Policy School. His main research area is carbon consumption and sustainable development.

Jun Liu, PhD is an associate professor in the Department of Media, Cognition and Communication at the Centre for Communication and Computing, University of Copenhagen, Denmark. His research areas cover political communication, information and communication technologies, and political sociology. He has articles that have been published in the fields of political science, communication, and sociology, including *Mass Communication & Society*, *Social Movement Studies*, *Acta Sociologica*, and *Television & New Media*, among others. His research has won several awards, including the Best Dissertation Award, the Information Technology and Politics Section of American Political Science Association (2014), and the Best Paper Award, the 2014 International Communication Association Mobile Preconference.

Li Liu is an associate professor at the Department of Geoscience and Natural Resource Management, University of Copenhagen. After her earlier education at Beijing Forestry University, she got her master's degree in landscape architecture from Danish Royal Veterinary and Agricultural University and PhD (2008) from the University of Copenhagen. Her current research focuses on how to achieve sustainable and climate-resilient city, especially sustainable urban water management, by urban green infrastructure and nature-based solutions. Her research includes interdisciplinary investigation of cities' planning and design practices on green infrastructure-based urban water management for climate resilience and sustainability transition, reviewing and assessing the roles of green infrastructure in improving urban water balance and providing multiple benefits to cities, and investigating new knowledge and methods needed for improving planning and design of urban landscape for climate-resilient cities. Liu's recent publications include 'Climate Resilience Strategies of Beijing and Copenhagen and Their Links to Sustainability', *Water Policy* 19, 997–1013, 2017; 'Potential Contributions to Beijing's Water Supply from Reuse of Storm- and Greywater', *Journal of Southeast University* (English Edition), 30: 150–157, 2014; and 'Multifunctional Green Infrastructure Planning to Promote Ecological Services in the City' by Niemela, J., Breuste, J.H., Guntenspergen, G., McIntyre, N.E., Elmqvist, T., and James, P. (eds.), *Urban Ecology: Patterns, Processes, and Applications*, Oxford University Press, 2011.

Dorthe Hedensted Lund is senior researcher at the University of Copenhagen, Department of Geosciences and Natural Resource Management, Section for Landscape Architecture and Planning. Her main field of research is governance within the fields of urban and rural planning. In later years, she has mainly worked with the governance of climate change adaptation at municipal and city levels in Denmark as well as African cities. She is deeply involved in an action research programme on Landscape Futures where the application of collaborative planning methods and strategy making is to improve rural landscapes and local development. She teaches a master's level course titled theories of urban governance and planning and performs numerous guest lectures in different contexts. Recent publications are Agger, A. and D.H. Lund (2017), 'Collaborative Innovation in the Public Sector – New Perspectives on the Role of Citizens', *Scandinavian Journal of Public Administration*, 21,3,17–37; Herslund, L., Lund, D.H., Jørgensen, G., Mguni, P., Kombe, W., and Yeshitela, K. (2015), 'Towards Climate Change Resilient Cities in Africa – Initiating Adaptation in Dar es Salaam and Addis Ababa'. Pauleit, S. et al. (eds.), *Urban Vulnerability and Climate Change in Africa*, Springer; Jørgensen, G., Herslund, L., Lund, D.H., Workneh, A., Kombe, W., and S. Gueye (2014), 'Climate Change Adaptation in Urban Planning in African Cities – The CLUVA-project', *Resilience and Sustainability in Relation to Natural Disasters: A Challenge for Future Cities*, P. Gasparini, G. Manfredi, and D. Asprone (eds.), Springer.

Outi Luova, PhD works as a university lecturer at the Centre for East Asian Studies at the University of Turku, Finland. She is vice director of the Centre and director of the Finnish University Network for Asian Studies. She holds an adjunct professorship (East Asian Studies) at the University of Helsinki and teaches frequently about sustainable urban development in Chinese universities. Her PhD thesis dealt with ethnicity-based economic cooperation between China's ethnic Korean areas and South Korea. Luova's current research focuses on China's urban governance especially with regard to environmental issues. Her recent publications include 'Community Volunteer Associations and the Diversification of Urban Governance in China: Comparison between Four Districts in Tianjin' [in Chinese] in *Societal Organizations and National Governance – Comparisons between China and Foreign Countries*, Peng Xizhe, Jude Howell, and Wang Chuanlan (eds.), Shanghai: Fudan University Press, 2016, and 'Charity Paradigm Change in Contemporary China: From Anti-socialist Activity to Civic Duty', *China Information*, 2017. She is one of the coordinators of the Sino-Nordic Urban Governance for Sustainable Cities Network (UGN).

Yuan Ren, PhD is professor of demography and urban studies in Fudan University, China. He is deputy director of the Institute of Population Research (IPR). He is fellow of New Century Excellent Talents awarded by Chinese National Education Commission (2012). He is one of the coordinators of the Sino-Nordic Urban Governance for Sustainable Cities Network (UGN). His recent publications include *Post-Demographic Transition* (Shanghai: 2016); *Reshaping Welfare Institutions in China and the Nordic Countries* (Helsinki & Shanghai: 2014); *Migration and Urbanization in Contemporary China* (Shanghai: 2013); and *Temporary Migrants' Living Patterns and Their Social Integration in Urban China* (Shanghai: 2012).

Jesper Schlæger, PhD is a distinguished professor in the School of Public Administration, Sichuan University. The chapter for this book was written during his previous position in the Centre for East and South-East Asian Studies at Lund University. His main field of research is Chinese public administration, in particular e-government. He is the author of *E-Government in China: Technology, Power and Local Government Reform*, which includes a chapter on digital urban management. His articles are published in *Journal of Chinese Governance*, *China Information*, *Chinese Journal of Communication*, and *Policy & Internet*. He is faculty member on a professional degree (MPA) in urban operations and management and has developed and taught classes on municipal administration, urbanization in a globalized world, sustainable traffic, and social innovation. He has also contributed case-study materials on digital urban management to the Chinese national database for professional education.

Jesper Svensson is reading for a DPhil at the Smith School of Enterprise and the Environment, University of Oxford. He holds a Master of Science in Asian Studies from Lund University and two bachelor's degrees in political science and environmental science from School of Business, Economics, and Law at the University of Gothenburg. Prior to joining the University of Oxford, Jesper worked as a research assistant for the Transboundary Rivers and Adaptation to Climate Extremes in North America (TRACE, North America) project under Dr Dustin Evan Garrick. He is currently the senior editor for Global Water Forum.

Jinnan Wang works as a vice director of Chinese Academy for Environmental Planning. Wang is also titled as the director of key laboratory of national environmental planning and policy simulation. Wang's research focus on environmental planning and policy. Wang has published 15 books and hundreds of journal articles. Wang's current publications include *Environmental Planning: Theory and Methodology*, Chinese Environmental Press (2014); 'The Challenge and Countermeasure for Developing Environmental Think Tanks in China', *Journal Environmental Protection* (in Chinese); and 'The Environmental Function Assessment and Zoning Scheme in China', *Acta Ecologica Sinica* (in Chinese).

Jiangli Wang, PhD is associate professor of international politics in the Department of Political Science, College of Public Affairs at Zhejiang University in Hangzhou, China. She is also a researcher at Zhejiang University's Center for Non-Traditional Security and Peaceful Development Studies. Her recent publications include 'The English and Chinese Schools of International Relations: Comparisons and Lessons' (with Barry Buzan, 2014); 'The Historical Turn of Contemporary IR Study in China' (2013); and 'Green Justice: Possibility and Necessity of Global Green Governance' (2012).

Biao Wei is senior engineer and certified planner of Hangzhou City Planning and Design Academy. His recent publications include Terry (2009), *Liverpool: Regeneration of a City Centre* (Taylor, David, and Davenport), translated by Wei

Biao (2016); 'Analysis and Implications of Public Participation in Urban Planning in China and UK: Case Studies of Hangzhou and London from a Comparative Respective' (with Dai Zhemin, 2015); and 'High-Speed Rail and Urban Transformation in China: The Case of Hangzhou East Rail Station' (with Chen, Chia-lin, 2013).

Changdong Zhang is an associate professor of political science at the School of Government and a research fellow of the Institute of State Governance and Center for Civil Society Studies, at Peking University. He received his PhD in political science from the University of Washington, Seattle. Zhang Changdong's research interests include comparative authoritarianism, taxation politics/fiscal sociology, state and society relationships, and institutionalism, with a regional focus of China and Asia. Prof Zhang has published about 20 journal articles (including *China Review, Sociological Theory*, and *Non-Profit and Voluntary Sector Quarterly*) and book chapters both in Chinese and English.

Xiaoxin Zhang works as a professorate senior engineer at Beijing Municipal Institute of City Planning (BMICP). He is director of Department of Municipal Plan at BMICP. His current research focuses on stormwater management, urban flooding control, and sustainable water resources management. He has completed more than 20 projects funded by Beijing Municipal Government. Recently he is the team leader of the Local Flooding Control Plan in Beijing Central Area. In this project, the project team set up a 2D hydraulic model and finished the flood map for Beijing Central Area. Zhang's recent publications include 'Overview of Urban Stormwater Runoff Management in USA', *Journal of Water & Wastewater Engineering China*, 2014:40, and 'Strategy of Local Flooding Control in Beijing Central Area', *Journal of Beijing Planning Review*, 2012:05.

Jingjing Zhou,LLM is a lecturer in the Department of Politics, Sichuan University Jinjiang College. She has undertaken extensive research on e-government and smart city development, as well as translation of academic literature. She has recently participated in writing a national-level textbook on e-government.

Jun Zhu works as research professor in the Research Centre of Ecological and Environmental Strategy in the Urban Planning and Architectural Institute of Fudan University. His research focuses on ecological planning, environmental planning, and strategic environmental assessment. His current research projects include 'Urban comprehensive traffic plan environmental impact assessment', supported by the Ministry of Environmental Protection, and 'The technology to ensure ecological security in the City group of Yangtze River Delta', supported by the Ministry of Science and Technology.

Abbreviations

AECRU	Association of Energy Conservation and Resource Utilization
BWA	Beijing Water Authority
CAP	Copenhagen Climate Adaptation Plan
CBD	Central Business Districts
CCSS	Center for Civil Society Studies of Peking University
CCTV	China Central Television
CDR	Committee of Development and Reform
CEMS	Continuous Emission-Monitoring Systems
CMP	Cloudburst Management Plan
CNNIC	China Internet Network Information Center
COD	Chemical Oxygen Demand
CPC	Communist Party of China
CSO	Combined sewer overflow
DRC	Development and Reform Commission
EB	Education Bureau
EIA	Environmental Impact Assessment
EIC	Economic and Information Commission
ENGO	Environmental non-governmental organization
EPB	Environmental Protection Bureau
EU	European Union
FYP	Five-Year Plan
GDP	Gross Domestic Product
GHV	Green Home Volunteers
GI	Green Infrastructure
GONGO	Government-owned/organized NGO
HOA	Homeowners Association
HOFOR	Greater Copenhagen Utility
ICT	Information and communication technology
IQR	Inter-Quartile Range
JEEP	Jiufeng Environmental Energy Project of Yuhang in Hangzhou (earlier called the Jiufeng Waste Incineration Power Plant Project)

KEPC	Key environmental-protection cities
KPI	Key Performance Indicators
LID	Low Impact Development
MEP	Ministry of Environmental Protection
MoHURD	Ministry of Housing and Urban-Rural Development
NDRC	National Development and Reform Commission
NESDP	National Economic and Social Development Plan
NG	Network Governance
NGO	Non-governmental organization
NIMBY	Not In My Backyard. It is usually used to describe the opposition by residents against building a potentially harmful facility near them.
NLGCC	National Leading Group on Climate Change
NMCEP	National Model City for Environmental Protection
NPG	New public governance
NPM	New public management
PES	Payments for ecosystem services
PITI	Pollution-Information Transparency Index
PX	Paraxylene
QESCCUE	Quantitative Examination System on Comprehensive Control of the Urban Environment
RCE	Residential energy consumption-related carbon emissions
RFD	Residents-factory dialogue platform
RMB	Renminbi
SARS	Severe acute respiratory syndrome
SDC	Sino-Danish Center for Education and Research
SEPA	State Environmental Protection Agency
SWOT analysis	Strengths-Weaknesses, Opportunities-Threats
TCE	Transportation-related carbon emission
TPA	Traditional public administration
TPM	Traditional public management
URDCE	Urban residents' direct carbon emission
US	United States of America
WTE	Waste to energy

Introduction: Getting to Grips with China's Emerging Green Urban Governance

This volume focuses on how governments and social stakeholders in urban China engage with major environmental risks and how they tackle the governance of the long-overdue transition to greener, more livable, and more sustainable cities. We draw inspiration in this endeavour from Henri Lefebvre's observation on the importance of the ecological question:

> …it is correct to assert that the milieu of life and the quality of the environment have acquired an urgent, politically central status. Inasmuch as one accepts such an analysis, the prospects for action are profoundly reformed. Several well-known but somewhat neglected forms—such as associative life or grassroots democracy (*autogestion*)—must be reinstated as key priorities; they assume new meanings when applied to the urban. The question then is to know if social and political action can be formulated and rearticulated in relation to specific problems, that, even if they are concrete, concern all dimensions of life. (Lefebvre 2014/1989)

The contributors to this volume argue that challenges to China's environmental and ecological sustainability in conjunction with popular demands for a higher quality of urban life have made a green turn in China's urban development necessary and that this need has already attained a central political status. In consequence, the prospects for political action have been transformed in recent years, and the Chinese Party-state has been forced to reformulate its approach to urban governance. Rapid demographic growth, massive industrialization, dramatic urbanization, and severe environmental degradation—with their devastating consequences—are the primary factors driving these developments. For many in China, however, the political, economic, social, and ecological inequality caused by these processes are of equally great concern. The changes in the political winds, as well as the newfound resolutions to tackle these challenges at the city level, are reflected in what we call emerging urban 'green' governance regimes. In this context, many of the authors in this volume have a special focus on the role of what Lefebvre called *autogestion* in the quotation above, i.e. on popular action and participation as a new driver in the green turn in urban governance in China.

The concept 'green' does not come with an established scientific or normative definition; nor have the authors of this volume attempted to arrive at one. But the general assumption underlying China's urban green politics is reflected in the idea that economic growth, which is still a top national priority, must be achieved in an environmentally sustainable manner, i.e. it must not come at the expense of the environment (Fay et al. 2014).

As a development principle, sustainable development was raised in *Our Common Future*, by the World Commission on Environment and Development (World Commission 1987), and affirmed by the UN General Assembly in 1987 (United Nations 1987). In October 2015, the UN General Assembly (United Nations 2015) drew up a new working agenda for sustainable development over the next 15 years. As cities are core intervention arenas when it comes to sustainable development in any society, sustainable cities or sustainable urban development has also become prominent on both national and local policy agendas in China (Qiu 2006). Related concepts and programmatic approaches (eco-city, low-carbon city, green city, livable city, and sustainable city) have been widely discussed and experimented with in practice over the last three decades (Delman 2014). However, while the Chinese leadership has thus embarked on a course to improve China's environment, and not least its urban environment, the pursuit of economic growth still has a negative environmental impact.

China's leaders have repeated and refined their vision of a greener country in a great number of political programmes in recent years. A particularly comprehensive development programme from 2015 called for the promotion of eco-civilization (*shengtai wenming* 生态文明) across the Chinese territory. As a national strategy for sustainable development, improved environmental management, and a new green governance regime, eco-civilization presents new values and norms for China's green transition. The principles of eco-sustainability are to be rooted in people's lifestyles, in the processes of industrial production, in the mechanisms of sociopolitical interaction, and in the methods and structures of urban development and governance. Moreover, Chinese leaders anticipate further institutional reforms in connection with environmental property institutions, market-based resource pricing, mechanisms for effective ecological compensation, and sanctions against noncompliers (Zhonggong Zhongyang 2015; Xinhua News Agency 2013; Hu 2012). These points were re-emphasized by Xi Jinping in his authoritative report to the 19th Communist Party of China Congress in October 2017 (Xi 2017).

Many cities in China developed their own approach to eco-civilization before it was fully developed as a national policy. In 2009, for example, Hangzhou[1] was chosen as an experimental site for urban eco-civilization by the Ministry of Environmental Protection. This was due to the city's considerable experience with a series of pilot environment and ecological development programmes since 1999. The main goal of Hangzhou's 10-year eco-civilization programme (2010–2020) is to develop and implement a morally informed new approach to comprehensive sustainable development, thereby ensuring continued economic growth while making the city more livable through a transition to greener production, transport, and daily life (Delman, Chap. 8; Delman 2016b; Delman 2014).

[1] Three chapters in this volume deal specifically with Hangzhou: Delman, Chap. 8, Burell and Almén, Chap. 10, and Wang et al. Chap. 11.

In China, as in other countries, these developments are taking place against the disturbing backdrop that the existing model of development has proven unsustainable (Kassiola and Guo 2010). China suffers from severe environmental degradation and from an increasing bio-capacity deficit. The country's per-capita ecological footprint—i.e. the demands put on its renewable resources—has far exceeded its ability to regenerate these resources. The size of the footprint is now 2.5 times that of the country's bio-capacity. The fastest growing individual component of China's ecological footprint is its carbon footprint, stemming from the use of fossil fuels in the economy at large, from the use of electricity by households, and from the consumption of carbon embodied in products for private, public, or commercial consumption (China Ecological 2012, 7–25). China's major cities now have larger per-capita carbon footprints than other major cities around the world (Baeumler et al. 2012, xlii; see also Ren and Liu, Chap. 6). These ecological and environmental problems do not just pose an important challenge to China's long-term development. Given the size of the country's population and the size and expected expansion of its economy, they also have a severe impact on global ecology and on the future sustainability of our planet.

The Chinese leadership's commitment to leading China through a green transition is reflected in a series of major policy documents, most recently in the 13th 5-year plan (Lianghui 2016), which covers the 2016–2020 period, as well as in recent urban development strategies (Zhonggong Zhongyang 2016; Guojia 2014). These policies aim to address the needs and the consequences of the fast and profound urbanization process that China has experienced since the 1980s. The rate of urbanization is already above 50%. The new policies also document that China's leadership is paying increasing attention to the well-being of the Chinese people. In fact, well-being has become one of the important targets of urbanization. China's most recent urbanization strategy from 2014, called 'new urbanization' (*xin chengzhenhua* 新型城镇化), depicts a new pattern of urbanization that emphasizes urban/rural integration, people-centred development, improving migrants' status as citizens, social integration, and environmental and ecological improvement (Guojia 2014). These ideas have now become basic guiding principles in China's urbanization and urban governance.

Most of China's cities have formulated a set of green policies like Hangzhou's; however, various local interests and stakeholders frequently delay or derail the process, preserving the status quo as much as possible in defence of their interests. They do not necessarily see any need to hurry; nor do they have the right incentives to change their behaviour (Kostka 2014; Ran 2013; Delman, Chap. 8). At the same time, many citizens seem impatient, and they are actively demanding improvements in the environment (J. Liu, Chap. 9; Burell and Almén, Chap. 10; Wang et al., Chap. 11; Zhang, Chap. 12).

It is thus timely and important to take the reader to the urban battlefronts, where the main fight to regain China's ecological balance is being waged. As Lund[2] notes

---

[2] Lund is the only contributor to this volume who has not worked with China. We have sought her contribution in order to develop a framework of commonly applicable theoretical concepts that can be used for reflecting on the findings of the case studies.

in her theoretical contribution on sustainability governance (Chap. 1), the conditions for human and non-human life are changing rapidly, and urban areas now have a dominant impact on global and local ecosystems. Cities increasingly embrace material flows, human activities, and governance structures that link them in complicated patterns both with their immediate surroundings and with global networks in which the challenges to and the conditions for effective governance are increasingly complex. Policy issues have become more intricate, and an increasing number of actors are gaining influence through global and local initiatives. As in urban zones elsewhere, great amounts of energy, food, materials, water, and land are consumed in China's urban areas, and huge volumes of waste, air pollutants, CO_2, and other greenhouse gases are produced there. As the nexus for economic growth, innovation, and employment in the national economy, urban areas necessarily and continuously develop new forms of governance to promote green and sustainable development.

As we see it, then, the urban green politics and action seen in China today have profound consequences for the future of China and of our planet itself. However, as J.L. Wang et al. point out in Chap. 11, there is also a need to take a step back—to consider justice in connection with these developments, i.e. principles that will lead to a fairer distribution of public environmental goods and a reduction or elimination of public bads. In Chap. 11, the authors argue that the earth's environment is an integral whole and that we cannot merely protect our own backyards. With the deepening ecological crisis, human beings need to reflect on their common history and their present predicament. It is time for a change of lifestyle and methods of production and to embrace a more holistic view of green justice. Humans, they state, must recalibrate their relationship with nature, with society, and with themselves. We are far from there as yet, however, and it is exactly the dynamics of China's green urban transition that have brought the contributors to this volume together.

The Purpose of This Volume

All the contributors here are interested in how China's environmental and sustainability challenges contribute to the transformation of urban governance and urban environments in China and how institutional reform, path dependency, and deepening globalization influence China's emerging urban green governance. The chapter authors examine the fight in China's cities *against* environmental degradation and *for* green change. While anchoring their studies empirically in the contemporary urban settings of the country, they contextualize their cases in theoretical, historical, institutional, and practical terms. They discuss whether and how new forms of urban green governance can help make China's cities greener, more sustainable, and more livable. Given the new Chinese framework for urban eco-civilization, the chapters also speak to international debates about ecological urbanism 'as a means of providing a set of sensibilities and practices that can help enhance our approaches to urban development', through the utilization of 'a multiplicity of old and new methods, tools, and techniques in a cross-disciplinary and collaborative approach toward urbanism developed through the lens of ecology' as noted by Mostafavi (2010).

Most of the chapters are based on fresh case studies that examine China's urban green governance from the inside out. All of them ask what deteriorating global and local environments do to China's system of urban governance, especially after national and local authorities in the country have realized that these challenges have to be dealt with decisively. Finally, the authors apply critical perspectives from the Chinese case to current international debates about urban governance in the face of the eco-sustainability challenges associated with urbanization.

Planet China and Urbanization Under Global Capitalism

China's urbanization over the last four decades has been rapid, dynamic, and environmentally devastating. The proportion of the population living in urban areas rose from around 20% in the 1980s to 56.1% in 2015. This means that about 770 million people lived in urban areas in the early 2010s, while about 600 million lived in rural ones. This trend is expected to continue during the coming decades, so that one billion of China's 1.4 billion people will be urban dwellers by 2050 (Ohshita et al. 2012). Such rapid urbanization poses huge environmental, ecological, and social challenges. Industrialization, the development of the nonagricultural economy, and urbanization are the driving forces behind it, and it is accompanied by accelerating changes in lifestyle and consumerism. In their study of carbon emissions by households in Shanghai, Y. Ren and D.S. Liu (Chap. 6) find that, although carbon emissions in China are still mainly caused by industrial production, the trend of development in China's cities is the same as that seen within cities in more developed countries. Large cities like Shanghai and Beijing are already well on their way to having an economy driven by services and consumption, and they will be the first to show how consumption is emerging as the main cause of overall carbon emissions in urban areas. Ren and Liu also find that different population groups are unequal in their direct carbon emissions. High emitters constitute 15.1% of the total population; however, their direct carbon emissions already account for 37% of overall carbon emissions. Low emitters, on the other hand, make up about 14% of the total population, but they produce only 4.4% of overall carbon emissions. In fact, the level of carbon emissions can already be regarded as an indicator of residents' socioeconomic status.

Of course, China's cities are comparable to cities around the world on a lot of different parameters. But as a relatively isolated political unit—first as the Chinese empire, subsequently as the Republic of China, and then as the People's Republic of China—China was in many ways a planet unto its own until it opened up to the outside world in earnest in the 1970s. This was due to the size of its territory, of its population, and of its economy; its long cultural and political history as a largely coherent, yet at times dramatically expanding political unit; its uniquely unifying language, which acts as an essential cultural glue; and its long history of large-scale urban settlements. However, China's most recent wave of modernization and reform, which has washed over the country since 1978, together with its interaction with a globalizing world, has thrust the country on to the global scene in ways that nobody had been able to predict, with the economic historian Mark Elvin as the eminent exception (Elvin 1973, 319).

Over the last half century, China has moved from being the empire at the centre of the Sinic world to becoming a powerful nation *inter pares*. Within the area of urban development and management, China basically followed its own practices until the late nineteenth century. In Elvin's words, the form of Chinese cities and the methods of their management, which had emerged over the course of more than two thousand years, marked '…[a] new civilization. The creation of the city began a crucial decoupling between the dominant, decision-making, part of the population, now living increasingly in a built environment, and the rest of the natural world. In other words, where and when a decision was made coincided less and less with where and when its environmental impact was felt….Increasing [decisional distance—*authors*] has progressively lessened the awareness of and sensitivity to the environmental effects of their policies among rulers and their advisers. This still holds today' (Elvin 2004, 94).

China's cities have always been large by the standards of their era. As the country evolved under its ruling dynasties, its systems for urban planning and management became more complex and sophisticated, in line with the changes taking place in each period. They also became more varied in their principles and approaches, but the urban hierarchy of rank and size dating back to imperial times still continues to matter today (Wu and Gaubatz 2013: 44, 49).

Against this backdrop, the term 'urban planning and management' is used here to signify the specific Chinese tradition under which cities were seen as political and military nodes in a thinly spread, hierarchical, and (mostly) effective imperial system of government that was developed in order to exploit and control the country's huge territory, its people, its resources, and its economic flows. The pattern whereby urban areas absorbed resources from rural society and control the national economy was strengthened during the period of the socialist planned economy, from the 1950s to the 1970s. This approach led to a rigid urban/rural binary structure, the legacy of which is still evident today, for example, through the particular type of Chinese urban residence registration (*hukou* 户口), which has privileged urban residents and made it very difficult for rural residents to obtain urban residence (i.e. full urban citizenship) since the 1950s. From the early 1980s, however, rural residents have been allowed to migrate to the cities in ever greater numbers to find jobs as 'peasant workers' (*nongmingong* 农民工). Still, these migrants with rural *hukou* largely remain second-class citizens—only now in the cities. Binary or segregated urban/rural systems of public finance and social management still exist, and the urban/rural binary continues to generate risks for urban governance in connection with resource management, inequitable development, and control over property (Yu et al. 2014). Effectively, only 35.3% of the total urban population had a local *hukou* registration in 2012, while 52.6% of the country's population lived in cities (Guojia 2014). The strong historical focus on privileged urban-centric development has never ceased, and the use of distinct approaches to political control founded on urban and urban/rural hierarchies has never been abandoned. As a result, in the view of Yao (2011), China's accelerated urbanization over the last few decades has 'resulted in extensive damage to its agricultural community, even leaving little time for adjustment. This seemingly irresistible force of modernization and urbanization is accelerating with an endless propelling, penetrating and destroying ability'.

Given the huge area of the country, more than twice that of the current European Union, there are considerable regional differences in styles of governance within China. However, under the hierarchical party-state system (on which more below), China's cities remain under the jurisdiction of higher levels of government. Still, urban planners, managers, and developers have naturally been forced to fit their ideas and governance patterns to local situations as local circumstances change. Since 1949, moreover, through policy transfers (cf. Peck and Theodore 2010), China's urban planners have liberally utilized ideas, experiences, and inputs from elsewhere—first from the Soviet Union and later from the West—in order to develop and modernize China's cities. They have done so to such an extent that Chinese urban leaders, according to Hsing You-tien, have now dropped industrialization as a distinctive marker for their efforts and replaced it with urban ideals. Not only have they adopted the jargon of 'city branding', 'urban marketing', and 'global cities', but they have also embarked on standard urban projects like the grand redevelopment of inner-city districts and the creation of central business districts (CBD) equipped with the infrastructure and services needed by head offices of large national and multinational corporations (Hsing 2010, 18–19). Over the last two decades, many of these iconic projects have been designed by world-class architects, some of them Chinese, and they are already textbook examples of urban projects. These developments testify to the openness of urban planners and developers in China; however, some claim that China has learnt little, if anything, from the negative experiences of the West during its urbanization (Bosker 2014). Indeed, 'weird' buildings are now being criticized and disallowed by the Chinese leadership (Rivers and Chung 2016). They are also often ridiculed by residents. The new China Central Television (CCTV) tower in Beijing, for example, has been nicknamed 'big pants' (*dakuzi* 大裤子) by locals.

As in other Asian countries, urban modernization and transformation in China embrace three historical processes. Taking place concurrently, but with diverse impacts at different times, they are globalization, urbanization, and the building of a modern nation-state. While globalization varies in its impact across Asia, it has prompted a liberalization of Asian economies and an embrace of various forms of capitalism (McKinnon 2011, 7–11). China's new 'socialist market economy' has been termed state or authoritarian capitalism (Witt and Redding 2012), inasmuch as it continues to be dominated for the most part by a hierarchical party-state system (described below) and by state-owned enterprises. It involves a particular variant of neoliberalism—one combining strong intervention by the local state with a radical market orientation. At the same time, the local state in China is highly entrepreneurial: it has used its privileged access to land to conduct lucrative land transactions, to promote the local generation of wealth, and to maintain its authoritarian rule. Local institutions have been changing rapidly to accommodate these processes, all within a context which was largely nondemocratic and mostly lacking in the rule of law. Finally (and somewhat paradoxically), local authorities have been increasingly endowed with powers, mandates, and assets, due to the wish of the central leadership to decentralize authority and power (Lin et al. 2014).

China's transformation from a traditional agricultural society into a modern industrialized one has brought about phenomenal changes in spatial terms, with a dramatic increase in the number of cities. In 2016, China had 626 cities, of which around 100 had more than 1 million inhabitants. The national and the local state has been at the core of this urbanization process: not only has it led it, but it has also formed an integral part of it. The local state has mobilized its resources to expand and strengthen cities, and the struggle over resources—their accumulation, their enjoyment, and their deployment to overcome barriers to urbanization—has come to define the local state and resulted in its consolidation. At the same time, urban expansion has created much larger urban regions, calling forth territorial dynamics which have provoked intrastate and territorial competition, and enabled the expansion and scaling of state power (Hsing 2010, 7–19).

Capitalism in Asia, including its authoritarian variant in China, has produced new levels of education, new kinds of occupation, and new opportunities for private space—not least for the new middle classes (McKinnon 2011; Ren and Liu, Chap. 6). Forces of individualization demand new arrangements for work, living, and transportation and lead to the contestation and reformulation of public and private norms and values (McKinnon 2011; Yan 2010). It is clearly necessary to study the confluence of all of these factors and processes when examining the challenges to urban sustainability in China, as the authors in this volume acknowledge and do in different ways.

For China's urban authorities, then, it has now become a concern to make the nation's cities more pleasant and viable to live and to work in. They have therefore engaged with academia to use modern environmental science in planning, governing, and managing urban processes and cities. At the same time, they have come to recognize that new methods of government are required if the challenges that cities face are to be addressed. Among these methods, the institutionalized participation of citizens and other societal stakeholders in policy development, in co-governing, in overseeing the government, and in conducting urban planning and decision-making have become prominent, not only outside China (cf. Lund, Chap. 1) but also to some extent within China. This is amply reflected in the strategies for urban development decided most recently by Chinese leaders (Zhonggong Zhongyang 2016; Guojia 2014). Furthermore, urban elite actors are increasingly inspired from outside through academic exchanges, study visits, and official programmes like city-to-city collaboration (Delman 2016a) which lead to new types of policy mobilization (cf. Peck and Theodore 2010) around China.

China's political and administrative authorities are indeed struggling today to tackle questions that have been debated in the academic literature for years (see Peck 2015): What, for example, is a city? Does 'the city' exist? Are current territorial and administrative borders and demarcations useful for dealing with the consequences of urban development and urban processes? How can multiple stakeholder interests be taken into account in urban development? How should broad coalitions and alliances for qualitative change—such as a green transition—be structured and built so as to facilitate not only policy development but also sustainable day-to-day management of environmental challenges, such as traffic congestion or the handling of garbage?

The Rulers on the Ground: China's Party-State System

The authors contributing to this volume hail from different disciplinary backgrounds, mainly in the social sciences and in urban studies. They are quite familiar with China's urban institutional environment and how it is governed. We recognize, however, that some readers may need guidance to fathom the political and institutional framework with which the authors are working academically and on the ground and to which they liberally refer. We shall therefore give a brief account of how China and Chinese cities are governed.

After the Communist Party of China (CPC) took power in 1949, it started to build a political system with parallel party and administrative branches. The Soviet Union served as the model in this regard, while China's imperial experience and the CPC's own experience with governing revolutionary base areas were also seen as crucial. The system that emerged operated with corresponding party and state organs that took responsibility for similar issues or sectors at all administrative levels: central, province/provincial city, prefecture/city, county/county city, and township. It is a hierarchical system in which the CPC maintains overall political, ideological, and administrative leadership and control over governments at all administrative levels, as well as over the people's congresses that elect them formally. As a consequence, the next highest level of party-state administration supervises and controls key political and administrative processes at the level immediately below.

For a long time, the party-state system was a world unto itself with no direct civil or societal constituency. It was largely driven by bureaucratic politics. The horizontal levels of party-state system at different administrative levels are known as 'areas' (*kuai* 块), i.e. territorial units of government; the vertical bureaucracies linked to ministries in Beijing are known as 'lines' (*tiao* 条), and they operate at all territorial administrative levels within nation-wide systems (x*itong*系统) with shared mandates and identity. But the complex hierarchical nature and the often partisan political interests of the constituent horizontal and vertical parts of the party-state system produce internal fragmentation and competition both horizontally and from top to bottom—between different sectors, between administrative levels, between vertical bureaucracies, and at times between party and state (Lieberthal and Oxenberg 1988; Brødsgaard 2016). Since the CPC has insisted on maintaining the basic structure of the parallel political-administrative system until now, political scientists use the term 'party-state' as shorthand for the conjoined systems of the party and the government.

From the outset in 1949, the party-state leadership recognized the need to mobilize, manage, and control the broader population. It established a number of so-called mass organizations for the purpose, such as trade unions, a women's federation, a communist youth league, and professional or trade associations—one for each sector. These bodies continue to exist, but the CPC has also come to accept that—with liberalization of the economy and society—more channels need to be opened up for representation by increasingly variegated and sophisticated social groups (see Burell and Almén, Chap. 10; Wang et al., Chap. 11), who have become well versed in using different channels, including the social media, to promote their interests (cf. Schlæger and Zhou, Chap. 7; J. Liu, Chap. 9). Over time, the CPC has proven itself to be adaptable vis-à-vis societal demands, and its flexibility allows it to continue governing. This modality of government has been called 'consultative authoritarianism' (He and Thøgersen 2010).

The CPC is a popular party and its membership continues to grow. In 2017, the Party had about 89.5 million members (China Daily 2017), i.e. about 6.6% of the entire population or some 8% of the population eligible for membership. While the CPC does not claim its mandate from direct elections, it does however assert its popular support based on the feedback it receives through its huge membership, through mass organizations, through organs of the party-state system, and more recently through social organizations and social media. Under the consultative authoritarian arrangement, social organizations have been able to gain some measure of operational autonomy (cf. Zhang, Chap. 12). Furthermore, as the channels of interaction between social organizations and the state have opened up, social actors and organizations have exploited new opportunities to impact policy-making (Teets 2014). But the CPC continues to insist on its right to co-opt social organizations, primarily to be able to control them, like it does in relation to all media and state-owned enterprises. Several authors in this volume explore the dynamics of the increasing interaction, tension, and/or collaboration between the party-state and societal interests and the consequences thereof for green urban governance (cf. Schlæger and Zhou, Chap. 7; Delman, Chap. 8; J. Liu, Chap. 9; Burell and Almén, Chap. 10; Wang et al., Chap. 11).

Despite its hierarchical organization and its internal fragmentation, the party-state has proved to be a dynamic political-administrative organization. Combined with the decentralization of decision-making powers to lower levels of administration—not least the provinces and the big municipalities—these dynamics have opened up space for wide-ranging local experiments, for varying speeds of reform at the regional level (Heilmann 2007, 2008), and for openness in many cities to establishing a dialogue and even collaboration with external stakeholders to address local issues, as several of the contributors to this volume document (Delman, Chap. 8; J. Liu, Chap. 9; Burrell and Almén, Chap. 10; Wang et al., Chap. 11; Zhang, Chap. 12).

In order to mitigate fragmentation and internal competition, which can be both productive and disruptive, the party-state has developed a number of sophisticated integrative mechanisms over time (Lieberthal and Oxenberg 1988). These aim to deal with increasingly complex political challenges. Among them, leading groups have become important. They cut across and combine the often contradictory interests of the various party and state organs to make them collaborate to ensure more effective policy implementation. They have thus become increasingly important for addressing complex or thorny policy issues (Miller 2008), such as environmental degradation, eco-civilization, and climate change. These issues involve coordination between a host of bureaucracies, and they demand the active participation of civil society, businesses, and citizens if solutions are to be found and implemented (Delman, Chap. 8; Delman 2016b).

The party-state system is highly meritocratic in its recruitment practices and in its approach to human resource development, with a strong focus on the careers of individual party-state leaders. The system is exceedingly professionalized and technocratized in the execution of its mandates. City officials are sufficiently well-educated to interact productively with experts in academia, society, and the business sector and to apply up-to-date national and international research and science in their work (Mol and Carter 2006).

The complexity of the party-state system and the absence of independent supervision and control of it have however resulted in widespread abuse of power and corruption. The problem is made yet worse by the continued dearth of transparency in the execution of government (cf. Wang et al., Chap. 11). This has given rise to far-reaching popular dissatisfaction and widespread distrust of party-state bodies. The current high-profile campaign against corruption under Xi Jinping's leadership is clearly meant not only to restore proper morals and behaviour within the party-state administration but also to garner new legitimacy for the CPC as the ruling party (Hewitt 2015).

Fundamental political reforms are not currently on the drawing board, but as noted above the party-state has proven agile and able to adapt to new circumstances. It has done so through continuous administrative reforms and though development of its consultative mechanisms. This is also the case within urban green governance, which has become an increasingly important political arena for the local party-state after many years of negligence. As this volume demonstrates, a shift is taking place from uniform top-down government to a more versatile form of governance, which combines traditional public administration, new public management, and network governance in a variety of ways (cf. Lund, Chap. 1). But it does not put the power monopoly of the party-state at risk. In fact, we would stress that the emerging urban green governance regimes are deeply influenced by the party-state system, and it will continue to exert a critical influence at all levels of operation on China's green development in the years to come.

China's Green Urban Governance and the Theoretical Landscape

Since the open-door policy was initiated in 1978, China's cities have been reconstructed and reinvented, in order to engage with modernization needs and globalization. The dynamics, trajectories, and systematics of this development are based significantly on a negotiation between the national heritage, the socialist practice, and the international experience, with the latter being absorbed into local urban design, planning, and management (cf. Peck and Theodore 2010; Peck 2015). This negotiation has been expressed through new urban and architectural designs (as noted above); through the establishment of special economic zones, export-processing zones, financial centres, central business districts, eco-cities, and similar models on the one hand; and through the incorporation of large numbers of rural migrants, not only as workers in industry, construction, and services but also as co-inhabitants of China's urban areas (Wu and Gaubatz 2013; Hsing 2010). Hsing argues that urbanization is at the centre of the new policy priorities and policy discourse of China's urban leaders. The focus is on urban development projects that determine the dynamics of the local state and its relations with the market and society (Hsing 2010, 114–5). At the same time, urban residents with *hukou* have seen a rapid upgrading of their living environments (Wu and Gaubatz 2013). This urbanism is distinctive in its privileging of place production over industrial production (Hsing 2010, 114–5) and as argued in this book: governance over government.

China's cities are rapidly growing beyond their traditional physical boundaries and their politically constructed administrative hierarchies and borders. New city-regions, by some called mega-cities, are emerging, creating a new logic of scale production and governance in China's urban landscape. In response to economic globalization and attempts at re-centralizing power at the central level, they aim to develop regional economies through fostering of regional competitiveness. This entails administrative annexation of smaller territorial units by bigger ones, new integrated spatial plans at the regional level, and regional institution building (Wu 2016). Thus, while China used to be a planet unto itself, with its own traditions and experience in urban development and management, by some called a cellular structure, the country's economic and urban processes have now come to mirror and to play into socio-spatial transformations at the global level which are challenging the categories once used to describe, analyse, and even manage China's cities. Theoretically speaking, these transformations embody, to cite Brenner and Schmid (2014), the blurring and re-articulation of urban territories at different scales that challenge prevalent epistemological assumptions, categories of analysis, and objects of investigation within urban studies. China's urban dynamics and processes cast doubt on what Brenner called established 'methodological city-ist' explanations and understandings of the city and of urban processes, i.e. perceptions of the urban as a bounded, nodal, and relatively self-enclosed socio-spatial construction (Brenner 2014a). In an effort to move beyond traditional 'methodological city-ism' in urban

Unit of analysis	Open, variegated, multi-scalar: the urban as an unevenly developed yet worldwide condition and socio-spatial transformation
Model of territorial organization	Processual, dialectical: agglomerations ('cities') relate dialectically to their ('non-city') operational landscapes, which are in turn continually transformed through their roles in supporting agglomeration
Understanding of territorial development	Mediated through capitalism, state strategies and sociopolitical struggle: worldwide implosion/explosion of capitalist socio-spatial organization, encompassing the evolving relations between agglomerations and their operational landscapes within a crisis-prone capitalist economy
Model of *longue durée* historical-geographical change	Discontinuous, uneven: socio-spatial configurations (including both agglomerations and their operational landscapes) are creatively destroyed through the crisis-tendencies of capital (mediated through state institutions and sociopolitical struggles), contributing to successive rounds of territorial differentiation and re-differentiation of various spatial scales

Fig. 1 Urban theory without an outside. (Source: Brenner (2014a, 22))

studies, Brenner proposes an urban theory 'without an outside' (shown in Fig. 1)—a new approach to understanding the phenomenon he calls 'planetary urbanization'.

Brenner's model is a time-relevant and insightful *moment of theory* (we know what was there before it—i.e. 'methodological city-ism'—but not yet what comes after it). It involves a dialectical reading of 'the urban' as a systemic, world-spanning, and contradictory social condition. Yet such scepticism might easily result in a swing of the theoretical pendulum from the one extreme to the other, i.e. from imposing existing theories onto new territories, despite their greatly differing experiences, to embracing an extreme urban positivism, in which every city is treated as a special case (Peck 2015).

While we recognize these prospects, which can be either constructive or destructive, we are acutely aware that urban studies must develop in line with shifts in the global terrains of urbanization—one of these being the massive urbanization seen in Asian countries, not least China, in recent decades (McKinnon 2012). In the past, Asia and the rest of the so-called South were basically left out of urban studies or reduced to places in deficit as compared with Western cities, i.e. places characterized by disorder, poverty, informality, and so forth (Robinson 2006, 117). There was apparently little desire to examine or to understand urban processes in a wider comparative perspective, including the world outside the privileged West. However, it is critical for urban studies to incorporate the examination of cities in the 'global South', the 'East', and the 'post-colonies' on their own terms and not as exceptions, exoticisms, or theoretical anomalies (Peck 2015). China's experiences are important in this context, and Brenner's framework is helpful for examining and explaining the processual, dialectical, and disruptive nature of urbanization in that country.

At the same time, we recall that China has developed its own approaches to contemporary urban management (as defined above) and governance based on its historical experiences. Therefore, in order to avoid depicting China as a theoretical 'anomaly' or an outlier (cf. Peck 2015, 161), this volume dissociates Chinese experiences with urban green governance from essentializing theoretical paradigms, while at the same time probing the applicability of urban governance theories to China. Our approach follows Peck's line of thinking that: 'It cannot be sufficient to hold singular cases at ambivalent distance from supposedly "offshore" theories' (Peck 2015, 162). Were we to focus on China's urban experience as a case unto itself, we would risk, as Schmid (2014, 2005) argues, failing to offer new insights into broader issues of global urbanization and its differentiated regional and local patterns. By focusing this volume on an understudied phenomenon—urban green governance in China—we hope to contribute to the global debate on sustainable urban development. Urban green governance, we suggest, is becoming an essential driver of urban processes globally, and what we observe in China is directly relevant elsewhere. It may also be productive to theorize new urban green governance structures, institutions, and processes that entail a multitude of political instruments in response to the behaviour of key stakeholder groups in urban regimes and processes at different scales (cf. Lund, Chap. 1) irrespective of the unique characteristics of China's party-state system.

Given the central role of the party-state in urban governance in China, it is important to examine the role it plays in the green urban transformation. To take one example, the party-state's massive planning system is still a key steering instrument in urban governance. The system dates back in part to the emergence of the socialist planned economy in the 1950s, but it is also rooted in China's imperial bureaucratic traditions. In combination, these two legacies make the planning system both pervasive and rigid. Structured along vertical lines, it stretches from the level of the State Council down to urban districts and rural counties. It also encompasses state-owned companies at various administrative levels (He et al., Chap. 3).

Furthermore, due to the complex nature of environmental issues, the effective implementation of policies to deal with them requires horizontal, cross-sectoral cooperation—and this has proven difficult within China's 'silo-ized' party-state system (cf. Delman, Chap. 8). China's local governments are facing huge challenges in integrating urban environmental plans with urban master plans and economic-development plans. In many cases the different plans are not coordinated (He et al., Chap. 3), and different city departments may take different or competing approaches to similar environmental challenges (Liu et al., Chap. 5). New ways of targeting policy interventions have therefore become important. Ren and Liu (Chap. 6) examine patterns of carbon consumption among different groups of citizens and argue strongly that a focus on high emitters will be necessary if public programmes to mitigate carbon emissions are to be implemented successfully.

In a wider perspective, this volume's focus on China's urban green governance also speaks directly to theoretical and policy debates outside China and within international organizations. Since the 1990s, Chinese specialists and leaders have recognized that cities are crucial for national and global sustainable development. The need for new governance regimes was initially anchored in a debate, originating within international development organizations, about 'good' governance. It was also recognized that the promotion of sustainable development and sustainability governance are intimately linked to the competitiveness of cities within the global economy (Qian 2008). In this sense, the most recent Chinese debate on urban development has engaged directly with the challenges and consequences of globalization for the cities of the world. The idea of the 'global city' has become a household paradigm in Chinese politics and academic debate on urban questions (Zhou 2014). These theories have since come under criticism by some urban-studies theorists for being (in Peck's words) '…a(nother) case of heterodox urban theory being captured, co-opted and corrupted by the very power structures that were the impetus for the original critique' (Peck 2015, 164). However, this criticism has yet to find a foothold in the Chinese debate.

There could of course be arguments for differentiating between Western and Chinese theory and practice. Qian (2008) finds that the Chinese literature on urban sustainability governance is influenced strongly—perhaps too strongly—by neoliberal theory and by ideas of new public management (cf. Lund, Chap. 1), and this leads him to argue that there is a need also to consider Chinese values, China's political economy, and the country's specific social and cultural traditions when engaging with its approaches to urban governance. Still, Qian argues, from the outset, there has also been a recognition in China that multiple stakeholder governance must be part of the solution to the country's urban environmental and sustainability challenges. Local governments continue to be the primary agents of change, but they could not continue to act solely as controlling agents. They need to understand the wishes of their societal constituencies better. They must coordinate with the business sector and with social organizations and engage in mutually beneficial partnerships. They need to become more professional, transparent, and efficient in handling the political challenges they face. Finally, they must eventually secure popular participation in governance (Qian 2008; cf. Lund, Chap. 1).

Furthermore, in Qian's (2008) view, sustainability governance is concerned with redesigning the urban setting to become more livable. There is a need to establish policies and institutional frameworks for stakeholder coordination, especially with regard to participation by businesses and the public. The local state should be 'competitive' in performing its tasks and in enlisting stakeholder engagement (Qian 2008; cf. Schlæger and Zhou, Chap. 7; Delman, Chap. 8). Efforts to achieve sustainable development must be guided by specific plans, and the different levels of government have to be held to account through proper indicator systems (Qian 2008; cf. He et al., Chap. 3; Delman, Chap. 8). Finally, policies must be enacted to ensure that the pursuit of sustainable development is just and fair and that it does not result in greater social inequity (Qian 2008; Wang et al., Chap. 11).

As noted above, Chinese cities are incorporated into hierarchies of administrative status, regional importance, and size. The different administrative tiers have traditionally been associated with differential privileges, with lower levels being subordinate to and less privileged than higher levels. Since the post-Mao reforms of public administration in China started in the 1980s, this hierarchical privileging or, at times, underprivileging of cities has led to inter-city competition for status and resources. Party-state leaders and their administrations at various levels of government have used these dynamics as a lever to stimulate urban development. Competition within the fragmented hierarchical party-state has thus become a key driver of urban development, to the extent that urban leaders actively compete to improve the standing of their city on hierarchical listings that focus on issues such as growth, branding, innovation, livability, and the environment (e.g. Chengshi paiheng, n.d.; Chengshi paiheng yu pingjia 2012). This competition is stimulated further by the party-state's system for assessing and controlling the performance of government organizations and their leaders (Delman, Chap. 8; Almén 2017). Hierarchization and comparison are also promoted by many academic studies to reflect and stimulate competitive thinking (e.g. Huan 2009). In Chap. 4, Luova analyses the implementation of environmental plans in two inner urban districts of Tianjin and shows how local characteristics lead to variation in policy implementation even at the level of urban districts due to such competition.

The 'procedures' or practices of modelling and inter-referencing thus become part of wider processes that entail national and international referencing and comparison. Ong (2011) refers to these phenomena as 'worlding' processes. They are distinctive, because they anchor globalization at the local level. City leaders look for antecedents to inspire them in their work with urban development (cf. Peck and Theodore 2010; Peck 2015), often with the assistance of experts or scholars. They internalize images and mappings of urban exemplars, thereby absorbing and then promoting the idea that it is possible and desirable to move up the status hierarchies (Bunnell 2015). These practices frequently involve policy mobility rather than direct policy transfer, through the creation of mental maps of 'best cities'. The policies embraced may inform future strategies (Bunnell 2015: 8) through the engagement of a range of 'agents, practices and performances involved in the social production and transformative circulation of forms of policy knowledge' (Bunnell 2015: 6). Practices of this kind resonate deeply with the Chinese tradition of policy experimentation and selective policy implementation under hierarchy (Heilmann 2007; Brehm and Svensson, Chap. 2; Luova, Chap. 4; Delman, Chap. 8). Urban experiments in environmental mitigation, for example, open up for unexpected opportunities which the pursuit of ecological practices affords to urban environments, spatial formations, and governance structures (cf. Mostafavi 2010).

However, while China's state-level green policies are increasingly ambitious and internationally referenced, local implementation often remains fragile or defiant, to the extent that it contradicts the much hyped experimental approach to policy implementation (Ran 2013; Kotska 2014). Brehm and Svensson, in a meso-level analysis of environmental model cities in Chap. 2, find that model-city programmes—as voluntary and incentive-based schemes for protecting and improving the urban environment—are exposed to system-inherent forces that lead to an increasing distance between programme objectives and local practices. With institutional elaboration and diffusion of the same model, the space for innovation in governance gradually narrows. Specific model schemes often come about in response to a top-down political paradigm, so it is difficult to uphold innovative impulses over time. In the end, a model scheme inevitably reaches a point of saturation. The end of a model scheme's lifecycle is reached once a competing programme offers new opportunities to gain political and economic rents (Brehm and Svensson, Chap. 2). Furthermore, many eco-city projects come under criticism for being diluted, due to compromises with entrepreneurial and technocentric local leaders (Joss and Molella

2013; de Jong et al. 2013); or else they are dismissed as a disguise for unsustainable land-development projects (Chien 2013).

In Jennifer Robinson's terms (2006, 1), this volume thus *takes the world of China's cities as its starting point*. We acknowledge the diversity and complexity of all cities, and therefore it could be perfectly possible to recognize or even theorize the city '…as a place of mobility, flow and everyday practices, and [to read] cities from their recurrent phenomenological patterns' (Amin and Thrift 2002, 7; cf. Lund, Chap. 1), but we also argue that urban processes need to be governed and that we find not only patterns of difference but also of similarity when we compare governance practices across cities. As Brenner notes (Fig. 1), the state is a strategic player. It mediates the (often disruptive) influences of capital, while simultaneously being engaged by sociopolitical struggles. The studies in this volume focus primarily on China's disrupted urban physical environments and on how the state and other actors act to mitigate the damage. This raises a series of questions regarding the execution of urban government and the governance of urban eco-sustainability challenges: Who governs the green transition? What is governed? How is it governed? What are the implications for China's urban green practices and for the established system of urban government? What are the implications of Chinese experiences for urbanism and urban studies? These are the questions that this book sets out to answer through its many case studies.

Structure of the Volume

The introduction above puts China's urban development into a historical and a theoretical perspective, arguing that China's urban development experience must be assessed on the basis of the country's own experiences with urban management. At the same time, China's engagement with the outside world, not least globalization since the late 1970s, has opened China's cities to international influence and global flows that allow us to apply contemporary urban theory to examining the governance of China's urban green transition. We also pinpoint the areas where China's experiences may help us redevelop and rearticulate the theories that we work with.

Section I Green Urban Governance: A Theoretical Perspective

In Section I of this volume, D. H. Lund (Chap. 1) contributes a guiding theoretical perspective on governance in relation to urban green and sustainable development. Based on her work on urban governance in the Nordic countries and with reference to the individual chapters in this volume, she contributes essential theoretical insights and substance as well as practical inspiration for the analysis of and reflection on urban green governance dynamics in China, and this thread is taken up and discussed in the Epilogue at the end of this volume. Lund notes that cities produce many of the environmental challenges that societies all over the world face. They are great consumers of energy, of food, of water, and of land. They are also great producers of waste, including air pollutants, CO_2, and other greenhouse gases. Therefore, they have a great impact on ecosystems and on the conditions for human and non-human life. At the same time, they are crucial for economic growth, for innovation, and for employment in national economies. It is therefore in cities that governance for sustainability may find its most fertile soil, and if successful will have its greatest effect, argues Lund. She proceeds to ask what type of governance can promote sustainable urban development? She first addresses some of the elements in an urban transition towards sustainability, after which she considers three governance paradigms—traditional public administration (TPA), new public management (NPM), and new public governance/network governance (NG)—and their advantages and disadvantages in terms of sustainability. She draws on the mainstream Western literature on governance and on Nordic examples which are within her field of expertise. She concludes, however, upon reading the case studies in this volume that they reflect commonly accepted understandings of governance theory and exemplify a variety of hybrid governance arrangements at different scales and seen from different perspectives which together provide a solid foundation by which to reflect on governance dynamics and their potentials in terms of promoting systemic sustainable development, in China and elsewhere. Lund argues that the cases included here also show that, despite the evident differences between China and most of the Western world, in terms of governance in general, there are similar trends in terms of more polycentric governance, an extension of actors beyond institutionalized power structures who gain power in the efforts to combat environmental degradation, and also in some of the governance technologies applied to increase sustainability in particular from the NPM toolbox such as performance assessments, voluntary incentive schemes, and benchmarking. Therefore, the prospects for sustainable urban governance, she concludes, will depend on the ability of cities to choose appropriate hybrid governance paradigms and associated tools within the context of the Chinese political-administrative system.

Section II Policy Mobilization, Planning, and Implementation

While many of the chapters in this volume touch upon policy mobilization that aims to find new ways to improve urban green governance, the chapters in this section examine specific types of policy mobilization and implementation, through policy schemes for greening cities and at the city level (education); through attempts to improve China's disjuncted urban planning practices, especially with regard to urban environmental planning; through more focus on bottom-up practices and new network governance approaches in storm water management; and finally through examining how to differentiate between target groups for specific green interventions, *in casu* reduction of carbon emissions.

In Chap. 2, Stefan Brehm and Jesper Svensson argue that, since the 1980s, China's central government has created various model and incentive schemes aimed at systematically and concurrently promoting innovative approaches for protecting the urban ecological environment. In this context, programmes such as the 'model city for protecting the environment', 'garden city', 'eco-city', or 'low-carbon city' have become an integral part of China's system for urban environmental governance. However, the role of these policy-incentive schemes for promoting best practice is only partly understood. The study contributes to this literature with a conceptualization of model cities as a dynamic governance instrument. The analysis suggests that the distance between programme objectives and local practices increases with programme maturity. Model-city schemes inevitably reach a point of saturation once a competing programme provides new opportunities to gain political and economic rents.

Jia He, Cunkuan Bao, Jun Zhu, and Jinnan Wang argue in Chap. 3 that traditionally different sectors in China have developed different types of plans with a variety of purposes and that they have been implemented by a highly fragmented planning system. Thus, due to the lack of an overarching design for the planning system in toto, inconsistency, redundancy, and conflict are often observable in the process of making and implementing plans at all levels of administration, including the city level. To some extent, multi-planning integration may help to improve the effectiveness of implementation and to resolve contradictions among different types of plan. However, within the context of such multi-planning integration, the environmental plan at city level faces not only new opportunities but also some new challenges. For example, although thousands of environmental plans have been issued in practice, the development of legislation and academic studies in this area have somehow slowed down, and few technical documents or guidelines have been issued by the environmental authorities to guide the implementation of them. The authors review the history of the environmental plan, discuss its relationship with other plans, and analyse the main challenges. They take Benxi City in Liaoning Province as a case showing how an integrated Urban Environmental Master Plan can be designed in practice. Finally, they offer some suggestions for future improvement.

Outi Luova examines implementation of environmental education policies in Tianjin in Chap. 4. She argues that environmental policies are cross-sectoral by nature and that during their implementation they have to be adapted to the differing contexts of different sectors. Still, she notes that there is variation in the implementation process between different geographical areas. Luova's study looks at how two inner urban districts in the city of Tianjin (one poor and one wealthy) implement their new environmental plans in the educational sector. Chinese urban districts have substantial freedom to adapt educational policies to fit local conditions, and schools themselves are encouraged to define and to emphasize their own specific priorities. Under these circumstances, how can environmental-policy goals be reached in an effective way, asks Luova? Her findings show that, with regard to numerical targets, the poor district did better than the rich one. This was because a green profile was cheap to acquire, it opened the door to substantial external funding, and it made possible an improved bargaining position vis-à-vis the city government. However, the rich district was able to provide solid resources for its schools, enabling them to excel in terms of qualitative objectives, i.e. nurturing green lifestyles and promoting respect for the environment. The study highlights (1) specific features of environmental-policy implementation in the educational sector, (2) variety in policy-implementation processes in inner urban districts, and (3) the importance of clear qualitative targets in the definition of policy goals.

In Chap. 5, Li Liu and her co-authors examine urban water management in the two capital cities, Beijing and Copenhagen. Their basic argument is that cities play a critical role for the sustainable management of planetary freshwater. At the same time, cities need to adapt to climate change. This offers cities an opportunity to improve freshwater management. The authors first describe the status of urban water management in the two capitals, from the standpoint of sustainable development and climate resilience. In particular, they look into the degree to which the local water balance has been displaced. They review key water challenges and instruments falling within the range of traditional public administration (TPA), new public management (NPM), and network governance (NG) in both cities. They then consider the potential impact of these instruments and the relevance of a green-infrastructure approach through improved urban sustainability governance. They find that neither Beijing nor Copenhagen can neglect to make some use of top-down approaches (TPA and NPM) with centralized hardware systems, strategic planning, and management. However, there is a great potential to inject more in the way of bottom-up practices (NG) into the mix of governance instruments, thereby improving water supply, wastewater management, and storm water management in a sustainable way. They also find that Copenhagen's successes in using NG approaches may serve to inspire Beijing. Further, both cities could seek a greater reliance on NG to find alternative approaches to managing floods and storm water through urban landscapes and develop associated mechanisms to engage a wider range of actors from both public and private sectors. The attraction of NG becomes yet more evident when the authors link issues of storm water management to questions of water supply and wastewater management, given that decentralized storm water facilities can provide water-supply resources and alleviate pressures on municipal wastewater treatment.

In Chap. 6, Yuan Ren and Liu Daisong divide urban residents' direct carbon emissions into two categories: residential energy consumption-related carbon emissions and transportation-related carbon emissions (TCEs). Based on a survey of urban residents' carbon consumption that the authors conducted in Shanghai in 2013, the authors analyse demographic and socioeconomic characteristics of high emitter and demonstrate the inequality of carbon emissions among different groups by income, consumption, and lifestyle. Ren and Liu conclude that high emitters should bear more social responsibility than others to help achieve a transition to a low-carbon society, and they discuss how to encourage changes in their lifestyle so as to facilitate this transition.

Section III The State's New Tools of Green Urban Governance

In recent years, local governments in China have become more sophisticated in adopting new and more diverse tools to improve urban green governance. While the development of these tools is often initiated and guided from the top, the tools may deliberately aim at engaging the public in ways that make local governance more transparent and accountable and at times also engage citizens in co-producing the information, the tools, and also the possible solutions.

In Chap. 7, Schlæger and Zhou explore new technology as an integrative mechanism in urban environmental governance in China. The added value of this approach lies in its focus on how ICT (information and communication technology) affects administration from both a top-down and a bottom-up perspective. Their main finding is that digital environmental monitoring serves to level out some of the information asymmetry between central government principals and local level agents, thus paving the way for a more centralized model of environmental urban governance in China that strengthens the power of the central state. Schlæger and Zhou also note, however, that ICT has facilitated a shift on to a more sustainable path of development, by making information about the ills of environmental degradation available to a broader group of citizens, In fact, citizens now enjoy increasing opportunities to participate in the co-production of basic public environmental services. In sum, the effect on environmental governance is to make local governments more open to control, both from above (through systems of centralized control) and from below (through citizen-activated fire alarms). Even so, many of the systems and software applications examined in their study are still in an early phase of adoption; therefore, their conclusions must be seen as indicative of the direction of development of use of information and communication technologies in urban green governance.

In Chap. 8, Jørgen Delman asks whether Hangzhou's system of public-sector performance reviews promotes sustainability governance in the city. First, he examines how ideas and concepts relating to sustainable urban development are incorporated into urban development programmes in Hangzhou. He then focuses on the design of the system for performance reviews of party-state organizations and their leaders. Performance reviews comprise a variety of measures for performance assessment and performance management. These procedures have increasingly come to involve social participation, in order to allow for a public voice and public participation in governance. His analysis demonstrates how the city's public performance review and management system incentivizes major government departments working with sustainability-related issues to toe the green Communist Party line. Even more, innovative provisions to enhance the public accountability of the local government may also help to promote green governance in Hangzhou. This may pave the way, he argues, for narrowing the well-known implementation gap in connection with implementation of urban environmental policies.

Section IV Society Knocking on the Door

While the Chinese Party-state is continuously experimenting with its approach to addressing China's environmental crisis at the local level of implementation, there is no denying that social or civil society activism has played a crucial role in forcing local governments to open up and become more inclusive in their approaches. Citizens contest local decisions, projects, or environmental incidents by taking to the streets and to social media. Local NGOs pursue policy lobbying. They may not always win their case, but they often manage to elicit concessions. The issue of proper or adequate justice is often at stake, and one of the chapters argues the need to focus on practising *green justice* with a long-term perspective to improve urban green governance in the interests of all the relevant stakeholders, not least citizens who are often driven by not-in-my-backyard considerations. Eventually, China's local state must change its governance structure, argues Zhang Changdong in Chap. 12, from a state-led model to one of state-society collaboration.

In Chap. 9, Jun Liu examines how online activism against a specific type of a potentially polluting industrial project travels across China from one urban region to another, thus linking urban communities of protest together across regions. Since the emergence of the environmental crisis in China in the early 1990s, Liu notes that environmental social activism has emerged as a key type of digital activism, as well as a major challenge to urban governance. While scholars have studied issues arising from digital activism in general and environmental activism in particular, most have failed to scrutinize the possible interconnections among different instances of digitally mediated political contention. To advance such an understanding, he employs the concept of 'cycles of contention' in his study—a concept helpful for investigating recurrent mechanisms of protest in contemporary society. His case comprises seven anti-PX (anti-petrochemical) environmental protests in China between 2007 and 2014. During this period, he conducted 54 in-depth interviews. He found that the coverage in the traditional media, on the one hand, served to legitimize and to modularize these anti-PX protests, thereby facilitating the adoption of digital media as part of the repertoire of contention, which helped making political contention sustainable over the long run. The use of digital media, on the other hand, enabled protestors to diffuse contention widely and quickly and to learn from past experiences. He concludes that, as a new challenge to urban governance, digitally mediated environmental activism is shaped by the specific communication ecology in China.

In Chap. 10, Burell and Almén examine the consequences of social protests for environmental governance in China. Environmental protection is a policy area where the party-state has allowed civil society to play an increasingly important role. Chinese citizens can take part in state-sanctioned forms of political participation, but one of the most important ways they can influence policy is by protesting.

If we are to understand urban environmental governance in China, the authors argue, we must study the impact of such protests. Students of social movements in China have made great efforts to understand what explains protest success in that country. But the authors contend that, in order to understand environmental governance, we must look at all the different outcomes of a social protest. In most cases a protest elicits at least some concessions from the authorities, even when the protesters themselves perceive it as a failure. The study examines environmental governance by looking at the varying outcomes of three different waves of environmental protest in Hangzhou. Protests may lead to the development of new institutional tools such as dialogue platforms or involve external actors like courts or NGOs. Through these mechanisms, citizen protests come to form a new and important element in China's fragmented system of governance.

In Chap. 11, Wang Jiang Li and her co-authors review a public protest against the construction of a waste incineration power plant in 2014 in Yuhang District, Hangzhou. The authors reflect on the absence of justice between the residents and the local government from the perspectives of procedural justice and spatial justice, respectively, which are the core dimensions of *green justice*, a holistic philosophical perspective on the relation between humans and nature. They argue that not-in-my-backyard (NIMBY) movements and closed local governmental decision-making processes are the major cause of such conflicts, which is a very common dilemma of environmental governance in the process of China's urbanization. To build a waste incineration power plant is not only a concrete measure of environmental governance but also a protection for good quality of civic lives, which reflects the overall need of urbanization. Therefore, the authors argue that the relevant parties, i.e. governments, residents, and interest groups, need to adhere to green justice and holism to deal with public bads like garbage and subsequently share the environmental public goods based on common and long-term interests.

In Chap. 12, Zhang Changdong argues that sustainable development requires that China's governance structure be changed: from a state-led model to one of state-society collaboration, in which both government initiatives and civic engagement are found. Over the last decade, which saw rapid economic growth together with rising concerns about unsustainability, the Chinese government started to address the problem of sustainability. At the same time, many non-governmental organizations (NGOs) have been trying—regardless of their varying degrees of dependence on the state—to promote urban sustainable development, among other things by lobbying the local (authoritarian) government to bring a variety of stakeholders into the policy-making process. Using comparative case studies and a quantitative analysis of policy lobbying by NGOs, Zhang found that many NGOs were able to influence government policy-making through various channels. At the same time, the quantitative evidence suggests that sustainability-related NGOs—i.e. NGOs dedicated to the defence of the environment and of vulnerable social groups—have a degree of policy influence.

Finally, in the Epilogue, we summarize the main findings of the volume and elicit how they contribute to our understanding of urban green governance in China and to what extent the Chinese experience can be helpful in refining our theoretical understanding of how urban development processes are governed.

Copenhagen, Turku, Uppsala and Shanghai, 13 December 2017

References

Almén, O. (2017). Participatory innovations under authoritarianism: Accountability and responsiveness in Hangzhou's social assessment of government performance. *Journal of Contemporary China*, publ.mon-line 28.10. https://doi.org/10.1080/10670564.2018.1389003

Amin, A., & Thrift, N. (2002). *Cities*. Cambridge: Polity.

Baeumler, A., Ijjasz-Vasquez, E., & Mehndiratta, S. (Eds.). (2012). *Sustainable low-carbon city development in China*. Washington, DC: World Bank.

Bosker, Bianca (2014). Why haven't China's cities learned from America's mistakes? *The Guardian*, 20 August.

Brenner, N. (2014a). Introduction: urban theory without an outside. Brenner (ed) (2014), pp. 14–30.

Brenner, N. (Ed.). (2014b). *Implosions/explosions. towards a study of planetary urbanization*. Berlin, jovis Verlag GMBH.

Brenner, N, & Schmid, C. (2014). Planetary urbanization. Brenner (ed.) (2014), pp. 160–163.

Brødsgaard, K. E. (ed.). (2016). *Earthquakes, energy, and the environment: Fragmented authoritarianism in China Today* (working title). Routledge.

Bunnell, T. (2015). Antecedent cities and inter-referencing effects: Learning from and extending beyond critiques of neoliberalisation. *Urban Studies, 52*(11), 1983–2000.

Chengshi paiheng yu pingjia. (2012). 城市排行与评价(南京专辑). 信 息 摘 编 (金陵图书馆编), 第1期 (总第112期),30 January.

Chengshi paiheng (城市排行) (n.d.). 中国城市环境宜居指数排名. *Chinacity.org.cn*, http://www.chinacity.org.cn/csph/csph/63949.html. Accessed 29 June 2016.

Chien, S. S. (2013). Chinese eco-cities: A perspective of land-speculation-oriented local entrepreneurialism. *China Information, 27*, 73–96.

China Daily. (2017). *CPC has nearly 89.5m members*. 30 June, http://www.chinadaily.com.cn/china/2017-06/30/content_29952238.htm. Accessed 20 Oct 2017.

China Ecological. (2012). China ecological footprint report 2012. Consumption, production and sustainable development. WWF, Institute of Geographic Sciences and Natural Resources Research (CAS), Global Footprint Network, Institute of Zoology (CAS), Zoological Society of London. http://awsassets.panda.org/downloads/china_ecological_footprint_report_2012_small.pdf. Accessed 3 Jan 2013.

de Jong, M., Wang, D., & Yu, C. (2013). Exploring the relevance of the eco-city concept in China: The case of Shenzhen Sino-Dutch Low Carbon City. *Journal of Urban Technology, 20*(1), 95–113.

Delman, J. (2014). Climate change politics and Hangzhou's green city making. In E. Björner & P. O. Berg (Eds.), *Branding Chinese mega-cities: Policies, practices, and positioning* (pp. 249–261). Cheltenham: Edgar Elgar Publishers.

Delman, J. (2016a). Ny motor i Danmarks forhold til Kina: Strategisk myndighedssamarbejde [New motor in Denmark's relations with China: Authority-to-authority collaboration]. Sørensen, Camilla og Delman, Jørgen (red.) (2016), Særnummer af *Økonomi og Politik*, 'Dansk Kina-politik—fra spørgsmål om eksport og danske arbejdspladser til ny verdensorden' [Danish China policy—from questions about exports and Danish jobs to a new world order], April, pp. 23–32.

Delman, J. (2016b). Urban climate change politics in China: Fragmented authoritarianism and governance innovations in Hangzhou. Brødsgaard, pp. 156–180.

Elvin, M. (1973). *The patterns of the Chinese past*. Stanford: Stanford University Press.

Elvin, M. (2004). *The retreat of the elephants—An environmental history of China*. New Haven/ London: Yale University Press.

Fay, M., Wang, J.-z., Draugelis, G., & Deichmann, U. (2014). Role of green governance in achieving sustainable urbanization in China. *China & World Economy, 22*(5), 19–36.

Guojia. (2014). 国家新型城镇化规划(2014—2020年). 人民网-人民日报, 17 March. http://house.people.com.cn/n/2014/0317/c164220-24650538.html. Accessed 25 Mar 2014.

He, B., & Thøgersen, S. (2010). Giving the people a voice? Experiments with consultative authoritarian institutions in China. *Journal of Contemporary China, 19*(66), 675–692.

Heilmann, S. (2007). Policy experimentation in China's economic rise. *Studies in Comparative International Development, 43*(1), 1–26.

Heilmann, S. (2008). From local experiments to national policy: The origins of China's distinctive policy process. *The China Journal, 59*, 1–30.

Hewitt, D. (2015). China's anti-corruption Campaign: Top Beijing and Shanghai officials under investigation as probe deepens. *International Business Times,* 11 November 2015. http://www.ibtimes.com/chinas-anti-corruption-campaign-top-beijing-shanghai-officials-under-investigation-2179252. Accessed 29 June 2016.

Hsing, Y.-t. (2010). *The great urban transformation. Politics of land and property in China*. Oxford/New York: Oxford University Press.

Hu, J. (2012). Firmly march on the path of socialism with Chinese characteristics and strive to complete the building of a moderately prosperous society in all respects: Report to the Eighteenth National Congress of the Communist Party of China on November 8, 2012, http://language.chinadaily.com.cn/news/2012-11/19/content_15941774.htm. Accessed 2 July 2016.

Huan, Qingzhi (郇庆治). (2009). 城市可持续性与生态文明: 对青岛、承德和苏州的比较. 马克思主义与现实, no. 2.

Joss, S., & Molella, A. P. (2013). The eco-city as urban technology: Perspectives on Caofeidian International Eco-City (China). *Journal of Urban Technology, 20*(1), 115–137.

Kassiola, Joel Jay, Sujian Guo (2010). China's environmental crisis: Domestic and global political impacts and responses. Houndmills, Basingstoke: Palgrave Macmillan.

Kostka, G. (2014). Barriers to the implementation of environmental policies at the local level in China. *World Bank Policy Research Working Paper*. In *WPS7016 (August)*.

Lefebvre, H. (2014/1989). Dissolving city, planetary metamorphosis. Brenner (ed.) (2014), pp. 566–570 (Translation by Laurent Corroyer, Marianne Potvin and Neil Brenner of: Lefebvre, Henri (1989). Quand la Ville se perd dans une metamorphose planétaire. *Le monde diplomatique*, May).

Lianghui. (2016). 两会授权发布中华人民共和国国民经济和社会发展第十三个五年规划纲要. 新华社, 17 March, http://news.xinhuanet.com/politics/2016lh/2016-03/17/c_1118366322.htm. Accessed 7 July 2016.

Lieberthal, K., & Oksenberg, M. (1988). Policy making in China. In *Leaders, structures and processes*. Princeton: Princeton University Press.

Lin, G. C. S., Li, X., Yang, F. F., & Hu, F. Z. Y. (2014). Strategizing urbanism in the era of neoliberalization: State power reshuffling, land development and municipal finance in urbanizing China. *Urban Studies, XX*(X), 1–21.

McKinnon, M. (2011). *Asian cities. Globalization, urbanization and nation-building*. Copenhagen: Nias Press.

Miller, A. (2008). The CCP Central Committee's leading small groups. *China Leadership Review, 26*(September), 1–21.

Mol, A. P. J., & Carter, N. T. (2006). China's environmental governance in transition. *Environmental Politics, 15*(2), 149–170.

Mostafavi, M. (2010). Why ecological urbanism? why now? *Harvard Design Magazine*, 32 (*Design Practices Now*), 1 Spring/Summer, online: http://www.harvarddesignmagazine.org/issues/32/why-ecological-urbanism-why-now. Accessed 29 June 2016.

Ohshita, S., et al. (2012). The role of Chinese cities in greenhouse gas emission reduction. In *Briefing on urban energy use ad greenhouse gas emissions*. Berkeley: Lawrence Berkeley National Laboratory.

Ong, A. (2011). Introduction. Worlding cities, or the art of being global. In A. Roy & A. Ong (Eds.), *Worlding cities. Asian experiments and the art of being global* (pp. 1–26). Chichester: Wiley.

Peck, J. (2015). Cities beyond compare? *Regional Studies, 49*(1), 160–182.

Peck, J., & Theodore, N. (2010). Mobilizing policy: Models, methods, and mutations. *Geoforum, 41*, 169–174.

Qian, Z (钱振明). (2008). 基于可持续发展的中国城市治理体系:理论阐释与行动分析.城市发展研究 (城市治理 section)15 卷, 3 期. Urban Studies, 15(3), pp. 150–155.

Qiu, B (仇保兴). (2006). 紧凑度和多样性——我国城市可持续发展的核心理念. 城市规划 30.11: 18–24.

Ran, R. (2013). Perverse incentive structure and policy implementation gap in China's local environmental politics. *Journal of Environmental Policy & Planning, 15*(1), 17–39.

Rivers, M., & Chung, S. (2016). Future Chinese skylines could look more uniform. *CNN Style*, 22 February. http://edition.cnn.com/2016/02/22/architecture/china-weird-buildings/. Accessed 26 June 2016.

Robinson, J. (2006). *Ordinary cities*. London: Routledge.

Schmid, C. (2014). Patterns and pathways of global urbanization: Towards comparative analysis. *Brenner and Schmid, 2014*, 203–217.

Teets, J. (2014). *Civil society under authoritarianism: The China model*. New York: Cambridge University Press.

United Nations. (1987). *Report of the world commission on environment and development*. General Assembly Resolution 42/187, 11 December.

United Nations. (2015). *Transforming our world: The 2030 agenda for sustainable development*. General Assembly Resolution, 25 September http://www.un.org/ga/search/view_doc.asp?symbol=A/RES/70/1&Lang=E. Accessed 11 Aug 2016.

Witt, M. A., & Redding, G. (2012). China: Authoritarian capitalism. In M. A. Witt & G. Redding (Eds.), *The Oxford handbook of Asian business systems* (pp. 11–32). Oxford: Oxford University Press.

World Commission on Environment and Development. (1987). *Our common future*. Oxford: Oxford University Press.

Wu, F. (2016). China's emergent city-region governance: A new form of state spatial selectivity through state-orchestrated rescaling. *International Journal of Urban and Regional Research, 40*(6), 1134–1151.

Wu, W., & Gaubatz, P. (2013). *The Chinese City*. London/New York: Routledge.

Xi, J. (2017). Xi Jinping report to 19th Congress (full text) (习近平十九大报告(全文)). *Xinhua News* (新华网), 18 October, http://finance.sina.com.cn/money/bank/bank_hydt/2017-10-18/doc-ifymviyp2268296.shtml. Accessed 18 Oct 2017.

Xinhua News Agency. (2013). The decision on major issues concerning comprehensively deepening reforms(adopted at the close of the third plenary session of the 18th CPC central committee on November 12, 2013 (中共中央关于全面深化改革若干重大问题的决定:2013年11月12日中国共产党第十八届中央委员会第三次全体会议通过),http://cpc.people.com.cn/n/2013/1115/c64094-23559163.html. Accessed 2 July 2016.

Yan, Y. (2010). The Chinese path to individualization. *The British Journal of Sociology, 61*(3), 489–512.

Yao, X. Y. (2011). Accelerating rural urbanization: Risk is higher than the opportunities. *Exploration and Free Views, 2*, 24–26.

Yu, A. T. W., Yuzhe, W., Zheng, B., Xiaoling, Z., & Liyin, S. (2014). Identifying risk factors of urban-rural conflict in urbanization: A case of China. *Habitat International, 44*, 177–185.

Zhonggong Zhongyang. (2015). 中共中央国务院印发《生态文明体制改革总体方案》，人民日报, 22 September. http://politics.people.com.cn/n/2015/0922/c1001-27616151.html. Accessed 5 Oct 2015.

Zhonggong Zhongyang. (2016). 中共中央 国务院关于进一步加强城市规划建设管理工作的若干意见. 21 February. http://politics.people.com.cn/n1/2016/0221/c1001-28137648.html?utm_source=The+Sinocism+China+Newsletter&utm_campaign=1be1ce814b-Sinocism02_21_162_21_2016&utm_medium=email&utm_term=0_171f237867-1be1ce814b-29620869&mc_cid=1be1ce814b&mc_eid=bf60dc29d9. Accessed 24 Feb 2016.

Zhou, Z. (2014). *A study on globalizing cities. Theoretical frameworks and China's modes*. Hackensack: World Century Publishing Corporation.

Xi, J. (2017). Xi Jinping: report to 19th Congress (full text) [illegible]. *Xinhua Wang* [illegible], 18 October. http://finance.sina.com.cn/money/bank/[illegible]-10-18-doc-ifymvriy2265390.shtml. Accessed 18 Oct 2017.
Xinhua News Agency. (2013). The decision on major issues concerning comprehensively deepening reforms (adopted at the close of the third plenary session of the 18th CPC central committee on November 12, 2013) [illegible] (2013年11月12日中国共产党第十八届中央委员会第三次全体会议通过) [illegible]. http://news.xinhuanet.com/politics/2013-11/15/c_118164235.htm. [illegible] cn/n/2013/1115/c64094-23559163.html. Accessed 23 July 2016.
Yan, Y. (2010). The Chinese path to individualization. *The British Journal of Sociology, 61*(3), 489–512.
Yao, X. Y. (2013). Accelerating rural urbanization: Risk is higher than the gain. *Contemporary Economics and Rural View, 2*, 24–26.
Yu, A. T. W., Yuzhe, W., Zhibo, B., Xiaoling, Z., & Ligin, S. (2014). Identifying key factors of urban-rural conflict in urbanization: A case of China. *Habitat International, 44*, 177–185.
Zhongguo Zhengyang. (2015). [illegible], 22 September. http://politics.people.com.cn/n/2015/0922/c1001-27616531.html. Accessed 5 Oct 2015.
Zhongguo Zhengyang. (2016). [illegible], 21 February. http://politics.people.com.cn/n1/2016/0221/c1001-28137165.html?utm_source=The+Sinocism+China+Newsletter&utm_campaign=[illegible]-Sinocism02_21_16&utm_medium=email&utm_term=[illegible]. Accessed 29 Feb 2016.
Zhou, Z. (2014). *A study on eco-cities: Theory of human settlements*. Hackensack: World Century Publishing Corporation.

Part I
Green Urban Governance – A Theoretical Perspective

Chapter 1
Wider Theoretical Debates on Urban Sustainability Governance

Dorthe Hedensted Lund

Abstract Cities produce many of the environmental challenges that societies all over the world face. They are great consumers of energy, of food, of water, and of land. They are also great producers of waste, including air pollutants, CO_2 and other greenhouse gases, and the like. As such they have a great impact on ecosystems and on the conditions for human and non-human life. But they are also crucial for economic growth, for innovation, and for employment in national economies. It is therefore in cities that governance for sustainability may find its most fertile soil and if successful will have its greatest effect. But what type of governance promotes sustainable urban development? This chapter addresses some of the elements in an urban transition towards sustainability first, after which three governance paradigms are considered – traditional public administration, new public management, and new public governance/network governance – and their advantages and disadvantages in terms of sustainability, drawing mainly on Nordic examples and the Western literature on governance. The point is to provide theoretical as well as practical inspiration for Chinese cities. The chapter concludes that the prospects for sustainable urban governance will depend on the ability of cities to choose appropriate governance paradigms and associated tools for the tasks involved.

Keywords Urban sustainability · City systems · Traditional public administration · New public management · New public governance

1.1 Introduction

Since the ancient civilizations of Mesopotamia, Egypt, Greece, and Rome, human beings have been concerned with the environmental impact of their use of raw materials. The term "sustainability", however, was not found in the *Oxford English*

D. H. Lund (✉)
Department of Geosciences and Natural Resource Management, Section for Landscape Architecture and Planning, University of Copenhagen, Copenhagen, Denmark
e-mail: dhlund@gmail.com

J. Delman et al. (eds.), *Greening China's Urban Governance*, ARI - Springer Asia Series 7, https://doi.org/10.1007/978-981-13-0740-9_1

Dictionary until the twentieth century (Du Pisani 2006). But the rapid environmental deterioration taking place in China due to urbanization and a strong focus on economic development (see Introduction and Brehm and Svensson Chap. 2) has made the idea of sustainability very salient in governance discussions today.

What is sustainable development, then? One of the most widely used definitions is given in the Brundtland Report to the effect that "sustainable development is development that meets the needs of the present without compromising the ability of future generations to meet their own needs" (WCDE 1987, 43). The Brundtland Report was followed by the Earth Summit in Rio in 1992. One of the outcomes of the Rio Summit was Agenda 21, an action plan to promote sustainability. It dealt with environmental protection and antipoverty issues, as well as with the role of women, of children, and of various societal organizations (trade unions, businesses, etc.). According to the Rio definition, sustainability is a matter of ensuring *environmental protection*, *social justice*, and *economic stability* – all at the same time. By definition, then, sustainability is cross-sectoral and relational. Hence, sustainable development is as much about changing politico-administrative systems as it is about changing material flows (e.g. so as to reduce carbon-based energy consumption) or encouraging pro-environmental behaviour (Astleithner and Hamedinger 2003). But the concept is also fluid and ambiguous. It essentially amounts to a compromise between two agendas, namely, conservation on the one hand and development on the other hand. As such, sustainable development is a contradiction in terms. It calls both for conserving natural resources and for exploiting them (Du Pisani 2006). Nor are there any fixed criteria for how sustainability can be achieved in practice. Many different interpretations of the concept are thus in play (Jabareen 2006).

Some authors argue that the Brundtland definition of sustainability is too weak. They call instead for "strong sustainability" (cf. Liu et al., Chap. 5), or for "green justice" (Wang et al., Chap. 11). Both concepts put environmental protection at the forefront, since they see ecological integrity and the preservation of non-substitutable natural capital as the foundation for both economic and social sustainability (Ekins et al. 2003; Hediger 2008; Neumayer 2003). In Chap. 11, Wang et al. elaborate on this discussion in relation to Chinese conditions, as seen from an ethical green justice perspective.

While the concept of sustainability is contested, there is widespread agreement that cities and urban governance are crucial for achieving sustainable development. This is because a majority of the world's growing population lives in cities, and urbanization is continuing apace. This means that cities produce many of the environmental and social challenges the world faces. Cities are great consumers of energy, food, materials, and water. They are also great producers of waste and pollution, including greenhouse gases (Grimm et al. 2008; Keivani 2010). Furthermore, as urbanization rates increase, many cities in developing countries find themselves unable to cater for the needs of their growing population in terms of housing, education, and sanitation services. Environmental and social impacts are thus aggravated, as is vulnerability to environmental degradation – especially among the urban poor (Bicknell et al. 2009). Still, cities are also the national and global

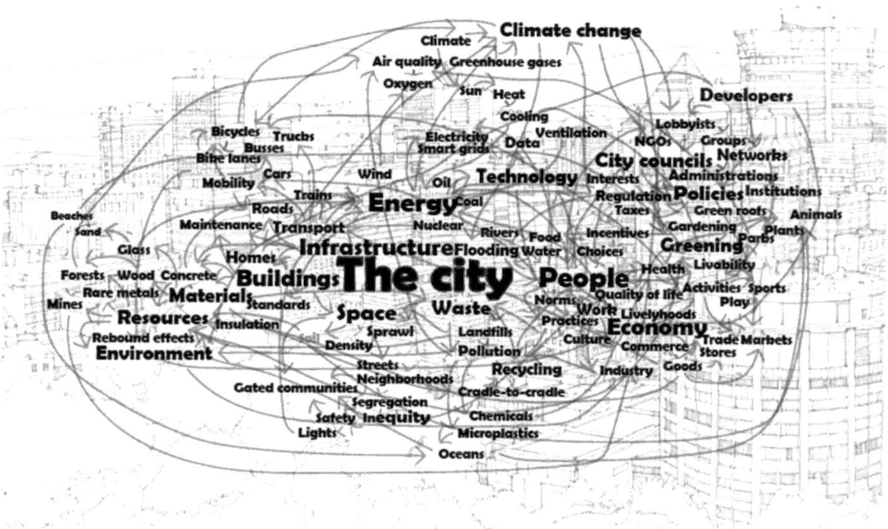

Fig. 1.1 The city as a complex system. In order to facilitate sustainable development, urban governance must be able to understand and address system linkages (Background drawing from: http://cdn26.us1.fansshare.com/photo/linedrawing/line-drawing-montreal-2060131755.jpg)

engines of growth and development, accounting for up to 85% of GNP in high-income countries and 55% in low-income ones. Their concentration of people and activities makes possible a better use of resources and a more efficient delivery of services than can be found in rural areas. Whether or not this potential is realized depends very much on how a given urban area is governed (Keivani 2010).

Urban governance for sustainability is a complex matter. The city is a complex system of material flows, human activities, and governance structures that link the city with its surroundings (Fig. 1.1). Conditions for effective governance are also changing, becoming more complex as greater numbers of actors gain influence (Meadowcroft 2010). This is also the case in China as discussed extensively in the following chapters.

In this chapter, I discuss the problem of governance for sustainable development, drawing mainly on Nordic examples and the Western literature on governance where the topic has a longer tradition than in China (cf. Introduction). Where relevant, I also cite chapters in this volume. The Nordic nations are known as frontrunners with respect to both environmental governance and social equity. It may thus be claimed they have moved further along the road to sustainable development than have most other countries. Much can be learned from the experience of the Nordic region when it comes to applying different governance technologies to promote sustainability; and such lessons may be useful in the Chinese context. Of course, the Nordic examples are embedded in a Western governance tradition and in more specific governance paradigms. Hence I do not argue that policies and governance models can be transferred directly to a Chinese context. They will and must inevitably "mutate" to fit the Chinese context (Peck and Theodore 2010). When governance technologies are transferred to another setting, it is important, however, that their normative foundation be understood and that they are adapted to the new context in which they are to function. The other chapters in this volume develop such a perspective in various ways. As explained in the Introduction in this book, some Western ideas on governance have been tried in China in particular related to new public management, and also aspects of new public governance are in hybrid forms becoming increasingly relevant as the following chapters show. This chapter will thus explain the international origins and normative preconditions for these governance paradigms to provide a framework which by to understand the dynamics and hybridization in China's emerging urban green governance.

I first address the city as an object of governance. How should we understand the city, if we are to promote sustainable development? The starting point is to understand the city as a complex, interrelated system and to consider some implications of this for efforts to promote sustainability through urban governance. Second, I review three different paradigms of governance which are prevalent in the Nordic region. My aim in so doing is to highlight the advantages and disadvantages of each model for sustainable development and to stimulate further discussion about their applicability in the Chinese context.

1.2 The City as a Complex and Interrelated System

Traditionally, cities have been defined according to spatial and social characteristics such as land use, concentration of economic activity, and density of inhabitants, buildings, and the like. Different authors have focussed on specific aspects of the urban fabric, but have failed to capture the complex urban processes that also define cities. These processes include flows of goods, materials, services, and people, as well as the relations among these different elements (da Silva et al. 2012). Over the latest decade or so, cities have also increasingly been conceptualized as forming (part of) dynamic socioecological systems (Gunderson and Holling 2002) or sociotechnical systems (da Silva et al. 2012; Grin et al. 2010). In this view, the dynamics of an urban system cannot be understood by looking at its individual elements in isolation. We must consider the system as a whole. One critique of this approach concerns the delimitation of the system. It is hard to analyse all possible relations in an urban system (including its many links beyond the spatial limits of the city itself), or to comprehend all interactions among such disparate elements as infrastructure, ecosystems, and human activities (da Silva et al. 2012). It may not be necessary, however, to be able to delimit the system before changes can be made, or to understand it in its entirety.

An advantage of the system approach is that changes towards sustainable development can take their starting point at many different points in the system. Take the goal of reducing carbon-based energy production. Progress can be made in this area by increasing the proportion of renewable energy, which may be seen as a sector-oriented solution. It can also be accomplished by instituting stricter standards for insulation in buildings; by developing and installing energy-saving technologies in home appliances, street lamps, etc.; by affecting consumption patterns through taxes and/or campaigns; by increasing plant cover to increase storage of carbon; and so on. If we remember all the relations that exist between energy production, consumption, and other elements in the city, an array of possible initiatives present themselves. Linkages within the system may also contribute to ripple effects, whereby the initial interventions or small niche inventions lead to system transitions due to relations within the system. There is also the risk that initiatives taken in other parts of the system will have unintended negative consequences.

A disadvantage of the system approach, from the standpoint of governance, is that attaining overall system sustainability is an immensely complex task, requiring comprehensive structural transitions. Furthermore, many of the persistent ecological and social problems we see in cities persist because they represent unintended side effects of co-evolutionary modernization processes that have fostered the structures we now have in terms of decision-making, the distribution of power, and the shape of institutions. Such structures are not, consequently, geared towards new developmental trajectories, and they inject inertia and resistance towards transitions into the system (Grin et al. 2010). A good example of this may be seen in Chap. 3, in which He et al. address the highly bureaucratic and sectoralized institutional structures in the planning system in China, and explains how the historical trajectory behind the different sector plans poses a challenge to achieving better integration.

The different chapters in this volume all take up different parts of specific Chinese urban systems and give examples of both actual and potential sites of intervention in Chinese governance to support transitions towards sustainability. Some of the chapters address institutional changes and their intended and unintended effects. In Chap. 4, for example, Luova looks at variations in policy implementation in different districts and considers the reasons for these variations. Delman (Chap. 8) and Schlæger and Zhou (Chap. 7) discuss the potentials and effects of new schemes for monitoring and assessing performance. Some chapters analyse efforts to induce pro-environmental behaviour among citizens through regulation. In Chap. 6, for instance, Ren and Liu analyse differences in carbon emissions along lines of social class, and they point to the importance of social norms and materialistic lifestyles in consumption patterns. Li et al. (Chap. 5), whose focus is on water management, ask what we can learn about regulating water consumption from the very different water-management approaches taken in Copenhagen and Beijing. Other chapters focus on the changing dynamics between citizens and the state resulting from engagement or contention (Delman, Chap. 8; J. Liu, Chap. 9; Burrell and Almén, Chap. 10; Zhang, Chap. 12). All of the chapters point to developments in China which may prove helpful in terms of moving towards greater sustainability in Chinese cities, but they also call attention to some of the barriers to such progress.

What is important to remember in a system perspective is that almost all changes in one subsystem will have unforeseen side effects – positive and/or negative – on other subsystems and thus on the urban system as such. This means that reflexiveness and learning should be an integral part of system governance, as uncertainty will always be part of any movement towards change. Any attempt at steering must accept "that the future is largely unknown and unknowable. … But even in the face of this radical uncertainty and indeterminacy human beings can try: 1) to orient society towards the attainment of desirable objectives and the avoidance of dangerous pitfalls; 2) to take action to protect groups that are especially vulnerable to the unfolding patterns of change; 3) and to re-order social institutions so that they are better placed to cope with whatever the future does bring" (Meadowcroft 2010, 302). For Grin et al., reflexiveness is the tool that can inform agents "how particular patterns are rooted in past structures, which reflect earlier orientations of societal development; and how other patterns, oriented to sustainable development, may contribute to and benefit from structural change" (Grin et al. 2010, 234). Where the reorientation of structures or social institutions is concerned, governance paradigms are right at the heart of the matter – both as a part of the structures and as the means to reorient them.

1.3 Governance Paradigms and Their Underlying Principles, Mechanisms, and Tools

If urban development is to be guided towards sustainability, steering and coordination will be essential. Due to the virtually all-encompassing nature of the efforts required, all of the different agents who influence the city will have to take part. As Meadowcroft argues: "In a fundamental sense, governance for sustainable development implies a process of 'societal self-steering': society as a whole is to be involved in the critical interrogation of existing practices, and to take up the conscious effort to bring about change." (Meadowcroft 2010, 302). This is not likely to happen, however, without some form of public governance and coordination.

What exactly do we mean by governance? In political science, "governance" can be understood in either one of two ways. It may denominate new forms of steering, encompassing public-private hybrids, greater pluri-centrism in the distribution of power, and the slipping of authority out of the hands of the nation-state (Rhodes 1997; Kjær 2004). Or it may denote in a more general sense all the ways in which societal coordination is achieved by public and private actors, including traditional bureaucratic methods as well as forms of coordination based on markets or networks (Pierre 2000; Kooiman 2003; Meadowcroft 2010). I employ the latter and more general definition in this chapter. When governance in the former and more specific sense is in question, I speak of *network governance* (NG) (Sørensen and Torfing 2007; Klijn and Koppenjan 2016) or *new public governance* (NPG) (Osborne 2010).

Osborne (2010) distinguishes three or more paradigms of public governance in the Western world, as do many other authors (Kjær 2004; Klijn and Koppenjan 2016; Koliba et al. 2010; Lerborg 2011). By a paradigm is meant the coherent patterns of thought – norms, basic assumptions, principles of governance, etc. –

which relate to practices of governing and which determine what is seen as legitimate decision-making, effective tools of governance, etc. (Lerborg 2011). Osborne terms the three main paradigms *traditional public administration*, *new public management*, and *new public governance.* These paradigms exist at one and the same time, as layered realities in the Western public sector, where each can be seen as a reaction to problems that plagued the preceding one (Hartley 2010). While the paradigms bear a relationship to existing practices, they should be understood first and foremost as analytical categories. In what follows, I present the three paradigms and their normative and institutional heritage. The aim is to provide an analytical framework by which to better understand the governance dynamics in China. In particular to understand the instances where parts of the paradigms have been transferred to the Chinese context to understand the ways in which the paradigms have been translated and altered during the transfer process (Peck and Theodore 2010).

1.4 Traditional Public Administration

The paradigm of traditional public administration (TPA) has its theoretical roots in political science and in the work of the German sociologist and economist Max Weber (1864–1920). Weber's account of bureaucracy puts professional expertise, objectivity, the rule of law, hierarchical organization, and clear lines of accountability at the centre. As such it builds, as an underlying basic assumption, on a modernist faith in science and rationality as the means to solve societal problems (Jensen et al. 2006). Bureaucratic organization can be seen, moreover, as a reaction to the unpredictable and arbitrary rule of kings, who claimed to derive their authority from God (Lerborg 2011). With the rise of modernity, science became the new source of authority for public administration, while liberalism made the aggregated will of the people the basis for political power. This paradigm has a long history, and it is thoroughly institutionalized today in Western systems of public administration.

The separation of politics from administration is immensely important in the TPA paradigm. The subjective realm, which is governed on the basis of power, is thus separated from the objective and professional realm, which follows standardized rules and procedures. The bureaucratic organization of public administration is considered to be effective, because specialization and sectoralized organization mean that educated specialists in the administration deal with the different issues. The administrators follow a public ethos of legality, impartiality, objectivity, and professionalism. Specific and personal interests have no legitimate place; serving the public good rather than partial interests is the goal of governance (Considine and Lewis 2003). The bureaucratic administration is seen as a neutral actor implementing laws on which parliament has decided (Kjær 2004).

In terms of normative content, equality and stability are central values under TPA. Everyone is equal under the law, with equal opportunities to influence law-making through elections. All are treated on the basis of objective rules and procedures. The associated understanding of democracy is that embodied in the procedural, liberal, representative model, in which equality is ensured through the principle of "one man, one vote" (Dahl 2006). Power is visible, and it is held accountable through the parliamentary chain of governance and civil rights (Considine and Lewis 2003; Considine 2002; Kjær 2004). The main tools of governance under TPA are rules, regulations, and the exercise of top-down power through command and control (Kjær 2004). Complex issues are dealt with through deconstruction – processed in specialized units with input from policy analysts, professionals, and scientific specialists (Klijn and Koppenjan 2016).

TPA has been criticized for being slow, inflexible, and ill-suited to solving complex and so-called "wicked" problems that call for innovation, cross-sectoral collaboration, and transdisciplinarity (Osborne 2010; Hartley et al. 2013). It has been criticized for not complying with its own ethos (Kjær 2004); for being machinist and out of sync with people's problems and needs (Lerborg 2011); for growing too big and encroaching on too many aspects of people's lives (Kjær 2004; Klijn and Koppenjan 2016); and for reducing citizens to voters and clients, i.e. passive subjects of governance rather than active co-producers of public value (Agger and Lund 2017; Aberbach and Christensen 2005). New ways of governing, making policy, and delivering services have therefore emerged.

1.5 New Public Management

New public management (NPM) grew out of the critique of bureaucracy. It spread from the 1980s onward, after a dramatic rise in public expenditures in the 1970s and associated problems of "big government" (Klijn and Koppenjan 2016; Lerborg 2011). It seeks to reduce the sphere of the state – i.e. of public bureaucracies – through deregulation, marketization, and the privatization of service delivery. It also attempts to improve state efficiency, by introducing decentralization, instituting performance management, establishing input and output control, and applying private-sector managerial principles in public organizations (Osborne 2010; Kjær 2004).

One aspect of NPM involves limiting the state to "steering", rather than "rowing". The role of governments is to set goals and to formulate policies. Implementation, on the other hand, is to be moved from public bureaucracies to separate public agencies or private organizations, which are to be held accountable through contracts, benchmarks, and performance measures. As such, the separation between politics and administration is even more pronounced under NPM than under TPA. Competition between service providers ensures effectiveness and efficiency. Due to the use of the market mechanism, only the best and most efficient providers hold their own (Hood 1991; Osborne and Gaebler 1993). To some extent, moreover, the market takes over the distribution of societal goods from the state.

The marketization of policy-making and service provision also affects the role of citizens. Under NPM, citizens are users and consumers in addition to being voters and clients, indicating a much more individualistic approach than is the case under TPA (Clarke 2006; Lucio 2009). NPM reforms have broadened citizens' opportunities for voice, as in the case of user boards in public institutions and local councils (Agger et al. 2010; Harris and Thomas 2011). User choice has been introduced in a number of public services, and new procedures for measuring user satisfaction and evaluating the quality of public services have been developed (Alford 2009; McLaughlin 2009).

In normative terms NPM reflects neo-liberalism and gives negative liberty (Berlin 2006) precedence over equality. Its basic assumption is that, if both public and private agents are freed from suffocating rules and regulations, they will be able to focus on producing positive outputs rather than on following rules and procedures – leading to higher quality and lower public spending (Lerborg 2011; Kjær 2004). In this understanding, there is no "public ethos" that suppresses private interests and furthers the public good thereby. Instead, private interests and power are checked through competition and public choice. Indeed, private interests are seen as desirable, because they create market demand and furnish information about public preferences to service providers. They are thus necessary for the market to function (Vaaben and Lund 2013).

NPM has been criticized for requiring excessive documentation. This has led, in the view of critics, to a lower quality of services due to poor contracts, to less time for public officials to provide services, and to an excessive focus on price as compared with quality. The last-mentioned failing results partly from the difficulty of measuring quality in a quantitative way. Markets also sometimes function badly in this area, because the "free" public choice – e.g. in relation to schools and hospitals – may not be that free after all because of externalities. Furthermore, the actions of elected representatives are anchored less effectively among those they represent. Finally, the focus under NPM undergoes a shift, thereby undermining the public ethos: formerly, a conscious choice was made about what the best solution for society is; now, the decision is left instead to economic dynamics (Lerborg 2011; Christensen and Laegreid 2008; Sørensen and Torfing 2007).

1.6 New Public Governance/Network Governance

In the 1990s, new public governance (NPG) or network governance (NG) rose to prominence. In theoretical terms, it is partly descriptive and partly prescriptive: the former regarding new governance practices and the latter regarding how wicked societal problems should be dealt with. In both cases, it is characterized by the contribution of multiple interdependent public and private actors in policy-making and service delivery. These actors participate in different types of networks, which assemble and function through negotiated interaction processes (Osborne 2010; Klijn and Koppenjan 2016).

As a prescription, NPG represents a reaction to the inability of the bureaucratic public sector to solve wicked problems that cut across sectoral boundaries. These wicked problems arise from structural constraints in bureaucratic organizations and from insufficient resources in terms of knowledge, finances, and governance tools.

Many large-scale societal problems – climate change, infertility, increased obesity, etc. – cannot be solved by means of regulation alone. Nor can pervasive poverty, increasing inequality, or environmental degradation, because the roots and causes of these problems are manifold and interrelated. For the same reason, market mechanisms also fall short. The answer seems to be to address these kinds of problems through various networks. Some of these networks may be collaborative policy networks which come together around a specific political cause; others may be publicly designated to address specific wicked issues, with the aid of a diversity of actors from both public and private organizations with different but complementary resources (Hajer 1995; Klijn and Koppenjan 2016). The idea is that bringing together people with different types of knowledge and different perspectives on the problem at hand can stimulate innovation and help to bring about better solutions (Sørensen and Torfing 2011; Blomqvist and Levy 2006).

In normative terms, this approach to public governance differs significantly from both TPA and NPM. First and foremost, the strict separation between politics and administration is abandoned. Both politics and administration may take place in the networks, which means that political power is being exercised by non-elected actors. Governance networks may operate in the "shadow of hierarchy" as Fritz Scharpf (1994) termed a persistent controlling power of traditional governments over the actions of networks. Nonetheless NPG constitutes a significant departure from the parliamentary chain of governance and the accountability and legitimacy this entails. Hence, ensuring democratic legitimacy and accountability in governance networks requires that said networks either be formally linked to the parliamentary chain and subjected to a more output-oriented logic (Sørensen and Torfing 2005) or that they be placed within a more deliberative model of democracy, in which case including affected parties within the network becomes imperative (Dryzek 2000). Also, accountability needs to be ensured by other means than through hierarchical authority and contractual relations, i.e. by means that guarantee transparency and which empower affected parties to impose sanctions on network decision-makers (Bovens 2007; Esmark 2007). Due to the fairly recent spread of network governance, there are no sedimented structures for ensuring accountability and legitimacy, such as those found under TPA. The role of citizens also undergoes a shift – from that of clients and consumers to that of co-creators of public policies and services (Agger and Lund 2017).

Invariably, such governance paradigms involve a simplification of the actual practice of governing. Many mixtures are practised, in both the norms and the tools of governing. Yet models serve several purposes: most importantly, they offer conceptualizations through which we may better understand different practices. For example, they can highlight some of the difficulties which arise from opposed normative content in the different models. They also lead us to expect inertia when governance practices are changed. Finally, models offer discourses and frameworks for action that may facilitate a more reflexive approach to governance in practice, especially for achieving sustainability.

Table 1.1 presents the main differences between the three paradigms with respect to important aspects of governance. The paradigms can be seen as ideal types.

Table 1.1 Three paradigms of governance

	Traditional public administration	New public management	New public governance/ network governance
Objectives	Delivering effective and uniform policies and services, in accordance with principles of equality, legitimacy, and the rule of law	Improving effectiveness and efficiency in public services and public organizations	Solving complex problems, improving quality of policies and services, facilitating inter-organizational coordination
Effectiveness and efficiency	Bureaucracy, specialization, expertise	Competition, market mechanisms	Inclusion of necessary knowledge, innovation
Democracy	Elected parliaments, separation of politics and administration	Aggregation of individual preferences, separation of politics and administration	Deliberation and partnerships, no separation of politics and administration
Decisions based on	Creation of political majorities	Procedures that guarantee free markets	Negotiations, leading to varying degrees of consensus
Tools/ technologies of governance	Laws, rules, regulations, loyalty of employees promoted based on merit	Contracts, benchmarks, performance measures, input and output control, public-private partnerships (purchaser-provider relationships)	Network management, network participation, network design, creating shared meanings, allocating resources, creating frameworks (economic, institutional, narrative), joint fact-finding, etc.
Compliance ensured through	Threat of legal sanctions	Threat of economic ruin	Trust and mutual obligations
Power	Visible, can be placed in government	Dispersed in the market	Shared in consensus-seeking networks
The role of the state	Control and steering	Steering, providing framework conditions for functioning markets	Facilitation of network governance through application of tools of governance
The role of citizens	Client, voter	Customer, user, consumer	Co-producer, co-creator
Number of actors	One dominant	Many buyers and sellers in the market	Some relevant, affected actors collaborate
Relation between actors	Hierarchy: authorities and subordinates	Principal-agent, independent, distant, impersonal relationships	Mutual dependence, close and trusting relationships
Central norms	Equality	Negative liberty	Community, trust
Theoretical roots	Political science and public administration and policy	Rational-choice theory and management studies	Institutional and network theory

(continued)

Table 1.1 (continued)

	Traditional public administration	New public management	New public governance/ network governance
Main problems	Slow and cumbersome administration, poor ability to deal with cross-sectoral problems	Issues in relation to non-marketed goods and services, excessive demands for documentation, loss of public ethos	Issues of democratic legitimacy and accountability

Sources: Kjær (2004); Osborne (2010); Klijn and Koppenjan (2016); Koliba et al. (2010)

The next question to be addressed is: what are the strengths and weaknesses of the different paradigms for promoting sustainable development in cities?

1.7 Advantages and Disadvantages of the Different Paradigms for Promoting Sustainability

If we take as a starting point that urban sustainability is about creating and supporting *environmental protection*, *social justice*, and *economic stability*, how can this be achieved? At first glance, it would seem that network governance/new public governance is the appropriate approach. We are dealing, after all, with complex problems that cut across sectors and involve non-marketed activities. Addressing them requires that citizens, public agencies, and private businesses and organizations play their part. This, however, is not the whole truth. No single approach suffices by itself. Instead, we must strike an appropriate balance between the different approaches and tools.

Cities must be understood as complex adaptive systems. Thus there are, as mentioned, many different approaches for achieving greater sustainability, including both straightforward initiatives and more complex ones. Take environmental protection. It can be accomplished by regulatory, market-based, or collaborative means; and it is conceivable that all three approaches will be needed if we are to achieve substantive change in how we live, work, produce, and consume. The challenge is to find the mixture which is appropriate for a given context and to avoid compromising either social justice or economic development in the process.

In the Nordic countries, traditional public administration has been successful in reducing industrial and agricultural pollution by means of regulation. For example, laws in the Nordic region safeguard forest cover, thereby protecting not just wood production but also groundwater, habitats, and biodiversity. TPA can also take credit for the famous bicycle infrastructure in Copenhagen, which has enabled a large part of the city's population to get by in daily life without a car. TPA can be extremely effective on issues where general laws and regulations are both politically and practically feasible. There is certainly no reason why the opportunities it opens up should not be exploited.

TPA also has some severe limitations. One is that it tends to be most effective on the level at which administrative structures for implementation and enforcement are present. In the case of Denmark, that would be the national or the municipal level. On global issues such as climate change, relying on TPA is less viable. For example, reducing the level of greenhouse gases in Denmark – where national regulations can be enforced – will be of very little help in solving the problem of climate change. It may instead lead to lower economic growth in Denmark, due to the competitive disadvantage in attracting investment that such regulations may produce. We do not have, of course, a global parliament or a world judicial system. Accordingly, international rules and regulations can only be used to solve global issues such as climate change *after* international networks have come to an agreement on regulations to be implemented – in all major greenhouse-gas-producing countries.

Another limitation is that TPA relies on sector-based initiatives that enhance the sustainability of the particular sector in question, but not necessarily that of the system as such. Such initiatives may also have unintended effects on other sectors. In Chap. 3, He et al. show how sectoral boundaries and power games hinder the integration of spatial plans in China, reducing the effectiveness of efforts to reach environmental goals.

NPM-based incentives can also be effective. They have been used successfully in the Nordic countries to reduce the negative environmental effects of motorized transportation. Taxes on fuel, water, and the discharge of waste water have resulted in lower consumption. Economic incentives to reduce energy consumption and increase energy efficiency have been introduced in connection with appliances. NPM may also involve the use of performance indicators to influence the behaviour of utility providers, as well as the marketization of environmental goods through economic schemes such as payments for ecosystem services (PES), whereby landowners are offered incentives to provide certain environmental benefits (such as conserving important habitats or maintaining forest cover on slopes to reduce the risk of landslides). The latter approach is now high on the international agenda, and it is being implemented through various programmes in many countries (Wunder 2005; FAO 2011). Such schemes are also conceivable in cities, where developers could be paid to provide ecosystem services. These schemes can only be implemented in cases where the provision of ecosystem services can be evaluated. The Chinese model city scheme presented in Chap. 2 by Brehm and Svensson is also an example of how benchmarking can be used as a driver for improving environmental performance. The analysis shows, however, that when taking the economic capacity of model cities into account, key environmental outcomes are not impressive. Hence revisions are necessary to improve environmental performance. In general, market-based incentives may be strong drivers of technological innovation which reduces the environmental impact of marketed goods, and economic incentives such as taxes and subsidies may be effective at encouraging consumption of a type friendlier to the environment. Copenhagen's greater success at reducing water consumption, as compared with Beijing, may perhaps be attributed to market-based regulation (cf. Li et al., Chap. 5). Performance-based assessments of public officials may also, to a certain degree, create effective incentives for public administration of an environmentally friendly kind, as Delman argues in Chap. 8. Here too, there is no reason why such tools should not be used where they are effective and appropriate.

NPM also has limitations. Market-based initiatives work best in areas where mature markets operate. Attempts to create new markets, such as for carbon emissions, can be difficult; and as the European Union Emissions Trading Scheme has shown, they can lead to perverse effects if they are not designed properly. In the EU case, the emissions trading scheme has been undermined by financial crisis, leading to a surplus of emission allowances and thus extremely low prices – and therefore no real incentives to reduce emissions (European Commission 2015). In Chap. 6, Ren and Liu moreover show that, in the case of middle- and higher-income households, social norms may be more important than prices for reducing carbon emissions. Furthermore, economic incentives only induce certain behaviours when such behaviours are viable options. Taxes on fuel, for example, only result in reduced car use when alternative modes of transportation are available. Like under TPA, moreover, initiatives taken under NPM tend to be sector-specific, which may lead to unintended effects in other sectors. Finally, initiatives based on competition run a high risk of creating winners and losers – thereby exacerbating economic and social inequality, undermining social justice, and resulting in reduced overall sustainability.

As for network governance, its main potential lies in its ability to transcend sector-based policies and to find more integrative solutions. This is particularly useful in areas where beneficial synergy effects are present. Cities may attempt, for example, to combine climate-change adaptation with urban recreation and urban greening (Wong and Brown 2009). Municipalities may enter into partnerships with utility companies and housing organizations to manage urban storm water in such a way as to provide recreational facilities and urban green areas (an option which is only possible through networks). Urban farming cooperatives may manage to increase food security, in the process both enhancing social capital and promoting urban biodiversity (Barthel and Isendahl 2013; Sharp and Smith 2003). As such, network governance is more likely to offer solutions through urban relations (see Fig. 1.1) than the other approaches are, and such solutions are consequently more likely to have a systemic impact. Integrative solutions can lead to necessary institutional changes, because the networks involved go beyond existing institutions and highlight barriers within them. Network governance is also more likely to spur civic engagement because citizens are given a much more active role as co-creators rather than passive clients. Civic engagement is in its own right an important aspect of sustainable development. Chapters 7, 8, 9, 10, 11, and 12 in this volume clearly show that civic engagement is becoming increasingly important in China, in the form of co-production of data collection or through Internet-based media, protests, and performance assessments of public services and their leaders, and increasingly puts pressure on public officials at local levels to govern in a fashion friendlier to the environment. Judging from the relevant chapters in this volume, the rise of civic engagement and contention in civil society against environmental degradation and reduced urban liveability may be one of the most important developments towards increased urban sustainability in China. This development has been made possible through the combination of changing the regulatory framework, new possibilities for citizens to partake in performance assessments (Delman, Chap. 8), and technological development which has reduced the risks of repercussions when engaging in civic activism and protests (e.g. Liu, Chap. 9).

Finally, network governance also has its challenges. One is to ensure accountability and legitimate decision-making, as already mentioned. Moreover, since network governance is the most recent, least institutionalized, and (normatively speaking) most divergent paradigm, it runs athwart existing structures – structures which embody TPA and NPM norms. For example, when Danish municipalities enter into network collaborations with utility companies, they quickly find themselves in situations where they are limited by regulations resulting from NPM reforms (more precisely, from efficiency benchmarks which the utility companies must follow). No collaborative project is allowed to be more expensive in establishment and operation than a "traditional project", and no added value from multifunctional approaches (such as recreational value, biodiversity conservation, etc.) counts in this calculation. The reason for this is that, according to NPM rationales, the sole function of the utility companies in question is to remove waste water from the city as efficiently as possible (Andersen 2013). Other restrictions may arise from norms derived from TPA. For example, when municipalities enter into innovation partnerships with companies in order to address societal problems through knowledge-sharing and collaboration, they may find themselves challenged by TPA norms to the effect that the public sector must keep its distance from the private sector and may not favour any one company over others. No guarantees can be given that a given company's products or services will be bought by the municipality. Such guarantees would be considered wrong from a public-ethos perspective, and they might give rise to a public scandal. This means that one of the most important incentives for companies to participate is removed. NPM too can obstruct trust-building and knowledge-sharing between companies and public organizations. It is extremely difficult to balance the distant, impersonal relationship of NPM with close and trusting relationship required of NPG (Vaaben and Lund 2013). As Grin et al. argue, challenges of this kind allow us to see governance not as a mode of control but as a mode of reflexivity, "in which established practices and structures – usually taken for granted – are being considered as open to change" (Grin et al. 2010, 231). As these examples suggest, strict regulatory or contract-based regimes may need to be loosened, if network governance is to be effective. This in turn requires that forms of accountability be developed that can substitute for contracts and legal rules. Perhaps assessments of the civic performance of networks will do the trick here.

1.8 Choosing the Appropriate Governance Hybrid for the Task at Hand

The starting point for this chapter has been that urban sustainability is a highly complex matter. Not only is the city a complex system of material flows, human activities, and governance structures that link the city with its immediate surroundings and with global networks. It is also the case that the conditions for effective urban governance are becoming more complex, as greater numbers of actors gain influence through both global and highly local initiatives. If we are to move towards sustainable urban development, steering and coordination will be essential. Due to the virtually all-encompassing nature of the efforts required, moreover, all of the different agents who influence the city will have to take part.

Governance through steering and coordination can take many shapes. It can be attempted by means of various tools, each with its own strengths and weaknesses. In this chapter, I have presented the main characteristics, rationales, and problems of the three paradigms of governance that prevail in the Nordic countries. Each has both potentials and pitfalls in relation to sustainable urban development. Each build on specific assumptions which must be taken into consideration before it is applied in different contexts. Reflexivity – assessing and possibly challenging existing norms and practices – is the keyword. Tools and approaches which have proven effective in one context may be useless in another. Regulation is effective when general laws and regulations are both politically and practically feasible. NPM is effective where relations of cause and effect are well understood, so that effective performance measures and incentive schemes can be designed. For example, regulating consumption by economic means alone requires that consumption be price sensitive, as well as independent of other social and material drivers. NPG may be effective at generating innovative, collaborative, and synergistic solutions to perceived problems; it requires that actors with the necessary complementary resources perceive a common problem, that they need each other in order to solve it, and that their institutional environment does not hinder collaboration. No single model is optimal in all situations.

The challenge for urban decision-makers and stakeholders is to give deep consideration to the urban system of which they are a part. What changes are needed? Which parts of the system can be altered in order to achieve such changes? Which instruments are fit for purpose? The task is not to choose *one* of the approaches reviewed: TPA or NPM or NPG. It is rather to find the appropriate tools for the task at hand, while explicitly taking into consideration the practices and norms which are fundamental to the different governance approaches. This will make it possible to create sufficient reflexivity to adapt and push existing institutions, structures, and practices towards a more sustainable future.

What follows in this book is a selection of case studies from China of approaches to promote sustainable urban development. The cases exemplify different hybrid governance arrangements at different scales and seen from different perspectives which together provide a solid foundation by which to reflect on governance dynamics and their potentials in terms of promoting systemic sustainable development. They also show that despite of the vast differences between China and the Western world in terms of governance in general, there are similar trends in terms of more polycentric governance, an extension of actors beyond institutionalized power structures who gain power in the efforts to combat environmental degradation, and also in some of the governance technologies applied to increase sustainability in particular from the NPM toolbox such as performance assessments, voluntary incentive schemes, and benchmarking.

References

Aberbach, J. D., & Christensen, T. (2005). Citizens and consumers. *Public Management Review, 7*, 225–246.

Agger, A., & Lund, D. H. (2017). Collaborative innovation in the public sector – New perspectives on the role of Citizens. *Scandinavian Journal of Public Administration, 21*(3), 17–37.

Agger, A., Löfgren, K., Torfing, J. (2010). *Nærdemokrati efter kommunalreformen: erfaringer med netværksbaseret borgerinddragelse* (Democracy after the municipal reform: Experiences with networkbased participation). Roskilde: Roskilde Universitetsforlag

Alford, J. (2009). *Engaging public sector clients. From service delivery to co-production.* Houndmills/New York: Palgrave Macmillan.

Andersen, C. (2013). *Rekreativ klimatilpasning. Samarbejde mellem kommune og spildevandsforsyninger* (Climate adaptation for recreation – collaboration between municipalities and utility companies). Master thesis, University of Copenhagen

Astleithner, F., & Hamedinger, A. (2003). Urban sustainability as a new form of governance: Obstacles and potentials in the case of Vienna 1. *Innovation: The European Journal of Social Science Research, 16*(1), 51–75.

Barthel, S., & Isendahl, C. (2013). Urban gardens, agriculture, and water management: Sources of resilience for long-term food security in cities. *Ecological Economics, 86*, 224–234.

Berlin, I. (2006). Two concepts of liberty. In R. E. Goodin & P. Pettit (Eds.), *Contemporary political philosophy* (2nd ed., pp. 369–386). Malden: Blackwell.

Bicknell, J., Dodman, D., & Satterthwaite, D. (2009). *Adapting cities to climate change. Understanding and addressing the development challenges.* Oxon/New York: Eartscan.

Blomqvist, K., & Levy, J. (2006). Collaboration capability – A focal concept in knowledge creation and collaborative innovation in networks. *International Journal of Management Concepts and Philosophy, 2*(1), 31–48.

Bovens, M. (2007). Analysing and assessing accountability: A conceptual framework. *European Law Journal, 13*(4), 447–468.

Christensen, T., & Laegreid, P. (2008). NPM and beyond – Structure, culture and demography. *International Review of Administrative Sciences, 74*, 7–23.

Clarke, J. (2006). Consumers, clients or citizens? Politics, policy and practice in the reform of social care. *European Societies, 8*(3), 423–442.

Considine, M. (2002). The end of the line? Accountable governance in the age of networks, partnerships, and joined-up services. *Governance-An International Journal of Policy and Administration, 15*(1), 21–40.

Considine, M., & Lewis, J. M. (2003). Bureaucracy, network, or enterprise? Comparing models of governance in Australia, Britain, the Netherlands, and New Zealand. *Public Administration Review, 63*, 131–140.

da Silva, J., Kernaghan, S., & Luque, A. (2012). A systems approach to meeting the challenges of urban climate change. *International Journal of Urban Sustainable Development, 4*(2), 125–145.

Dahl, R. A. (2006). Procedural Democracy. In R. E. Goodin & P. Pettit (Eds.), *Contemporary political philosophy* (2nd ed., pp. 107–125). Malden: Blackwell.

Dryzek, J. S. (2000). *Deliberative democracy and beyond. Liberals, critics, contestations.* Oxford: Oxford University Press.

Du Pisani, J. A. (2006). Sustainable development – Historical roots of the concept. *Environmental Sciences, 3*(2), 83–96.

Ekins, P., Simon, S., Deutsch, L., et al. (2003). A framework for the practical application of the concepts of critical natural capital and strong sustainability. *Ecological Economics, 44*(2–3), 165–185.

Esmark, A. (2007). Democratic accountability and network governance – Problems and potentials. In E. Sørensen & J. Torfing (Eds.), *Theories of democratic network governance* (pp. 274–296). New York: Palgrave Macmillan.

European Commission. (2015). *The EU Emissions Trading System (EU ETS) – European Commission*. http://www.cifor.org/publications/pdf_files/OccPapers/OP-42.pdf. Accessed 13 Oct 2015.
FAO. (2011) *Payments for ecosystem services and food security*. http://www.fao.org/docrep/014/i2100e/i2100e.PDF. Accessed 13 Oct 2015.
Grimm, N. B., Faeth, S. H., Golubiewski, N. E., et al. (2008). Global change and the ecology of cities. *Science, 319*, 756–760.
Grin, J., Rotmans, J., & Schot, J. (2010). *Transitions to sustainable development. New directions in the study of long term transformative change*. New York: Routledge.
Gunderson, L. H., & Holling, C. S. (2002). *Panarchy. Understanding transformations in human and natural systems*. Washington: Island Press.
Hajer, M. (1995). *The Politics of Environmental Discourse*. Oxford: Clarendon Press.
Harris, N., & Thomas, H. (2011). Clients, customers and consumers. A framework for exploring the user-experience of the planning service. *Planning Theory & Practice, 12*(2), 249–268.
Hartley, J. (2010). Innovation in governance and public services: Past and present. *Public Money and Management, 25*(1), 27–34.
Hartley, J., Sørensen, E., & Torfing, J. (2013). Collaborative innovation: A viable alternative to market competition and organizational entrepreneurship. *Public Administration Review, 73*, 821–830.
Hediger, W. (2008). Weak and strong sustainability, environmental conservation and economic growth. *Natural Resource Modeling, 19*(3), 359–394.
Hood, C. (1991). A public management for all seasons? *Public Administration, 69*, 3–19.
Jabareen, Y. (2006). A new conceptual framework for sustainable development. *Environment, Development and Sustainability, 10*(2), 179–192.
Jensen, H. S., Knudsen, O., & Stjernfelt, F. (2006). *Tankens magt. Vestens idehistorie (The power of thought. A history of ideas of the West)*. Copenhagen: Lindhardt & Ringhof.
Keivani, R. (2010). A review of the main challenges to urban sustainability. *International Journal of Urban Sustainable Development, 1*, 5–16.
Kjær, A. M. (2004). *Governance*. Cambridge: Polity Press.
Klijn, E., & Koppenjan, J. F. M. (2016). *Governance networks in the public sector*. Oxon/New York: Routledge.
Koliba, C., Meek, J. W., & Zia, A. (2010). *Governance networks in public administration and public policy*. Boca Raton: CRC Press.
Kooiman, J. (2003). *Governing as governance*. London: Sage.
Lerborg, L. (2011). *Styringsparadigmer i den offentlige sektor* (Paradigms of governance in the public sector). Copenhagen: Jurist- og Økonomforbundets forlag.
Lucio, J. (2009). Customers, citizens, and residents. *Administration & Society, 41*(7), 878–899.
McLaughlin, H. (2009). What's in a name: "client", "patient", "customer", "consumer", "expert by experience", "service user"—what's next? *British Journal of Social Work, 39*(6), 1101–1117.
Meadowcroft, J. (2010). Who is in charge here? Governance for sustainable development in a complex world. *Journal of Environmental Policy & Planning, 9*(3–4), 299–314.
Neumayer, E. (2003). *Weak versus strong sustainability: Exploring the limits of two opposing paradigms*. Cheltenham: Edward Elgar Publishing.
Osborne, S. P. (2010). *The new public governance? Emerging perspectives on the theory and practice of public governance*. London/New York: Routledge.
Osborne, D., & Gaebler, T. (1993). *Reinventing government: How the entrepreneurial spirit is transforming the public sector*. New York: Plume, Penguin Group.
Peck, J., & Theodore, N. (2010). Mobilizing policy: Models, methods, and mutations. *Geoforum, 41*, 169–174.
Pierre, J. (2000). *Debating governance. Authority, steering, and democracy*. Oxford: Oxford University Press.
Rhodes, R. A. W. (1997). *Understanding governance. Policy networks, governance, reflexivity and accountability*. Buckingham: Open University Press.

Scharpf, F. (1994). Games real actors could play: positive and negative coordination in embedded negotiations. *Journal of Theoretical Politics, 6*(1), 27–53.

Sharp, J. S., & Smith, M. B. (2003). Social capital and farming at the rural-urban interface: the importance of nonfarmer and farmer relations. *Agricultural Systems, 76*(3), 913–927.

Sørensen, E., & Torfing, J. (2005). The democratic anchorage of governance networks. *Scandinavian Political Studies, 28*(3), 195–218.

Sørensen, E., & Torfing, J. (2007). *Theories of Democratic Network Governance*. Palgrave Macmillan, New York.

Sørensen, E., & Torfing, J. (2011). Enhancing collaborative innovation in the public sector. *Administration & Society, 43*(8), 842–868.

Vaaben, N., & Lund, D. H. (2013). Bygge bro og bygge mure: paradokser i offentligt-privat innovationssamarbejde. *Økonomi & Politik, 2*, 67–79.

Wong, T. H. F., & Brown, R. R. (2009). The water sensitive city; principles for practice. *Water Science & Technology, 60*(3), 673–682.

World Commission on Development and Environment. (1987). *Our common future*. New York: United Nations.

Wunder, S. (2005). *Payments for environmental services: Some nuts and bolts*. Jakarta: CIFOR http://www.cifor.org/publications/pdf_files/OccPapers/OP-42.pdf, Accessed 13 Oct 2015.

Part II
Policy Mobilization, Planning, and Implementation

Chapter 2
Are Model Cities an Effective Instrument for Urban Environmental Governance?

Stefan Brehm and Jesper Svensson

Abstract Since the 1980s, China's central government has created various model and incentive schemes aimed at systematically and concurrently promoting innovative approaches for protecting the urban ecological environment. In this context, programmes such as the 'model city for protecting the environment', 'garden city', 'eco-city', or 'low-carbon city' have become an integral part of China's system for urban environmental governance. However, the role of these policy-incentive schemes for promoting best practice is only partly understood. This chapter contributes to the literature with a conceptualization of model cities as a dynamic governance instrument. The analysis suggests that the distance between programme objectives and local practices increases with programme maturity. Model-city schemes inevitably reach a point of saturation once a competing programme provides new opportunities to gain political and economic rents.

Keywords National Model City for Environmental Protection programme · Model-city programmes · Incentive schemes · Urban project's life-cycle perspective

2.1 Introduction

Why are some cities more successful in reconciling economic development goals and environmental preservation? And what can other municipalities learn from them? These questions are at the heart of Chinese political experimentalism in the field of green or greening urban governance. The search for answers starts with the

S. Brehm (✉)
Centre for East and South-East Asian Studies, Lund University, Lund, Sweden
e-mail: stefan.brehm@ace.lu.se

J. Svensson
Smith School of Enterprise and the Environment, University of Oxford, Oxford, UK

J. Delman et al. (eds.), *Greening China's Urban Governance*, ARI - Springer Asia Series 7, https://doi.org/10.1007/978-981-13-0740-9_2

identification of good examples followed by a systematized approach to evaluate and transfer best practices. Yet governments and cities can be fast or slow learners, creative or formalistic adopters, leaders or followers. Thus in order to explain varieties, success and failure, it is necessary to extend the above questions by asking 'who can learn from them?' In this chapter we seek to provide some answers in the context of China's National Model City for Environmental Protection programme. The aim is to highlight a particular aspect of policy mobility (Peck and Theodore 2010), namely, the life cycle of model policies with distinct modes of learning and emulation. Life cycles of the exemplary, we argue, shed light onto one of the mechanisms that render policy mobility a complex, dynamic, and non-linear process. In the remainder of this section, we are going to situate our research in the field of environmental governance in China and then detail the succeeding research programme of this chapter.

Periods of high economic growth often come at the price of environmental degradation. In China, this tension has long been part and parcel of state planning. The Maoist period featured the 'struggle of men against nature', emphasizing 'permanent revolution' and hence dialectical conflict (Murphey 1967). Policies aimed mainly at exploiting resources and converting natural landscapes into cultivated areas. There was little concern for the consequences of environmental degradation or for the loss of biodiversity. A Reform and Opening process started in 1978, marking the beginning of a new, export-driven economic regime that once more sought to accelerate the process of accumulation. Natural-resource extraction accelerated, and productive capacity increased at an unprecedented pace. At the same time, the social and economic costs resulting from pollution increased at a rate never seen before. And while prevailing policies in the Maoist era and under Deng Xiaoping differed, the driving force behind mounting pollution and intensified exploitation of resources was and still is the aim to promote economic development as the fundament and legitimation of political and military power.

In recent years, China's modern environmental crisis has added a twist to the resource-power nexus, because the rapid deterioration of the natural environment compromises human health and living conditions in such a comprehensive way that sustainability emerges as a new determinant for regime stability (Heberer and Schubert 2008; Francesch-Huidobro et al. 2012). Policies aimed at establishing and negotiating a better balance between environmental and economic interests have been seen as critical among leaders of the central government in particular (Eaton and Kostka 2014). Local governments, by contrast, are often less enthusiastic. They need, after all, to find practical solutions for various problems; and it is they who shoulder the financial burden for effective implementation (Ran 2013). What is more, dense networks of local businesses and political elites impose formidable challenges to the emergence of new regimes characterized by environmental transparency and accountability (Hillman 2014). Conflicts of interest between central planners and local governments play out at the intersection between, on the one hand, a vertical command line (*tiao* 条) stretching from the Ministry of Environmental Protection (MEP) in Beijing down to environmental-protection bureaus (EPBs) at the county level and, on the other hand, the horizontal organization (*kuai* 快) of government bodies at provincial, prefectural, county, town, and village levels (Liu et al. 2012). EPBs at all administrative levels are responsible for overseeing and implementing environmental policies and regulations handed down from the MEP. Yet local governments provide the lion's share of resources for the EPBs and have a say in appointing their leadership. There is thus ample space to protect local vested interests against regulatory requirements for protection of the environment (Van Rooij et al. 2013).

Grassroots behaviour often seems to be at odds with the intended outcomes of environmental rules and regulations. This nurtures scepticism that the current set of economic incentives and authoritarian top-down monitoring will be effective in tackling the challenges ahead. Wang (2013b) describes the perverted behavioural patterns of local governments as meeting the 'letter' without coming close to the 'spirit'. Local cadres, he states, are often willing to invest in environmental infrastructure as a means for stimulating growth, promoting political patronage, and pursuing personal economic gain. But such expenditures rarely improve environmental performance, because political credit and personal financial gain are correlated with money spent, rather than with environmental outcomes achieved.

However, the institutional development of environmental-governance mechanisms does not simply end there. In fact, a series of case studies gives reason to suppose that local practices are much more diverse than can be explained with the stylized dichotomy between *tiao* and *kuai* (Li et al. 2011). Recently, therefore, researchers have taken a fresh look at the meso-level, which connects macro-environmental outcomes with micro-practices at the grassroots. In this context, Li and Higgins (2013) posit that the central government deploys environmental-management programmes like the 'national model city for protecting the environment' in order to accommodate diversity and encourage local innovations.

The existing literature on environmental model cities can be divided into two fields. A macro-normative strand evaluates the criteria for model cities in terms of the stated aims of policy (Yu 2014; Chen and Gao 2011; Chen et al. 2010; Li and Higgins 2013). Works in this vein provide recommendations on how better to align requirements, monitoring, and selection processes with the overall aim of the policy in question. A second branch of the literature deals with micro-level evidence for political entrepreneurialism as a driving force behind the environmental commitments of municipalities. Case studies highlight the role of local political leaders who mobilize time and resources when they sense opportunities for economic kickbacks and career advancement (Li et al. 2011; Van Rooij et al. 2013).

Previous research provides little insight into the model-city programme as a tool for bargaining between central and local interests. Apparently model cities sometimes act as true change-makers (Li and Higgins 2013); in other cases they represent not much more than a half-hearted accounting exercise where the problems are relocated a few kilometres away (Liu 2013). The current body of literature attributes this variation to specific circumstances that emerge sporadically rather than systematically. Factors mentioned in this vein as conducive to success are political leadership, intensified civic engagement in environmental matters, more effective institutional arrangements for decentred governance, a stronger emphasis on the rule of law, and revised criteria for cadre evaluation (Kostka and Mol 2013; Li and Higgins 2013; Li et al. 2011; Liu et al. 2013).

In this chapter, we argue that current research has a blind spot. It is reasonable to assume that local leaders appreciate the potential economic and political benefits of becoming a model city. It is also intuitive that some cities enjoy better preconditions for attaining the title. But why do management quality and the environmental outcomes of one and the same management programme vary so greatly, as case studies suggest? In order to solve this puzzle, we propose a life-cycle perspective on environmental-management programmes. Our framework highlights changing incentives over the course of a scheme's development, i.e. from its inception until its phasing-out. We show in particular that criteria, management, and as a consequence incentives change over time. Performance disparities are thus not solely due to the idiosyncratic characteristics of individual leaders or localities; they are also the result of systematic factors inherent in the management of the environmental-incentive scheme.

We set out to analyse, in five steps, the life cycle of the National Model City for Environmental Protection (NMCEP) programme. We first provide a brief introduction of the origins and current institutional setting of environmental model cities. In Sect. 2.3, we use descriptive statistics to compare the performance of environmental model cities over time and in comparison with other municipalities. This quantitative analysis provides the empirical background for the following two sections. In Sect. 2.4, we seek to explain the emergence and development of the model-city scheme from the supply side (local government). In Sect. 2.5 we do the same from the demand side (the central state). The logic of this analytical structure follows from understanding policy mobility as a distinct approach to policy studies. Peck and Theodore (2010) emphasize in this context: 'Models that travel therefore reveal at least as much about "demand-side" needs, imperatives, and anxieties as they do about supply-side inventiveness'. Our analysis deploys the 'experimentation under hierarchy' framework (Heilmann 2008) and introduces the concept of paradigm-steered technological change in order to integrate a life-cycle perspective on model-city schemes. In Sect. 2.6 we conclude.

2.2 National Model City for Environmental Protection Programme

China's first and most prominent incentive scheme for promoting economic development in tandem with higher environmental standards is the National Model City for Environmental Protection (NMCEP) programme. The first municipality to acquire model-city status was Zhangjiagang, in Jiangsu Province. Qin Zhenhua, the city's party secretary between 1992 and 1998, was inspired by Singapore's 'Garden City' approach and decided to make Zhangjiagang the cleanest city in China. Qin's foremost aim was to attract foreign firms, as well as to profile himself as a progressive and visionary leader. His efforts were backed and promoted by Xie Zhenhua, who was director of the State Environmental Protection Agency (SEPA) at the time.

Xie lauded Zhangjiagang's achievements during a national meeting on environmental protection held in that city in 1995. One year later, after further lobbying activities, the fourth national meeting on environmental protection named Zhangjiagang the first model city for protecting the environment (Li et al. 2011). Kunshan, the second NMCEP city, is located about 100 kilometres south east of Zhangjiagang. Officials there hesitated to recognize Zhangjiagang's influence on their own policy choices, instead highlighting their independent achievements. Two additional co-located cities, Taicang and Wujiang, joined the model cluster in 2001 and 2003, respectively. In these cases, however, policy decisions were driven by the ambition of the provincial government to establish a model-city cluster (Li et al. 2011). As of April 2012, 78 cities and 6 urban districts across China had been awarded the title. More than two thirds of these (58/84) are located in developed regions along China's east coast, with 21 in Jiangsu, 20 in Shandong, 10 in Guangdong, and 7 in Zhejiang. Figure 2.1 shows that environmental model cities cluster within three regions, namely, Shandong, Jiangsu/Zhejiang, and Guangdong.

With its conversion from a pioneering experiment into a national incentive scheme, the NMCEP became more formalized and standardized. The programme now builds upon the Quantitative Examination System on Comprehensive Control of the Urban Environment (QESCCUE). The latter involves a compulsory examination of the environmental performance of all key environmental cities in China. The

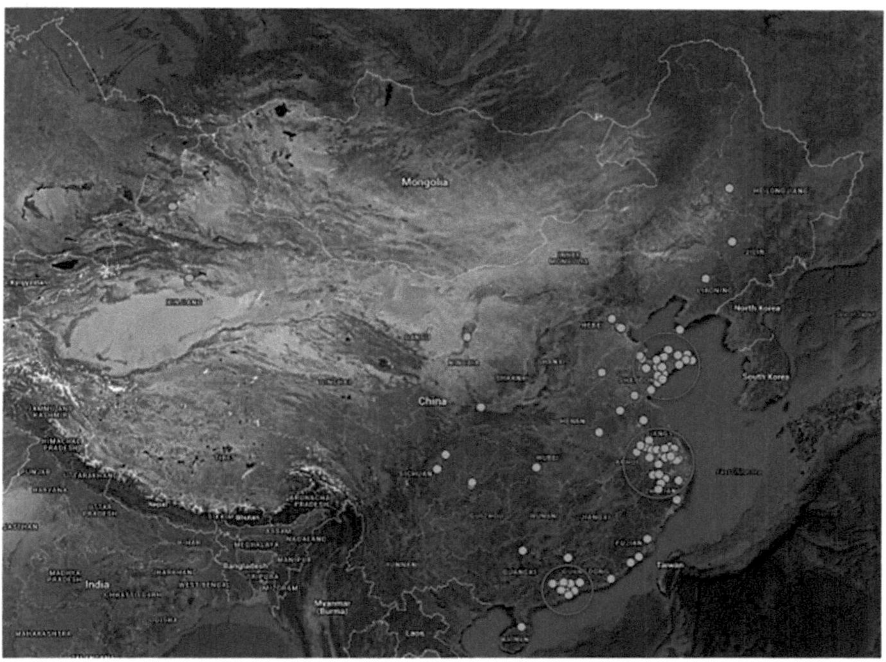

Fig. 2.1 Spatial distribution of environmental model cities

NMCEP scheme differs from the QESCCUE in that its evaluation criteria cover not only environmental issues but questions of socio-economic development as well—a particularity with the purpose of motivating cities listed at the higher end of the QESCCUE rankings (Li and Higgins 2013). The NMCEP programme is voluntary and open to all cities and districts. The assessment comprises 3 basic conditions and 23 specific targets under 4 categories: socio-economic standards, environmental quality, environmental construction, and environmental management. To attain national model-city status, a municipality must conduct self-assessments and implement a local plan in line with the MEP's requirements.

Before a city can initiate a 'model-creation process', the following preconditions must be met. The city needs to:

- Perform national and provincial tasks of pollutant control.
- Be among the top-ranking cities in its province in overall environmental performance for 3 consecutive years.
- Have no serious environmental incidents in its urban districts during the past 3 years and no violations of environmental regulations the past year.
- Develop an environmental emergency plan (MEP 2005).

Moreover, if a city meets the MEP's requirements and submits its report to the MEP and to the provincial environmental-protection bureau (EPB), the city needs to follow a public-notification rule. During the process the MEP conducts evaluation and site visits, after which assessment reports are made available to the public for feedback. The process of becoming a national model city takes 3–4 years (Li et al. 2011, 120–121). In order to maintain their status, model cities have to report their progress to the provincial EPB each year and to be re-evaluated by the MEP every 3 years. Failing to meet the benchmarks of the NMCEP programme can mean losing the title. Performance criteria for becoming an environmental model city are defined in detail; and, as seen in Table 2.1, they often specify numerical benchmark values. Thus, the programme rewards cities for measurable outcomes but does not evaluate innovative performance, i.e. the process by which benchmarks are achieved. This leads to a paradox that environmental targets need to be met even at the expense of the environment.

2.3 Model Cities in Comparative Perspective

Section 2.2 demonstrated that model cities are subjected to a rigorous and comprehensive assessment process and that successful candidates for the title have to demonstrate outstanding management capabilities in combination with high economic performance. It is therefore not surprising that we find the lion's share of award winners along the economically advanced east coast. Yet the intimate connection between economic development and environmental performance raises a question: are model cities committed to environmental objectives beyond such improvements as a result from technological progress and shifting industrial structures—i.e.

Table 2.1 Assessment indicators for National Model City for Environmental Protection programme

Basic requirements
1. Fulfilment of main national and provincial tasks of pollutant control
2. Absence of serious environmental incidents in urban districts during the past 3 years and of violations of environmental regulations in the past year and development of environmental-emergency plan
3. Placement among top-ranking cities in province in overall environmental performance for 3 consecutive years
Economic and social
4. Annual per-capita disposable income in the past 3 years of at least 10,000 RMB for urban residents and of at least 8500 RMB for cities in western China; environmental spending >1.7% GDP
5. Year-by-year decline in energy consumption per unit of industrial added value
6. Declining water consumption per unit of GDP
7. Year-by-year decline in emission intensity of major industrial pollutants per unit of industrial added value
Environmental quality
8. Satisfaction of national standards for annual average concentration of major pollutants in urban air; satisfaction of secondary standard in 85% of total number of days per year for daily average concentration of pollutants
9. Compliance with central quality standards in source areas for drinking water
10. Fulfilment of environmental functional requirements for water quality within jurisdiction and fulfilment of transregional water-quality requirements at cross section of city's domain exit
11. Urban environment noise level < 60 dB(A)
12. Traffic noise level < 70 dB(A)
Environment construction
13. >35% of total area in green area (per-capita green area above national level for cities in western China)
14. >80% of urban waste water treated, >20% of urban sewage water recycled
15. Stabilization of emissions standards for key industrial enterprises
16. >50% of total energy consumption is clean energy use
17. >80% of motor vehicles inspected in accordance with environmental standards
18. >85% of urban garbage treated in accordance with environmental standards
19. >90% of industrial solid waste treated or reused
20. Safe disposal of hazardous waste in accordance with law
Environmental management
21. Target-responsibility system for environmental protection in place. Environmental indicators included in performance evaluation of leading party and government cadres. Government officials assigned responsibility for implementing public-announcement system for environmental quality
22. Implementation of EIA law for new projects
23. The Environmental Protection Agency is an independent institution. Environmental capacity-building to meet standardized national building requirements
24. Public-satisfaction rating for urban environmental protection >80%
25. Environmental education integrated into primary- and secondary-school education >85%
26. Implementation of urban environmental-sanitation provisions and fulfilment of requirements for environmental management of surrounding areas

Source: Compiled by the authors from MEP (2005)

beyond those produced by economic modernization itself? Also, are model cities as a group more homogeneous than other localities with respect to key indicators of environmental performance? That is what we will expect to see if the programme effectively aligns policy goals of the central state with the interests of local officials and hence serves as a meso-level governance instrument harmonizing local idiosyncrasies with national strategies.

To get a better understanding of the particular achievements of model cities, we compare them with other municipalities. Economic disparities in China are substantial, so it seems sensible to juxtapose model cities with jurisdictions which are similar in terms of administrative rank and economic development. To this end, we collected information on cities listed by the MEP as 'key environmental-protection cities' (KEPCs). KEPCs have a special obligation to take the lead in implementing green policies, and they are subject to comprehensive environmental monitoring (Wang et al. 2013a; Liu and Wang 2013; Marquis et al. 2011). Getting listed as a key environmental-protection city is not something done at the discretion of a municipality; by contrast, participating in a model-city programme is voluntary. Thus, requirements for model cities usually go beyond the benchmarks that apply to KEPCs. The following descriptive statistics are based on about 400 observations for 128 cities during the 2008–2011 period; 40 jurisdictions in this sample hold the title of 'National Model City for Environmental Protection'. Statistical information was taken from three sources: the China City Statistical Yearbook (various issues), which includes all key economic and environmental indicators; a list of environmental model cities provided by the Ministry of Environmental Protection[1]; and the regular reports issued by the Institute of Public and Environmental Affairs[2] on the pollution-information transparency index.

We present key performance indicators (KPIs) for environmental and economic development in relation to the qualification criteria for environmental model cities listed in Table 2.1. In particular, we deploy publicly available indicators and statistics on environmental management, energy consumption, pollution intensity, and the treatment of discharge water and of solid waste. All performance indicators are measured relative to a municipality's GDP. This allows us to mitigate the performance bias arising from economic modernization rather than local environmental commitment. The variations within and between each of the two groups (model cities and other cities) are depicted as boxplots, which are a standardized way of displaying the distribution of data. The 'whiskers' above and below the box show the locations of the minimum and maximum. The central rectangle spans the range from the first to the third quartile (interquartile range, IQR), and the horizontal line within the box marks the median, i.e. the typical value for a representative of each group. The dots, finally, depict outliers, which are defined as values three times (or more) above or below the IQR.

Figure 2.2 compares municipalities in terms of economic and demographic factors. As expected, model cities have a relatively high per-capita GDP, and a typical

[1] Available at http://english.mep.gov.cn/inventory/Model_cities/. Accessed on 2016-02-11

[2] Available at http://www.ipe.org.cn/en/about/report.aspx. Accessed on 2016-02-11

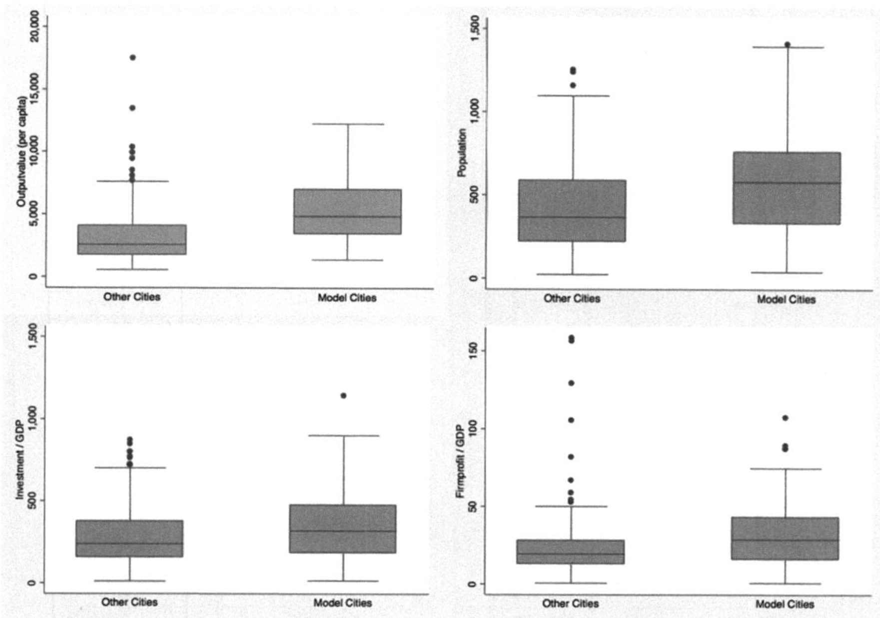

Fig. 2.2 Economic and demographic characteristics of sample cities

model city tends to be larger than a typical key environmental-protection city without model-city status. What is more, model cities attract more investments and their firms are more profitable. This is most probably the result of a larger concentration of industries with advanced production content and of service industries with high value added. Variance within the model-city group is also greater, as compared to that within the group of other municipalities. Model cities, then, are more likely to be economic high performers; however, the variation among them in this regard is significant.

In the next step, we look at the potential environmental-management capacity of our sample cities. The boxplot in the upper-left portion of Fig. 2.3 suggests that model cities mobilize more budgetary resources for protecting the environment than do other cities. This gives us reason to suppose that model cities have larger environmental bureaucracies, which ideally translates into enhanced management capacity. The pollution-information transparency index (PITI) in the upper-right box supports this conjecture. The PITI measures a municipality's compliance with environmental regulations that aim to render local governments more accountable. The total score is an aggregate of eight sub-indicators relating to discharge-fee data, clean-production audits, environmental-impact assessment reports, enterprise environmental-performance ratings, responses to public information requests, information disclosure on records of enterprise violations, disposition of verified petitions and complaints, and results of enforcement campaigns against polluting enterprises.

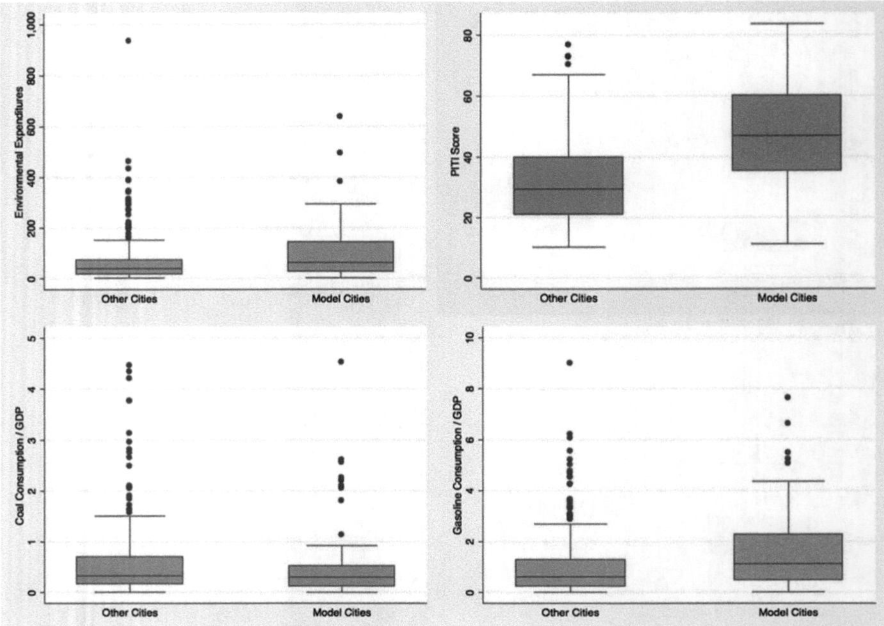

Fig. 2.3 Potential environmental-management capacity of sample cities

The boxplots at the bottom of Fig. 2.3 show coal and gasoline consumption relative to a municipality's GDP. Model cities seem to use less coal and more gasoline. The reduction in coal consumption is in line with the energy-consumption patterns of advanced economies, where production facilities are more efficient and/or less reliant on energy, due to the shift to the service sector. The demand for fuel, on the other hand, increases with a rapidly expanding fleet of vehicles—the status symbol of the rising middle class. All in all, model cities seem to possess a higher environmental-management capacity; moreover, the comparatively low number of outliers implies that this is a universal characteristic of this group, despite substantial within-group variation. The data on energy use, however, point to different consumption patterns rather than to efforts to render production more sustainable.

In Fig. 2.2, finally, we look at environmental-performance indicators found in official statistical yearbooks at the city level. In three out of four cases—i.e. treatment of solid waste, of sulphur dioxide, and of coal dioxide—model cities are underperforming relative to their economic weight. In the case of discharge-water treatment, there is no obvious difference in performance. Still, the model cities appear to be more homogeneous as a group than the other cities, since the number of outliers in their case is relatively small (Fig. 2.4).

In Fig. 2.5, we visualize the dynamic properties of model-city performance. To this end we graph kernel densities of the four environmental indicators for model

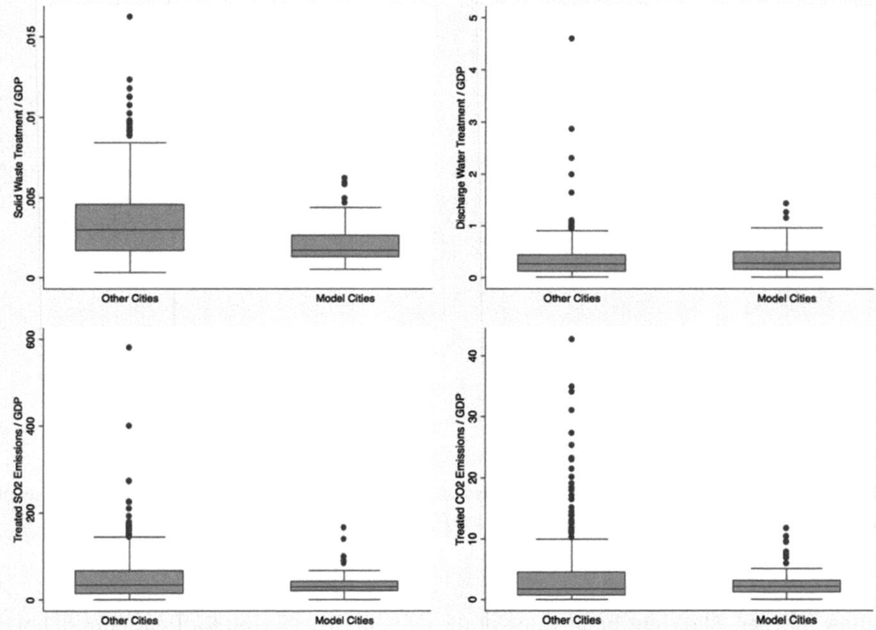

Fig. 2.4 Environmental performance indicators for sample cities

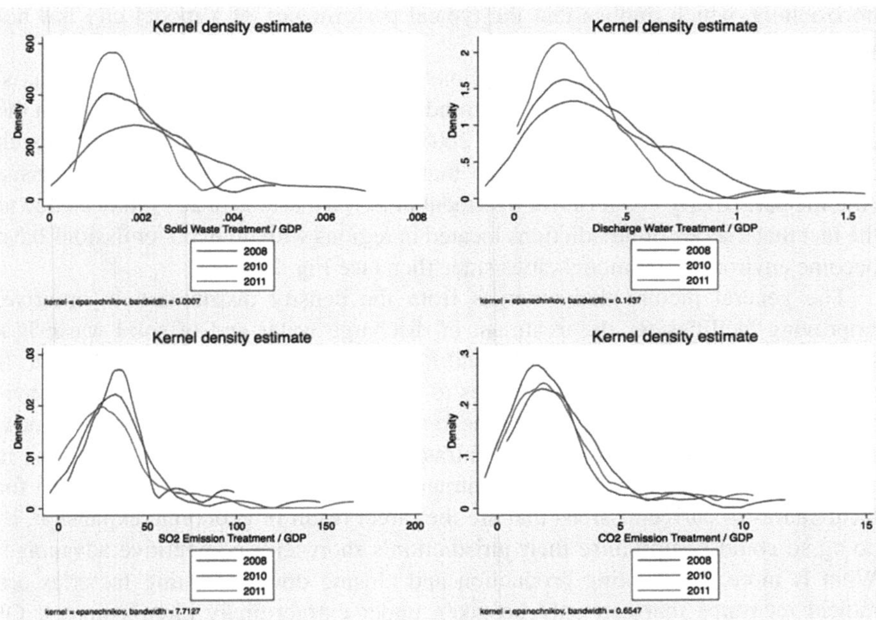

Fig. 2.5 Kernel densities of model cities' environmental performance by year

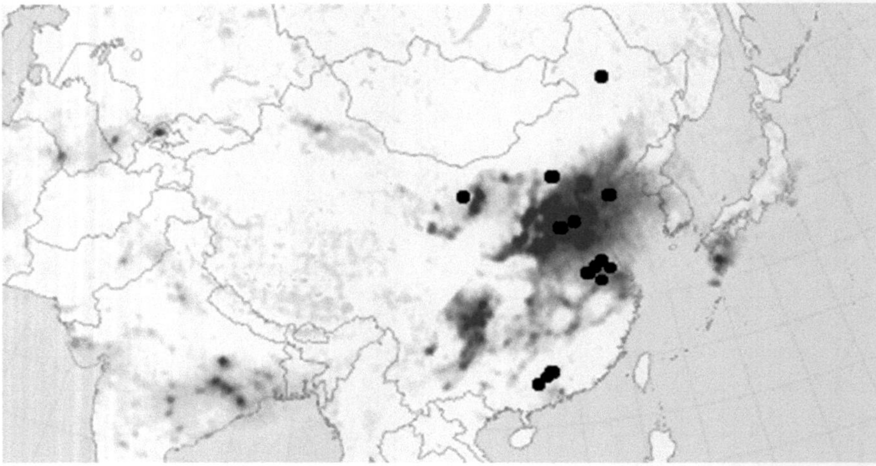

Fig. 2.6 NASA satellite image of SO2 emissions (2011–2014) and location of model cities as of 2011/2012 (Source: http://visibleearth.nasa.gov/view.php?id=87154 (accessed 2016-11-04))

cities by year. The right-hand slope of the density distribution for treatment of solid waste and of discharge water is getting steeper, and the right-hand tail thinner. This indicates that model cities increasingly centre on a (minimum) benchmark value, i.e. heterogeneity is decreasing over time. The density peak does not shift horizontally, which implies that the typical performance of a model city has not improved over the sample period.

Treatment of SO_2 and of CO_2 emissions is similarly static on the horizontal axis. Yet in neither case can we find a clear trend towards decreasing heterogeneity. In the case of CO_2 emissions, the values for 2008 may be extraordinarily high, inasmuch as the denominator (GDP) contracted that year due to the global financial crisis. And the particularly low level for treatment of SO_2 emissions in 2011 may be due to the fact that a series of jurisdictions located in regions with high SO_2 emissions have become environmental model cities since then (see Fig. 2.6).

The general picture that emerges from the density distribution is intuitive. Improving facilities for the treatment of discharge water and of solid waste is a straightforward investment process, and it is comparatively easy to implement. It also serves the interests of local cadres to establish visible manifestations of environmental concern and at the same time to nurture clientelist ties through the allocation of capital (Wang 2013a). By contrast, reducing SO2 and CO2 emissions is more complex, and it involves commitments by many stakeholders. It is hard for local cadres to curb emissions that are the direct result of economic expansion, as doing so could compromise their jurisdiction's short-term competitive advantage. What is more, interrupting production and closing down polluting factories are radical measures that can only be taken under extraordinary circumstances. Of course, one could also try to convince citizens to make less use of private cars, or to encourage enterprises to invest in more filters and in energy-efficient equipment. Such things are difficult, however, because they require incentives and political support which exceed the potential benefits of a model-city scheme.

On a general note, it bears stressing that the comparison and analysis presented here are not exhaustive, due to the limited availability of statistical material. Still, some preliminary conclusions can be drawn. Environmental model cities are economically more advanced than are other key environmental-protection cities, and they possess a relatively high capacity to implement environmental policies. Yet, in terms of energy consumption and environmental outcomes, model cities do not have a particularly good track record relative to their economic weight. From this perspective, model cities seem to thrive on environmental improvements that result from economic modernization, in combination with their own enhanced efforts to demonstrate environmental leadership. There is no obvious connection between leadership and environmental performance. At least for the 2008–2011 period, model cities show no significant progress towards improving their average achievements, although they do display a trend towards converging on (minimum) benchmark criteria.

2.4 Supply of Model Cities

The development of the environmental model-city programme follows a pattern that Heilmann (2008) calls 'experimentation under hierarchy'. In order to hedge against the political risks of policy innovations, local cadres tend to seek the support and encouragement of higher-level political sponsors (Heilmann 2009, 2008). Then, after a powerful policymaker has endorsed the results of a successful experiment, campaigns and lobbying are needed in order to broaden support for the initiative and eventually to diffuse the policy innovation as a model for a wider region or even the whole nation (Heilmann 2008). This concept has gained wide currency in recent years, and various scholars have successfully applied it to explain policy outcomes within a range of contexts (Tsai and Dean 2014; Wang 2009; Ahlers and Schubert 2013; Xia and Pahl-Wostl 2012).

The environmental model-city programme was inspired by the clean-city approach of Zhangjiagang. A major difficulty in developing such a political innovation is managing the operational risks involved. Thus, realizing a 'grand idea' such as 'turning Zhangjiagang into China's cleanest city' required commitments from various departments and the involvement of many stakeholders. A willingness to cooperate cannot, however, be taken for granted, because Chinese bureaucracies are staffed and led by cadres that compete for political and professional recognition of their individual performance rather than their collective achievements (Eaton and Kostka 2014). Qin Zhenhua was the highest-ranking official in the municipality; hence he possessed the political clout to bring officials into line. Zweig (1997) describes Qin's management style as authoritarian and as strongly focused on ideology, intimidation, and a development-at-all-costs mentality. He actively monitors local activities and after recommending changes, he sends his staff to follow up. Those who have not responded to his orders are routinely moved aside, and their power is passed on to someone who listens to him. While many intellectuals and local government officials in Jiangsu Province felt that this heavily authoritarian, almost Maoist, leadership style was no longer appropriate in China, Qin has been remarkably successful (Zweig 1997, 245).

The elevation of Zhangjiagang's initiative from a local vision into the prototype for a national model-city scheme would have been impossible without a patron in the central government. To this end, Qin aggressively forged clientelist ties with Beijing throughout his period of service in Zhangjiagang. In the early 1990s, for instance, he campaigned in the capital for free-trade zone status. In order to gain this approval, he sent young girls from the city to serve as maids and housekeepers in the homes of several central-government officials—and Zhangjiagang's municipal government paid their wages (Li 1997). In 1994, Qin embarked on his mission to make Zhangjiagang the cleanest city in China, as a means of attracting foreign investors. Public toilets were kept clean, garbage collection was made comprehensive, and citizens were fined for throwing litter on to the streets. Car horns were forbidden; polluting enterprises had to move out. In 1995, the provincial propaganda department used Zhangjiagang as a model unit for General Secretary Jiang Zemin's promotion of the 'two civilizations' idea. This concept had emerged as early as under Deng Xiaoping and referred to material and spiritual civilization, respectively. The former concerned China's economic progress; the latter encompassed progress in morality, ethics, and ideology. Jiang began to emphasize spiritual civilization in 1996. To this end he established the Party's Central Committee Spiritual Civilization Steering Committee, which oversaw the activities of spiritual-civilization offices at various levels. Schools, work units, neighbourhood committees, and the like promulgated public morality and civilization guidelines. Ratings and award systems were established to promote individuals exhibiting exemplary behaviour (Dynon 2014). Qin Zhenhua's initiative to clean Zhangjiagang was certainly designed to demonstrate this spirit of morality and ethical progress. A provincial meeting was held in the city, and word was spread to Beijing. Jiang himself visited Zhangjiagang and wrote 16 characters known as the 'Zhangjiagang spirit' (Zweig 1997).

Jiang's campaign for ideological and ethical renewal focused on behavioural change. Consequently, neither provincial leaders nor their central-government counterparts were concerned that China's first environmental model city showed very little interest in protecting the natural environment or maintaining biodiversity. The ultimate aim was to generate a clean and tidy urban space, in order to make the city more attractive to foreign investors. In fact, a good deal of imagination is needed in order to rediscover Zhangjiagang as the model archetype for environmental model cities today. This highlights Zhangjiagang's role as an ideological or spiritual innovator, rather than a technological leader.

In order to convert the 'clean-city spirit' of Zhangjiagang into a manageable incentive scheme, central planners had to define the character and conditions of an environmental model city. The main source of content for the programme was the 'Japan-China Environmental Development Model City Scheme'. Enacted in 1997, this programme benefitted from foreign funds and technologies, and it sought to boost the environmental performance of three Chinese cities: Dalian, Chongqing, and Guiyang (Baker 2015). Transnational cooperation, in tandem with the efforts of international organizations to document and benchmark the environmental-protection measures taken, shaped the assessment criteria. These mechanisms, the purpose of which was to render the state of the environment quantifiable, became a way to demonstrate status (Hoffman 2009), thereby linking environmental governance on the one hand with political/economic incentives on the other.

The measures taken by environmental model cities follow a fairly uniform pattern: relocating polluting industries, greening urban spaces with grass and trees, and improving infrastructure for waste-water treatment and garbage collection. Two brief accounts of Shenyang and Dalian illustrate this: In 1984, for example, Shenyang was ranked one of the 10 most polluted cities in China. In 2001, the city's leaders declared that they wanted the city to achieve model environmental status. During the subsequent 3 years, Shenyang closed down over 600 factories, upgraded 300 more, removed industries from the downtown—replacing them with residential and business districts—and increased the number of wastewater treatment plants (Economy 2006, 179).

Using the experience of Kitakyushu city in Japan as a guide, the city government in Dalian championed a strategy of 'greening' the city with the planting of trees and grass, 'rationalizing' city planning, and relocating industry out of the city centre to the suburbs (Hoffman 2011, 113).

Model cities awarded the title do not have to demonstrate a high level of innovativeness, but rather comprehensive and sophisticated methods for monitoring, measuring, and benchmarking progress and political efficacy. Thus in contrast to the Zhangjiagang spirit, which is characterized by a qualitative way of knowing—i.e. a visionary leader's interpretation of a problem and its solution—the emerging programme deploys numbers and figures in order to construct and define the issue at hand. Statistics, in this context, are used as purveyors of 'truth' and a means of 'knowing' what is wrong (Greenhalgh 2008).

Figure 2.7 summarizes and systematizes the development of the environmental model-city programme. In this context Qin Zhenhua, the party secretary and mayor of Zhangjiagang, represents a political entrepreneur in the Schumpeterian sense, i.e. an individual possessed of the spirit and energy to create and supply an innovative good which establishes a new market and generates handsome political and economic profits. Yet the origins of the model-city programme can be traced back not just to Zhangjiagang, but to at least three other cities as well: Dalian, Chongqing, and Guiyang. With the help of foreign funds and technologies, these municipalities sought to emulate successful experiences from other cities around the globe. The archetypal model city draws, then, from two different sources: political entrepreneurialism and qualitative/experimental content development. It is important to note that different organizations supplied these components. In Fig. 2.7, the early phase of the programme is located in the bottom-left rectangle. The circle symbolizes the relatively large returns that a successful political innovation generates.

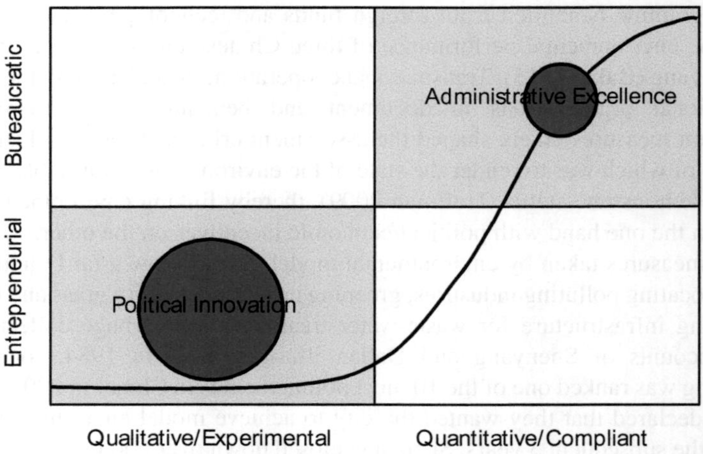

Fig. 2.7 Stylized facts of a model-city incentive scheme

When we compare the early phase of the model-city programme with the assessment process and related criteria at a later stage, as outlined in Sect. 2.2, it is easy to see that control and monitoring of performance indicators play an increasingly important role, while room for individual creativity narrows down. Aspiring model cities are still encouraged to adopt the programme to local circumstances.[3] Yet the quantitative approach of a mature incentive scheme does not encourage fundamental innovations, because the statistical way of knowing predefines the problem and its solutions. Creativity is replaced by bureaucratic processes of measuring, ranking, benchmarking, and adjusting. Thus imitating model cities are not rewarded for political vision, but rather for administrative excellence and an ability to comply with high standards. We illustrate the mature phase of the model-city incentive scheme in the top-right rectangle of Fig. 2.7. The s-shaped line stylizes the number of model adopters. Compared to the early phase, the mature phase is characterized by a large number of municipalities offering the same original political innovation. This implies that profits shrink and that incentives to join the programme shrink too.

2.5 Demand for Model Cities

Experimentation under hierarchy requires that a powerful patron endorse a successful political innovation and that he/she organize lobbying activities as a means of increasing the number of model-city supporters within upper government/CCP

[3] Available at http://env.shqp.gov.cn/gb/content/2007-06/27/content_143812.htm. Accessed on 2016-04-12

echelons (Heilmann 2008). The patron's role is best described as that of a broker, and the reward is a 'finder's fee'. But where does the demand come from? In the case of Zhangjiagang, it seems that Jiang Zemin created a political market with his promulgation of the idea of 'two civilizations'. Abstracting from this observation, we contend that political programmes or paradigms create political markets. Hence, in an effort to understand the demand side of model-city schemes, we extend the experimentation-under-hierarchy approach in this section. To this end, we make use of Kuhn's concept of the scientific paradigm (Kuhn 1962), which posits that scientists in a field are first attracted to a certain methodological approach, but eventually discard it in favour of another paradigm. A paradigm, in Kuhn's understanding, is a scientific approach to some phenomenon which provides model problems and solutions for a community of scholars.

Scientific advances do not just take the form of small incremental steps along the trajectory of a given paradigm; they can also take the form of major jumps and starts. Creative space along the trajectory of a paradigm is small, since the latter structures the formation of new knowledge, while embarking on a new paradigm opens vast creative space, as the direction of knowledge formation is not yet prestructured. Dosi (1982) extends Kuhn's scientific paradigm to a theory of technological change. He differentiates between market-pull factors and technology-push factors for innovation. In his view, incremental change along the trajectory of a given technological paradigm springs mainly from market-feedback mechanisms, in which the search for new products or processes focuses efforts in predefined directions. By contrast, radical inventions require more autonomy from short-term feedback mechanisms, because departing from a technological paradigm results in a lack of structures and thus more uncertainty, in terms of the road to take and ultimately the outcome of innovative efforts.

Kuhn's concept of the scientific paradigm provides us with the tools to understand the entrepreneurial aspect of political innovation, whereby an entrepreneur makes a political programme persuasive by solving a paradigm-relevant problem. Dosi, by contrast, helps us to grasp the convergence of model cities over time in terms of policy content. In this respect, the emergence of experimental and trial regions is driven by technology-push factors within a politically funded space. Then, once technological and organizational experiences have been turned into measurable processes, innovative activity is confined to incremental steps, i.e. the sort produced by market-pull factors. This perspective meshes well with our earlier observation that model cities tend to become more homogenous as a group, but without continuing to push the technological frontier.

A paradigm-driven demand function implies that a model-city incentive scheme develops along a life cycle that comes to an end when a new paradigm draws entrepreneurial attention and resources into a new, relatively unexploited political market offering the prospect of higher returns. Illustrating the truth of this claim for the environmental model-city programme, Fig. 2.8 shows the incentive scheme's pattern of diffusion in relation to a subsequently emerging model—that of the low-carbon city. The political trigger for the model-city programme was Jiang Zemin's campaign in 1996 to consolidate the concept of 'socialist spiritual civilization' as a

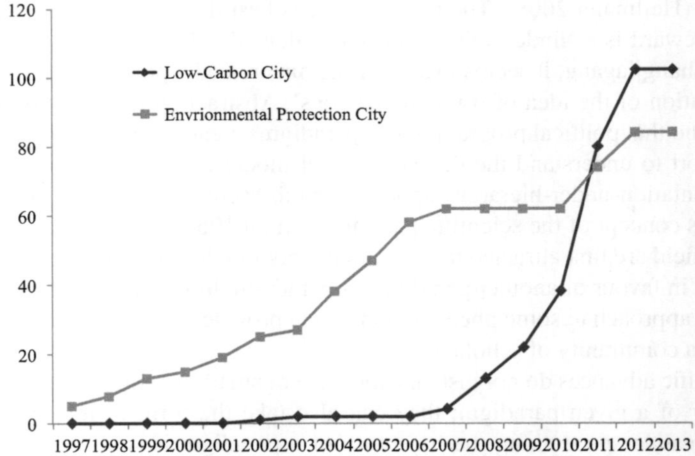

Fig. 2.8 Diffusion of model-city schemes (Source: http://english.mep.gov.cn/inventory/Model_cities/ (accessed 2016-11-04))

tool for re-legitimating party leadership (Cooke 2009). In this context, the environmental model-city programme sought to provide examples of where economic and environmental performance can go hand in hand. As we see in Fig. 2.8, adoption speed in the model-city programme accelerated after 2002, with the promulgation of the Cleaner Production Law and its paradigmatic switch from pollution control to pollution prevention (Mol and Liu 2005). Since then, more and more cities have incorporated ecological objectives into the core of their local policy agendas. Model diffusion stagnated from 2007 until 2011, when ten more cities won the title. There have been no additional nominations since 2012 (Table 2.1).

The stagnation of the NMCEP programme coincides with the emergence of low-carbon cities as a newly emerging model type. The drive for low-carbon innovations took off with China's 11th five-year plan, which aimed at reducing emissions by 20 per cent by 2010. Low-carbon model cities, as a catalyst, began to mushroom in 2007, when China's leaders propagated the 'low-carbon economy' as a new political vision for the country. At that time the central government established the National Leading Group on Climate Change (NLGCC) and presented China's national plan for responding to climate change. The National Leading Group was headed by China's premier and was tasked with coordinating national action on climate change (Qi et al. 2008). Within this framework, a series of cities started developing programmes to save energy and reduce emissions. Like their environmental model-city counterparts, low-carbon cities relied on foreign assistance and were linked to international diplomacy (Bulkeley et al. 2012; Economy 2006). Hence, content generation in the low-carbon model also rests on experiments and trial zones.

Also like their environmental model-city counterparts, low-carbon cities are associated with an archetypal entrepreneurial model—in their case that of Baoding city—which provides a spiritual narrative but little significant technological content. In 2008, the Ministry of Construction and the World Wildlife Fund jointly chose Shanghai and Baoding as two pilot cities for the construction of a low-carbon city. Baoding was also among the first low-carbon pilot cities selected by the State Council, due to its important role in providing an example for other cities (Oberheitmann and Ruan 2013). The city was chosen because of its mayor's vision of promoting economic development by developing a low-carbon industrial sector (Qi et al. 2008). Like many other cities, Baoding established a high-tech industrial zone (in 1992). The local government started focusing on low-carbon industries early on. It invested in several corporations, established contacts with investors, and promoted research links with Tsinghua University. Since then the city has become a centre for new energy technology, even winning the title 'National New Energy and Energy Equipment Industrial Base' (Martinot et al. 2007). Baoding's low-carbon ambitions were clearly driven by economic interests and hence provided a good example of how to achieve economic success by promoting the new paradigm of a low-carbon economy. Reports like the following one nurtured the narrative of the Baoding low-carbon spirit:

> In 2007, Baoding proposed to become a solar-powered city in three years. Through the effort of the Baoding government about 105 of Baoding's residential districts have installed or upgraded solar power facilities and about 101 traffic lights at the crossroads have started to use solar power. The world's first 'power valley' building, integrating photovoltaic power generation and a five-star hotel, was formally put into use in Baoding. China's 'power valley' and 'solar power city' now enjoys a worldwide reputation. (Wang et al. 2013b, p. 217)

It's a model other cities in China hope to replicate. By applying some of the same advantages that have made China the world's factory in other industries—economies of scale and cheap labour—Baoding-based companies such as Yingli Group and Baoding Tianwei have rapidly made themselves into global players in the solar- and wind-energy markets (Mackinnon 2009).

Yet, even as Baoding is successfully developing and exporting low-carbon technologies, the city today remains China's third most polluted municipality, mainly because the local government has focused on low-carbon equipment but neglected carbon emissions as a result of increased production (Stanway 2014). It was clearly an example of a political innovator without content leadership. The latter is mainly being developed in five provinces (Guangdong, Liaoning, Hubei, Shaanxi, and Yunnan) and seven cities (Chongqing, Xiamen, Shenzhen, Guiyang, Nanchang, Tianjin, and Hangzhou). These are the local governments which the National Development and Reform Commission (NDRC), as the agency primarily responsible for formulating China's policy on climate change, selected with the aim of developing best practices (Zhou and Williams 2012). The NDRC's low-carbon demonstration sites, which encourage innovative best practice, are open in terms of their objectives, and they are evaluated qualitatively in terms of their success in establishing effective technologies and institutions for reducing carbon emissions (Khanna et al. 2014). At a later stage, their experiences will be gradually transformed

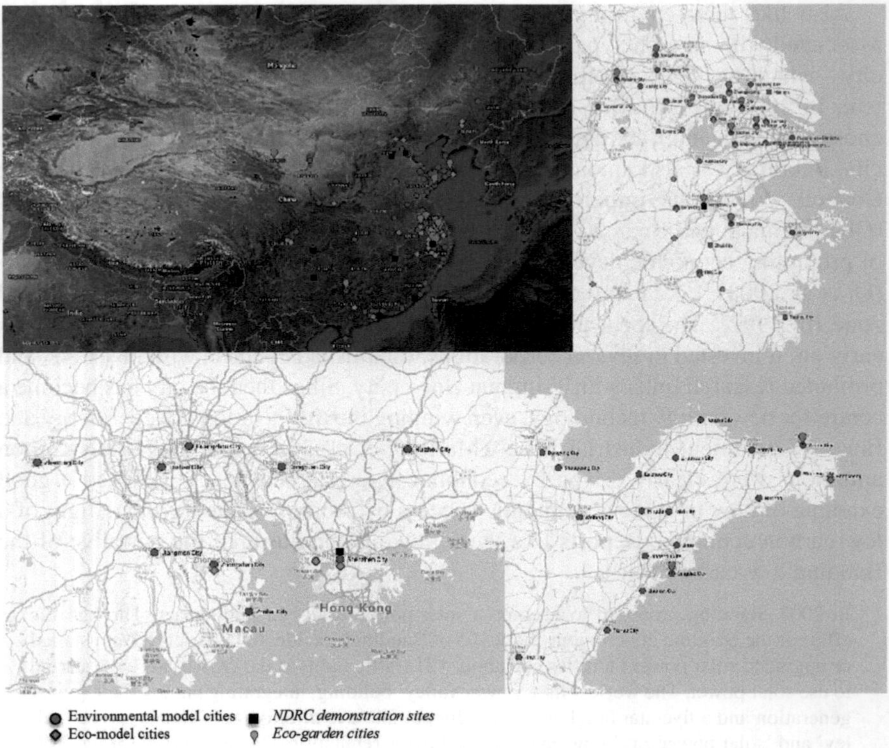

Fig. 2.9 Spatial distribution of various model-city schemes (Source: https://china.lbl.gov/sites/all/files/china_eco-cities_indicator_systems.pdf (accessed 2016-11-04))

into quantifiable processes, which will mark the second stage in the life cycle of this incentive scheme.

The emergence of a new political paradigm is not good news for the established environmental model-city scheme. In order to safeguard the value of the model-city programme, the Ministry of Environmental Protection sought to recalibrate its assessment criteria by adding a module on low-carbon technology adoption (Zhou and Williams 2012). Since 2007, accordingly, environmental model cities have been able to apply for an 'eco-model city' upgrade. (However, a final attempt to relaunch the model only took off with the adoption of the 12th 5-year plan in 2011, which strengthened political efforts to reduce carbon emissions and to institute stronger environmental protection.) In addition, the Ministry of Housing and Urban-Rural Development (MoHURD) restructured its garden-city programme, promoting the 'eco-garden city' as a new label.

Looking at the geographical distribution of the various model-city initiatives (see Fig. 2.9), we find that the NDRC's demonstration sites display a distinctly dispersed

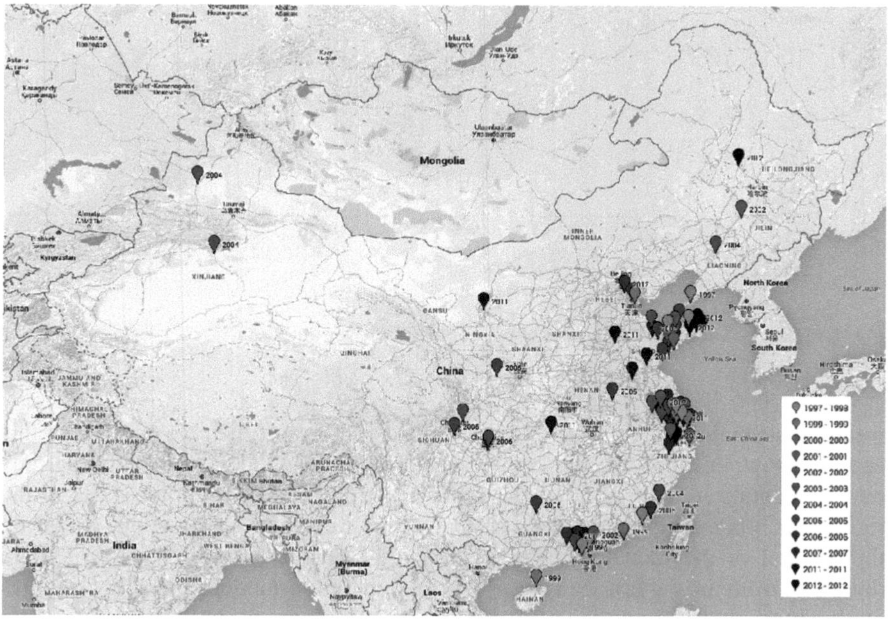

Fig. 2.10 Spatio-temporal distribution of environmental model cities (Source: http://english.mep.gov.cn/inventory/Model_cities/ (accessed 2016-11-04))

pattern. However, the eco-garden city and the environmental-protection model city/eco-model city appear to cluster in Jiangsu, while other provinces show relatively little enthusiasm. A possible explanation for this spatial pattern is that the original environmental model cities have a particular interest in keeping the established scheme alive, because a competing programme will inevitably reduce the political and economic value of the title. Later adopters of the NMCEP programme, however, do not have much to gain by complying with the low-carbon technology module, because the costs are relatively high compared with the potential political rewards. The cities that did not acquire environmental model-city status benefit from no lock-in effects and hence have no reason to take the risk of investing in a 'dying model'. The point of saturation is thus reached, and the environmental model city gradually loses its political relevance.

An alternative explanation would be that a model-city programme develops along several different cyclical courses, where less developed jurisdictions join in available at a lower cost. We consider this possibility from two perspectives. Firstly, if model-city adoption follows economic development, we may be able to detect a spatial diffusion pattern that starts out along the east coast and then spreads to the central and western provinces. Yet Fig. 2.10, which displays the spatio-temporal pattern of adoption, shows no obvious evidence of an east-to-west shift. Secondly, an adoption pattern driven by economic development should lead to changes in the

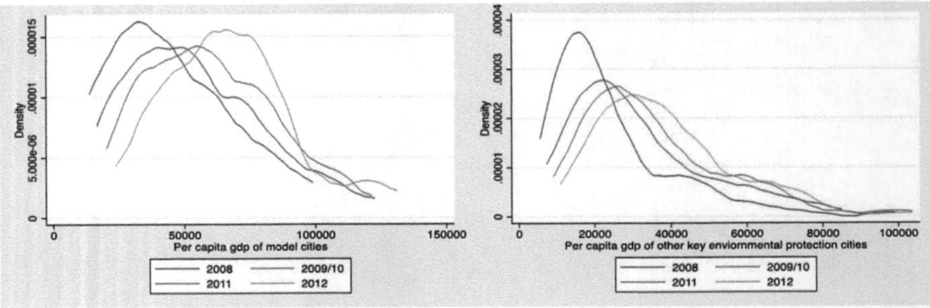

Fig. 2.11 Kernel-density distribution of per-capita GDP over time

wealth distribution among model cities towards the end of the programme. To this end we show, in Fig. 2.11, the kernel-density plots for model cities and other key environmental cities throughout our sample period. The distribution shifts to the right, which is mainly due to the fact that China's economy is growing. However, the left tail of the model-city distribution is getting fatter over time—a feature diverging from the pattern seen in the distribution of non-model cities. This supplies some support for the multicycle hypothesis. Still, if we are to substantiate this claim further, we will have to wait and see when and where new environmental model cities emerge.

2.6 Conclusion

With China's environmental crisis, the demand for knowledge about the role and effectiveness of governance approaches has increased. Much of the mainstream literature centres on the hierarchical command lines of government and party organizations and the incentives these create for protecting or neglecting the environment. More recent research points to a wide variation in local practices and policies that cannot be explained fully with a macroscopic model. Microstudies in this vein emphasize how individual motivations and the intrinsic dispositions of leaders and organizations can trigger productive or counterproductive outcomes. The meso-level—found in the space between one-size-fits-all explanations and atomic observations of individuals and places—still remains under-researched. Yet it is in this space that we find the greatest potential for developing new insights for effective governance instruments.

This chapter represents an effort to fill this research gap. We focus on model cities as a voluntary incentive scheme for protecting and improving the urban environment. Our analysis suggests that factors inherent in the system lead to a growing distance between programme objectives and local practices. We show that, with institutional elaboration and progressive model adoption, the space for creativity gradually narrows. Given that a model scheme is a response to a political paradigm, it is hard to sustain innovative impulses over time. Innovation also becomes less rewarding, as benchmarking criteria are getting unified and standardized. A model scheme will thus inevitably reach a point of saturation. The end of a model scheme's life cycle is reached once a competing programme provides new opportunities to gain political and economic rents.

This research speaks to policy mobility as a process transforming the original 'model DNA' and reproducing a variety of model mutations. We find that particular modes of emulation are shaped by a model scheme life cycle. On a general note, our analysis suggests that model-city schemes represent a valuable contribution to promoting best practice. However, in order to mitigate the negative effects of counterproductive behaviour, central planners need to assess carefully when to initiate a competing scheme. They must ensure that cities participating in an established scheme receive sufficient returns on their investment. Otherwise put, the lifetime of a model must not be too short. On the other hand, planners need to discourage the continuation of a scheme when its marginal social contribution slides below its social cost.

References

Ahlers, A. L., & Schubert, G. (2013). Strategic modelling: 'Building a new socialist countryside' in three Chinese counties. *The China Quarterly, 216*, 831–849.

Baker, S. (2015). *Sustainable development*. New York: Taylor & Francis.

Bulkeley, H., Broto, V. C., & Edwards, G. (2012). Bringing climate change to the city: Towards low carbon urbanism? *Local Environment, 17*(5), 545–551.

Chen, A., & Gao, J. (2011). Urbanization in China and the coordinated development model—The case of Chengdu. *The Social Science Journal, 48*(3), 500–513.

Chen, X., Geng, Y., & Fujita, T. (2010). An overview of municipal solid waste management in China. *Waste Management, 30*(4), 716–724.

Cooke, S. (2009). 7 'Religious work'. In *China's governmentalities: Governing change, changing government* (Vol. 33, pp. 125–150). Hoboken: Taylor & Francis.

Dosi, G. (1982). Technological paradigms and technological trajectories: a suggested interpretation of the determinants and directions of technical change. *Research policy, 11*(3), 147–162.

Dynon, N. (2014). Civilisation-state: Modernising the past to civilise the future in Jiang Zemin's China. *China: An International Journal, 12*(1), 22–42.

Eaton, S., & Kostka, G. (2014). Authoritarian environmentalism undermined? Local leaders' time horizons and environmental policy implementation in China. *The China Quarterly, 218*, 359–380.

Economy, E. (2006). Environmental governance: The emerging economic dimension. *Environmental Politics, 15*(02), 171–189.

Francesch-Huidobro, M., Lo, C. W.-H., & Tang, S.-Y. (2012). The local environmental regulatory regime in China: Changes in pro-environment orientation, institutional capacity, and external political support in Guangzhou. *Environment and Planning-Part A, 44*(10), 2493.

Greenhalgh, S. (2008). *Globalization and population governance in China* (pp. 354–372). In: Global assemblages. Blackwell Publishing Ltd.

Heberer, T., & Schubert, G. (2008). *Regime legitimacy in contemporary China: Institutional change and stability*. London: Routledge.

Heilmann, S. (2008). Policy experimentation in China's economic rise. *Studies in Comparative International Development, 43*(1), 1–26.
Heilmann, S. (2009). Maximum tinkering under uncertainty unorthodox lessons from China. *Modern China, 35*(4), 450–462.
Hillman, B. (2014). *Patronage and power: Local state networks and party-state resilience in rural China*. Stanford: Stanford University Press.
Hoffman, L. (2009). Governmental rationalities of environmental city-building in contemporary China. In *China's governmentalities: governing change, changing government* (pp. 107–124). Oxon: Routledge.
Hoffman, L. (2011). Urban modeling and contemporary technologies of city-building in China: The production of regimes of green urbanisms. In *Worlding cities* (pp. 55–76). West Sussex: Wiley-Blackwell.
Khanna, N., Fridley, D., & Hong, L. (2014). China's pilot low-carbon city initiative: A comparative assessment of national goals and local plans. *Sustainable Cities and Society, 12*, 110–121.
Kostka, G., & Mol, A. P. J. (2013). Implementation and participation in China's local environmental politics: Challenges and innovations. *Journal of Environmental Policy & Planning, 15*(1), 3–16.
Kuhn's, T. (1962). *The structure of scientific revolutions*. Chicago: University of Chicago Press.
Li, C. (1997). *Rediscovering China: Dynamics and dilemmas of reform*. Lanham: Rowman & Littlefield.
Li, W., & Higgins, P. (2013). Controlling local environmental performance: An analysis of three national environmental management programs in the context of regional disparities in China. *Journal of Contemporary China, 22*(81), 409–427.
Li, Y.-w., Miao, B., & Lang, G. (2011). The local environmental state in China: A study of county-level cities in Suzhou. *The China Quarterly, 205*, 115–132.
Liu, L. (2013). Chinese model cities and cancer villages: Where environmental policy is social policy. In I. Wallimann (Ed.), *Environmental policy is social policy—social policy is environmental policy* (pp. 121–134). New York: Springer.
Liu, Q., & Wang, Q. (2013). Pathways to SO 2 emissions reduction in China for 1995–2010: Based on decomposition analysis. *Environmental Science & Policy, 33*, 405–415.
Liu, L., Zhang, B., & Bi, J. (2012). Reforming China's multi-level environmental governance: Lessons from the 11th five-year plan. *Environmental Science & Policy, 21*, 106–111.
Liu, N., Lo, CW.-H., & Zhan. X. (2013). *Collaborative governance and corporate environmental compliance in China*. Conference paper.
Mackinnon, M. (2009). *The people's republic of green*. The Globe and Mail.
Marquis, C., Zhang, J., & Zhou, Y. (2011). Regulatory uncertainty and corporate response: How China's environmental enforcement is catching up to regulation and how business can keep up. *California Management Review, 54*, 39.
Martinot, E., Li, J., & Mastny, L. (2007). *Powering China's development: The role of renewable energy*. DC: Worldwatch Institute Washington.
MEP. (2005). *Evaluation criteria and implementation guidelines for national model city for protecting the environment during the eleventh five-year plan period (Revised)* [关于印发 "十一五"国家环境保护模范城市考核指 标及其实施细则(修订)的通知]. Available at http://www.zhb.gov.cn/gkml/hbb/bgt/200910/t20091022_174766.htm. Accessed on 10 June 2018.
Mol, A. P., & Liu, Y. (2005). Institutionalising cleaner production in China: The cleaner production promotion law. *International Journal of Environment and Sustainable Development, 4*(3), 227–245.
Murphey, R. (1967). Man and nature in China. *Modern Asian Studies, 1*(4), 313–333.
Oberheitmann, A., & Ruan, X. (2013). Low carbon city planning in China. In F. Urban & J. Nordensvärd (Eds.), *Low carbon development: Key issues* (pp. 270–283). London: Taylor & Francis.
Peck, J., & Theodore, N. (2010). Mobilizing policy: Models, methods, and mutations. *Geoforum, 41*, 169–174.

Qi, Y., Ma, L., Zhang, H., & Li, H. (2008). Translating a global issue into local priority China's local government response to climate change. *The Journal of Environment & Development, 17*(4), 379–400.

Ran, R. (2013). Perverse incentive structure and policy implementation gap in China's local environmental politics. *Journal of Environmental Policy & Planning, 15*(1), 17–39.

Stanway, D. (2014). *Pollution plagues China's Baoding despite bid to clean up act*. Reuters.

Tsai, W.-H., & Dean, N. (2014). Experimentation under hierarchy in local conditions: Cases of political reform in Guangdong and Sichuan, China. *The China Quarterly, 218*, 339–358.

Van Rooij, B., Fryxell, G. E., Lo, C. W.-H., & Wang, W. (2013). From support to pressure: The dynamics of social and governmental influences on environmental law enforcement in Guangzhou City, China. *Regulation & Governance, 7*(3), 321–347.

Wang, S. (2009). Adapting by learning: The evolution of China's rural health care financing. *Modern China, 35*(4), 370–404.

Wang, A. (2013a). The search for sustainable legitimacy: Environmental law and bureaucracy in China. *Harvard Environmental Law Review, 37*, 365–440.

Wang, A. L. (2013b). *Search for sustainable legitimacy: Environmental law and bureaucracy in China.*

Wang, L., Zhang, P., Tan, S., Zhao, X., Cheng, D., Wei, W., Su, J., & Pan, X. (2013a). Assessment of urban air quality in China using air pollution indices (APIs). *Journal of the Air & Waste Management Association, 63*(2), 170–178.

Wang, W., Zheng, G., & Pan, J. (2013b). *China's climate change policies*. Hoboken: Taylor & Francis.

Xia, C., & Pahl-Wostl, C. (2012). The process of innovation during transition to a water saving society in China. *Water Policy, 14*(3), 447.

Yu, L. (2014). Low carbon eco-city: New approach for Chinese urbanisation. *Habitat International, 44*(0), 102–110.

Zhou, N., He, G., & Williams, C. (2012). *China's development of low-carbon eco-cities and associated indicator systems (No. LBNL–5873E). Lawrence Berkeley National Lab. (LBNL), Berkeley, CA (United States)*. Available at https://escholarship.org/uc/item/0f4967nd. Accessed on 06 June 2018.

Zweig, D. (1997). Institutional constraints, path dependence, and entrepreneurship: Comparing Nantong and Zhangjiagang, 1984–1996. In J. H. Chung (Ed.), *Agents of development: sub-provincial cities in Post-Mao China* (pp. 213–253).

Chapter 3
Environmental Planning and "Multi-Planning Integration" in China

Jia He, Cunkuan Bao, Jun Zhu, and Jinnan Wang

Abstract Different sectors in China have developed different plans for a variety of purposes. Due to the lack of an overarching design for the whole planning system, inconsistencies, redundancy, and conflicts are often revealed in the process of making and implementing urban plans. From a normative perspective, multi-planning integration may help to improve the effectiveness of implementation and to resolve contradictions amongst different types of plan. Within the context of multi-planning integration, the Environmental Plan faces new challenges as well as opportunities. In this chapter, the authors examine the current status of the Environmental Plan and attempt to discern future trends in connection with multi-planning integration. They review the history of the Environmental Plan and carry out a SWOT analysis of it. They discuss the role of Environmental Plan by analysing its relationship to other plans. They take Benxi City in Liaoning Province as a case showing how an integrated Urban Environmental Master Plan is devised in practice. Finally, they offer some suggestions for future improvement.

Keywords Urban planning · Urban multi-planning integration · Urban Environmental Plan

J. He (✉) · C. Bao
Department of Environmental Sciences and Engineering, Fudan University, Shanghai, People's Republic of China

J. Zhu
Research Centre of Ecological and Environmental Strategy in the Urban Planning and Architectural Institute, Fudan University, Shanghai, People's Republic of China

J. Wang
Chinese Academy for Environmental Planning, Beijing, People's Republic of China

J. Delman et al. (eds.), *Greening China's Urban Governance*, ARI - Springer Asia Series 7, https://doi.org/10.1007/978-981-13-0740-9_3

3.1 Introduction

Planning plays a strategic and leading role in the Chinese development process (Zhang 2014b). In other words, the objectives, targets, and methods of development are determined by planning. However, conflicts and redundancies have revealed themselves during the making and implementation of plans in China. Different planning departments have acted separately to protect their own interests in the planning process (Wang et al. 2008), and a dispersal of authority has led to overlapping areas of competence that frequently has hampered cooperation and coordination. Irrespective of this, as reforms of the planning system have progressed, efforts have been made to coordinate different sectors and to integrate multiple plans (Fang 2007; Hu 2002). This has especially been the case ever since President Xi Jinping called for "integrating three or more plans at city and county level to form one plan and one blueprint" at a central working conference on urbanization in 2013 (Guo 2014).

The Environmental Protection Plan is a special plan for the use of the country's departments of environmental protection (Bao et al. 2013). It contains diverse provisions for protecting the environment, controlling pollution, treating effluents, and saving resources. According to the Ministry of Environmental Protection (MEP), the Environmental Protection Plan must incorporate three red lines: ecological conservation, controls on the emission of environmental pollutants, and efficient use of energy and resources. The consequence of these new provisions has been that the technical methods of environmental planning thus needed upgrading, in order to meet the requirements of defining the three red lines (Tang et al. 2015). Another and equally important issue in the effort to achieve multi-planning integration is how to deal with the relationship between the Environmental Protection Plan on the one hand and other types of plans on the other, such as the Urban Master Plan and the National Economic and Social Development Plan (NESDP) (Hu 2011; Huang 2012).

These are typical challenges when initiating governance interventions for urban sustainability. As Smith and Wiek (2012) have observed, administrative structures critically influence how environmental policies are implemented. They specifically address three traditional features of government that pose challenges to transition towards more sustainable governance: departmentalism, self-reliance, and incrementalism. Departmentalism refers to the absence of horizontal cooperation and coordination between different departments. Due to self-reliance, knowledge and resources from other departments and various non-governmental actors, such as researchers and companies, are not utilized in planning and implementation. Furthermore, incremental changes are seen as problematic, because mitigation of current environmental problems would require fast and drastic transformations instead of slow stepwise policies.

These challenges, especially departmentalism and self-reliance, are at the centre as China tries to reform its environmental planning system. This chapter addresses the attempts to integrate the multiple plans and to overcome the aforementioned challenges in the transformation towards greener governance. Because of the centrality of the planning system to China's environmental governance, it is essential to analyse the recent attempts to improve the coordination and integration of environmental planning in China. A study of China's planning system also generates much-needed empirical evidence for comparisons amongst cities: how to set environmental targets and design plan actions are critical questions in cities everywhere. In this chapter, we first describe the current status of environmental planning in China within the context of multi-planning integration. Our discussion includes a review of the developmental history of environmental planning and an analysis of the relationship between the Environmental Plan and other kinds of plans. It then studies the main challenges and recent developments that support the transformation towards effective multi-planning integration. We then take Benxi City in Liaoning Province as a case showing how an integrated Urban Environmental Master Plan is devised in practice. Finally, we present the main findings.

3.1.1 How to Understand Multi-Planning Integration in the Context of China

Multi-planning integration can be understood as a process of consolidating multiple plans into one integrated plan, so as to reduce overlap and enhance intra-plan coordination (Wei et al. 2012). Before analysing multi-planning integration further, however, we need to think about some related questions, such as what the multiple plans stand for (the subject of the plans), how such plans are to be integrated (the method of integration), and what the single integrated plan is to look like (the form of the one plan).

First, let us look at the "multiple plans" in order to understand their respective subjects (see also Fig. 3.1). The Land Resource Law of China, promulgated in 1999, clearly required that the Land-Use Plan be integrated with the Urban Master Plan. This requirement was strengthened in the amended Urban and Rural Plan Law of PR China, dated 2008. After the 2008 law came into force, for example, Shanghai merged the agencies responsible for each respective plan within the city, thereby playing a leading role in efforts to integrate the two types of plans (Yao 2010). Another example was the pilot reform project launched in October 2003 by the National Development and Reform Commission (NDRC). This involved the integration of three different plans – the National Economic and Social Development Plan, the Urban Master Plan, and the Land-Use Plan – in six cities or counties, including Suzhou City, Anxi County, Qinzhou City, Yibin City, Ningbo City, and Zhuanghe City. Unfortunately, this pilot project was unsuccessful due to various operational deficiencies, such as unmatched basic data, insufficient cross-departmental communication, conflicts of interest amongst different departments, and a failure to set environmental capacity requirements as a limiting condition. By contrast, the attempt at dual-planning integration in Shanghai went well, thanks to provisions for consolidated data, shared land classification, and standardized technical support.

Fig. 3.1 The practice of multi-planning integration in China

Having said that, it also bears stressing that, whilst some attempts at dual-planning or integration of these three types of plans have been made, they have not yet been able to take environmental capacity or ecological conservation into account. They may not help, therefore, to address existing land-use problems in connection with environmental damage, unplanned land development, or excessive resource consumption. It is increasingly important in China today that environmental protection be integrated into economic development and urban planning, so that land/spatial utilization can be harnessed. This need has naturally led to the formulation of the "multi-planning integration" concept, with environmental planning as a necessary component. In 2004, the NDRC and the MEP launched a pilot "four-plan integration" programme to plan reforms proactively, to establish mechanisms of coordination, and to implement policies of spatial restriction.

Secondly, let us look at the outcome of attempts at integration in terms of expectations for the "one plan". In 2013, when the requirement of multi-planning integration was initially put forward by the central government, there was a debate about the definition of the "one" plan. Was this "one" plan an existing plan as defined by law, such as Shanghai's dual plan, which actually used the Land-Use Plan as the basis for incorporating the Urban Master Plan? Or was the "one plan" a newly created plan that consolidated different plans? And if an existing legal plan serves as the basis, which plan shall it be? As per the Notice Letter on Pilot Implementation of the Urban Master Plan in a Tri-planning Integration Context, issued by the Ministry of Housing and Urban-Rural Development in January 2014, the Urban Master Plan was to serve as the "one plan" integrating the National Economic and Social Development Plan, the Urban Master Plan, and the Land-Use

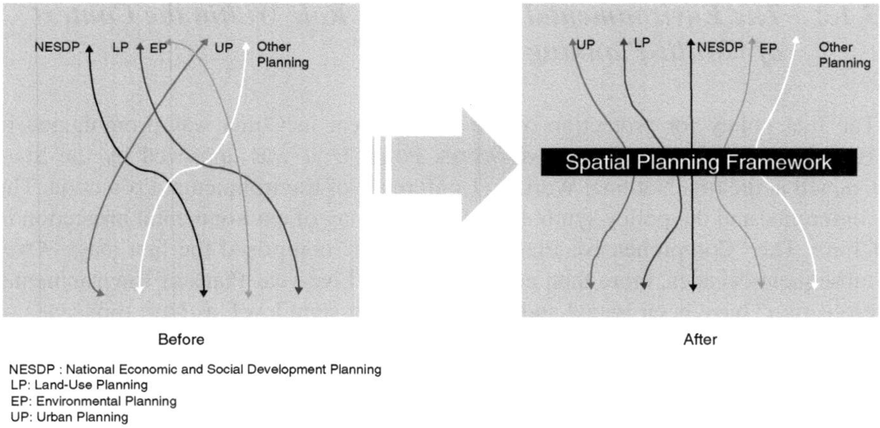

Fig. 3.2 Understanding multi-planning integration in China

Plan. However, the Notice Letter on Pilot Implementation of Multi-Planning Integration, issued collectively by four ministries in August 2014, did not clearly define what the "one plan" should be like. Instead it simply emphasized the importance of establishing a proper index system, in order to ensure that the integrated plan meets core development targets whilst maximizing local advantages. Typical for departmentalism and self-reliance, the situation as of 2016 is that different planners tend to regard plans of their own type as best suited to serve as the "one" plan. For example, urban planners think an urban plan should serve as the "one plan," whilst development planners think a development plan would be best. Land-use planners call for a Land-Use Plan; environmental planners support an Environmental Plan. Moreover, if a new integrated plan is formulated, what will be the relationship between this new plan and existing plans? And how are subsequent concerns – regarding responsible departments, functional boundaries, development methodologies, the scope of the plan, etc. – to be resolved?

The outcome of multi-planning integration largely depends on the correct understanding of "integration"(Zhang 2014a, b). The common understanding amongst scholars in China today is that "multi-planning integration" does not necessarily refer to the development of a single planning document (Wang 2011; Hu 2011; Wang 2009). It can refer instead to the formation of an integrated and systematic planning mechanism or platform, in the context of one city, in which the departments responsible for various plans coordinate their actions on behalf of a common vision (target) (see Fig. 3.2).

3.1.2 The Environmental Plan and Its Role Within the Context of "Multi-Planning Integration"

The first policy for protection of the environment in China was promulgated in 1973. Known as "The Thirtytwo Words Policy",[1] it was approved by the State Council at the first National Working Conference of Environmental Protection. The conference and the policy symbolized the beginning of environmental protection in China. The "Comprehensive Plan Running First" comprised the first stage. Over subsequent decades, more than eight rounds of "Five-year Plans of Environmental Protection" have been issued and implemented at state level, as have thousands of plans at city or county level. The Environmental Plan Department has been upgraded from a subordinate office to a ministerial department, and great progress has been achieved. Generally speaking, the Environmental Plan in China has been upgraded from being just a chapter in the NESDP to being an independent special plan. Furthermore, the governing concept of the Environmental Plan has gone through three stages of evolution. The stress in the first stage was on "controlling three types of pollutants, with end-pipe treatment". The stress in the second was on "spatial regulation and controls on the emission of pollutants". In the third and current stage, finally, the emphasis is on "sustainable development and overall optimization".

Within the context of the construction of an ecological civilization (cf. Introduction; Delman, Chap. 8), there is a close relationship between the Environmental Plan and other plans that steer economic development. Over many years, a planning system has been developed in China, with the NESDP, the Land-Use Plan, and the Urban Master Plan serving as its "three pillars". If we are to understand the role of environmental planning within the multi-planning integration framework, we must first analyse the relationship between the Environmental Plan and other plans, such as the NESDP, the Land-Use Plan, and the Urban Master Plan.

Firstly, let us discuss the relationship between the Environmental Plan and the NESDP. China has a tradition of issuing plans for economic and social development. According to China's Constitution, the role of the NESDP is to set out the overarching strategy for economic and social development in the nation (or a region) during a given period. It lays down the overall tasks and objectives of development, and the policies, measures, and strategies for meeting them. Maintaining ecological balance and preventing and controlling environmental pollution are amongst the key goals set out in the NESDP. As an essential part of the NESDP, the Environmental Plan has to harmonize these objectives with those of national economic development. The process of making the Environmental Plan and that of making the NESDP must therefore be synchronized. The main tasks and objectives of the Environmental Plan should be incorporated into the NEDSP, and the latter must involve the rational use of natural resources and must respect the carrying capacity of the environment,

[1] "The Thirty-two Words Policy", rendered in English, stands for "protecting the environment for the benefit of the people through overall planning, rational distribution, comprehensive utilization, public involvement, and converting waste into resources".

thereby maintaining ecological balance for the sake of sustainable development. In reality, however, the Chinese government has paid much more attention to economic growth ever since it embarked on its path of reform and development, in 1978. The NEDSP has had the leading role in the planning system, and the Environmental Plan has played a supplementary part, rather than forming the basis for the NEDSP. However, the situation has been changing since 2007, when the central government laid out its strategy for building an ecological civilization. More and more local governments have recognized the importance of environmental protection and have realized that the NEDSP must be formulated on the basis of a prior consideration of resource capacity and environmental security.

In addition to the NEDSP, the Urban Master Plan and the Land-Use Plan play an important part in the planning system as statutory plans. The Urban Master Plan manages the process of urbanization and sets out the direction of urban development, including the scale, spatial layout, and functional orientation of the overall construction. The relationship between the Urban Master Plan and the Environmental Plan is mainly reflected in the size and layout of the urban population, which determines the size and layout of the environmental infrastructure. The structure and layout of urban industry determines the functional zones, ecological red lines, and pollution control objectives of the Environmental Plan. In practice, the traditional planning process entails developing the Urban Master Plan first, after which the Environmental Plan is drawn up – in accordance with the Urban Master Plan. However, since local governments are eagerly pursuing urbanization and industrialization, this traditional approach has failed to ensure that the restrictions placed by the Environmental Plan on urban economic and social development are respected. This has jeopardized sustainable development in the urbanization process.

For a long time, as compared with the statutory plan mentioned above, the Environmental Plan has played a subordinate and supporting role in the planning system. However, with the government's decision to build an ecological civilization, the relationship between the Environmental Plan and related plans is changing (Lü et al. 2016). The goal of the NESDP has also evolved, from the simple pursuit of speedy economic growth to the pursuit of development quality. This shift – from the single pursuit of economic development to an attempt to achieve a balanced and coordinated development of environmental, social, and economic systems – leads to the necessity of multi-planning integration. The respective roles of the NESDP, the Land-Use Plan, the Urban Master Plan, and the Environmental Plan may be described as "Leadership, Enforcement, Platform, and Mitigant". In Fig. 3.3, we analyse this integration from a SWOT perspective.

The NESDP is "the leader" in the system: it lays down the overall objectives of development and serves as the basis for the other plans. The role of the Land-Use Plan and of the Urban Plan is to ensure that an appropriate balance is struck between the population, the economy, the environment, and the resource base. In so doing, they serve as the enforcement and platform for the NESDP. Then last, but not least, the role of the Environmental Plan is that of a mitigant.

How are we to understand the concept of "mitigant" (*zuoyao*佐药), and what are the tasks and functions of the Environmental Plan? Mitigant is a factor which "supports"

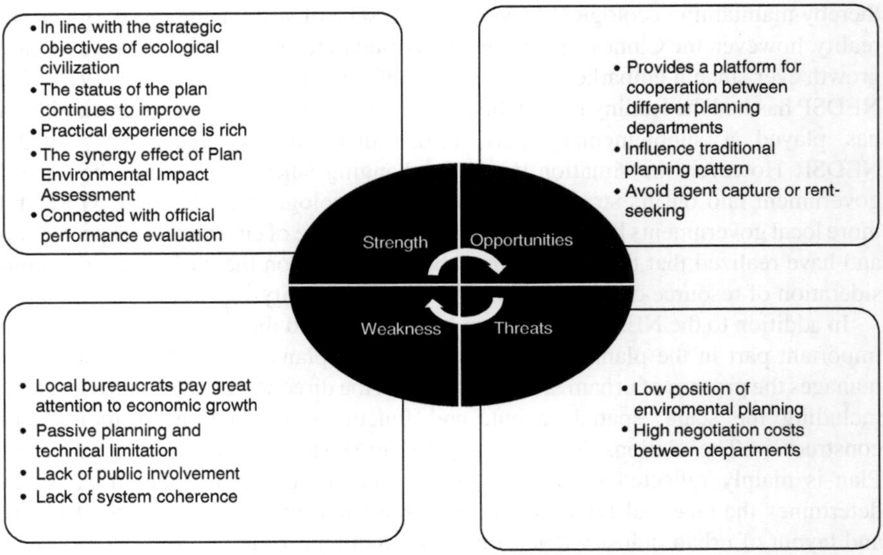

Fig. 3.3 The relationships involved in multi-planning

or "moderates". The tasks and functions of the Environmental Plan within the overall planning system are shown thusly in Box 3.1. The role of the Environmental Plan is to ensure the bottom line of environmental and ecological security. It does this by working as a basis for, and restraint upon, the NESDP, the Land-Use Plan, the Urban Plan, and other plans.

Box 3.1 The Role and Function of the Environmental Plan in the Multi-Planning System

How to Understand the "Mitigant"

- To support the implementation of the NESDP, the Urban Master Plan, the Land-Use Plan, and other plans
- To mitigate adverse impacts arising from the NESDP, the Urban Master Plan, the Land-Use Plan, and other plans
- To hold the bottom line of environmental security, thereby furthering the construction of an ecological civilization

Detailed Functions of the Environmental Plan in a Context of Multi-Planning Integration

- To achieve environmental targets and objectives as defined in the NESDP, the Urban Master Plan, the Land-Use Plan, and other plans

(continued)

Box 3.1 (continued)

- To formulate plans for pollution control and to implement measures to protect the environment against adverse impacts generated by the NESDP, the Urban Master Plan, the Land-Use Plan, and other plans
- To establish environmental restrictions or limitations on development plans and to optimize development layouts and facilitate the upgrading of industrial structure/composition, by instituting stringent ecological red lines, streamlining cleaner production standards, and promoting the saving of energy and the reduction of waste.

Tasks for the Environmental Plan

- To prevent and mitigate negative environmental impacts, to determine the carrying capacity of the environment, and to ensure the bottom line of environmental protection
- To implement high-level strategies for protecting the environment
- To coordinate with the NESDP, the Urban Master Plan, the Land-Use Plan, and other plans by providing spatial regulation to ensure ecological safety
- To balance conflicts of interest amongst government, market, and public actors by encouraging public involvement

3.1.3 Current Challenges and Positive Trends

Overall, during the 2010s, the policy environment has become more supportive for transition towards sustainability because the central government pays now more attention to people's increasing demands on environmental protection, democracy, legislation, fairness, and justice (cf. Wang et al., Chap. 11) in addition to economic development. Specifically, the aim of building an ecological civilization, set out by the Central Committee of the Communist Party of China (CPC) (cf. Introduction, Delman, Chap. 8), is an indicator of this change as it provides a supportive platform for the development of a strong Environmental Plan (see Fig. 3.3). As mentioned in the report of the 18th National Congress of the CPC in 2012, the construction of an ecological civilization should take a prominent place in all aspects of economic construction, political construction, cultural construction, and social construction – for the long-term benefit and sustainable development of China. Furthermore, in 2017, the 19th National Congress of Communist Party of China (CPC) defined comprehensive and synchronized development as a main tool to address the national key challenge of unbalanced regional development. In other words, environmental protection now plays as important a role as economic and social development. Therefore, the Environmental Plan, which provides the blueprint for achieving an ecological civilization, should also be given a prominent place, and it must be integrated fully into other types of plan – i.e. the NESDP, the Urban Master Plan, and so on (Wang et al. 2013).

In addition, it is not just the central leadership that attaches importance to environmental protection; local governments also pay more attention to the issue now. This is particularly the case since their record in reaching environmental targets – saving energy, controlling the emission of pollutants, etc. – plays an important part now in how their political performance is evaluated. Thus local governments make great efforts to set and to achieve appropriate environmental targets. On the one hand, if the targets are too high to reach, the performance of the local government officers will be assessed unfavourably, and their political careers will be jeopardized; on the other hand, if the targets are too low, environmental quality will not be improved enough to satisfy either the public or these officers' superiors (Delman, Chap. 8).

Furthermore, as mentioned above, thousands of environmental plans have been implemented in practice. Methods and techniques have evolved; knowledge and understanding have increased. The number of environmental experts and professionals is growing year after year, showing that the Environmental Plan has led to great expansion of environmental governance capacity.

Challenges have also revealed themselves in practice. The first issue here is the lack of coordination with other plans, which may result in low support from other departments and high negotiation costs – or even to overruling by other departments with greater power. In most cases, for example, the Environmental Plan has had to yield to the NESDP or the Urban Master Plan, with their focus on economic benefits. The second issue is the poor interlinkage within the Environmental Plan system itself. In other words, the Environmental Plan system lacks a comprehensive design linking the overall plan to special plans, i.e. from the state-level plan over provincial and urban plans to county-level plans. In some cases, the objectives and targets at state level could not be fully interpreted or implemented by the subordinate plan.

Last but not the least is the bad positioning of the Environmental Plan within the overall Chinese planning system. The Environmental Plan is only a special plan within the larger planning system, and the impact of the latter may be to sabotage the function of environmental protection for sustainable development. Although thousands of environmental plans have been issued over the years, the development of theoretical study and legislation in this area has somehow slowed down. For example, only the concept and general principles of environmental protection are mentioned in the introductory part of the Environmental Protection Law; detailed information in terms of content and working procedures is lacking. Furthermore, whilst thousands of environmental plans have been implemented in practice, few technical documents or guidelines have been issued by the environmental authorities.

3.1.4 The Urban Environmental Master Plan in Benxi City

In order to adapt to the requirements of the New Urbanization Plan issued in March 2014, the Ministry of Environmental Protection is promoting pilot practice of the Urban Environmental Master Plan. Twenty-four cities have been chosen for a pilot project, including Dalian, Guangzhou, Fuzhou, Haikou, and Benxi. Benxi, a city of more than 1.7 million, is located in the mid-area of Liaoning Province in northeastern China. It is known as the “steel capital of China” Although Benxi, as one of the old industrial cities of northeastern China, has made important contributions to the process of economic development, the city is also confronted with severe environmental problems. These include air pollution from steel factories and ecological degradation caused by iron mining. Benxi also faces the challenge of the new-type urbanization advocated by the central government since 2014. According to the main function, in the area plan of Liaoning Province (2014), shown in Fig. 3.4 and Fig. 3.5, Benxi will be the priority area of development within the province, with an important part to play in an urban strategy for building up the Shenyang City Cluster.

During the 5-year plan period for 2016–2020, Benxi will face double pressure. On the one hand, the city suffers from problems caused by an environmentally unfriendly industrial structure and physical layout created during its period of heavy industrialization, as well as by the fast urbanization of previous years. These issues will need to be addressed in order to meet public demands. On the other hand, subsequent negative environmental impacts arising from new-type urbanization must be avoided by means of the Environmental Plan. Benxi will need to integrate the Urban Master Plan and the Environmental Plan in order to address the environmental problems caused by the industrial structure and physical layout that arose during the process of urbanization and industrialization. The MEP therefore chose Benxi as a pilot city for integrating the Environmental Plan into a multi-planning system.

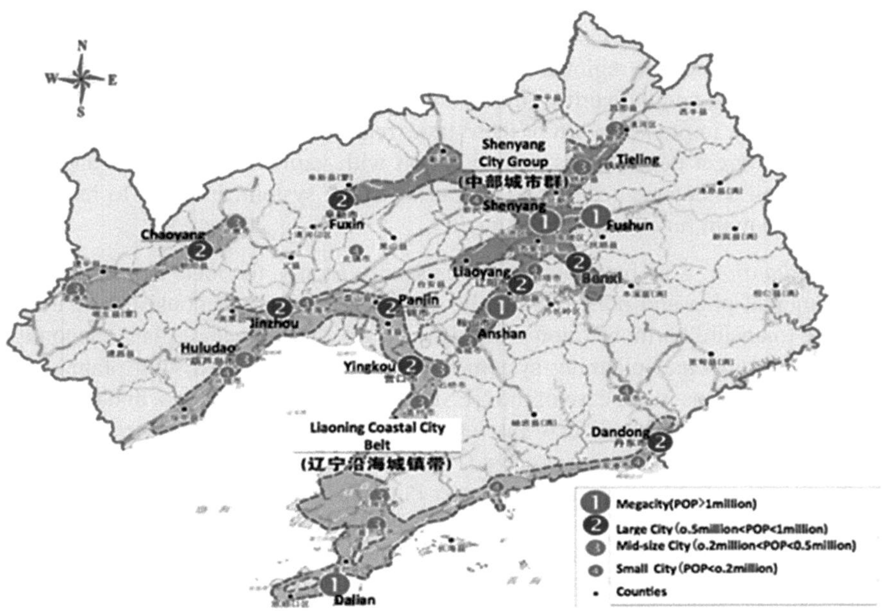

Fig. 3.4 Spatial layout of the Liaoning urbanization strategy in an urban strategy for building up the Shenyang city cluster

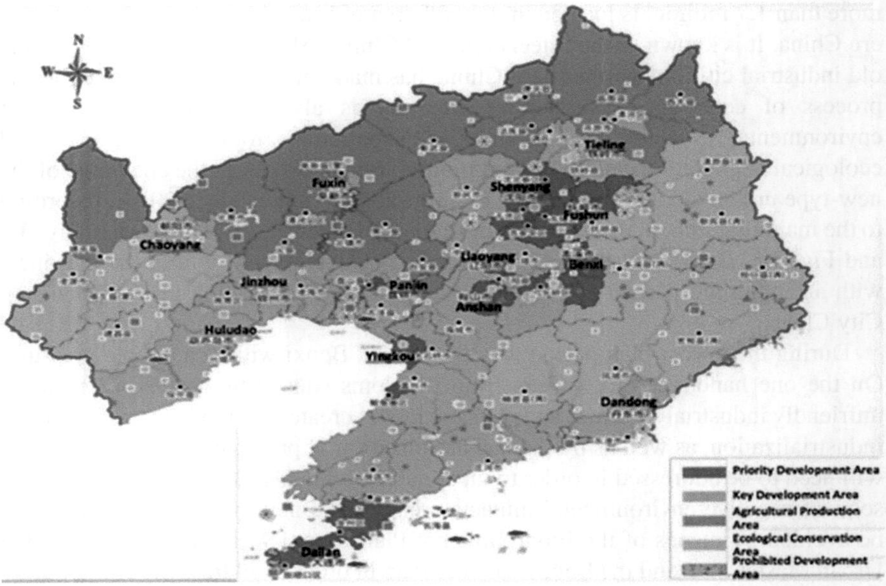

Fig. 3.5 Four main function areas of Liaoning

Unlike the traditional Environmental Plan, the Urban Environmental Master Plan is not limited in its content to setting targets and measures for the control and treatment of environmental pollution (Bao et al. 2013). Besides combatting environmental pollution, the Urban Master Plan is charged with such tasks as determining the carrying capacity of the regional environment, defining spatial control through ecological red lines, and delivering environmental justice through public involvement. The successful implementation of these tasks requires close horizontal cooperation amongst different departments, collaboration with various stakeholders and partners, as well as determined efforts to transform the system. In order to reach the goals, three main functions were defined:

1. *Coordination*: This function has three aspects. Firstly, the local system of environmental plan and policies must be coordinated and kept consistent, particularly to ensure harmonization with higher-level environmental plans. Secondly, the environmental planning system needs to be coordinated with planning in other sectors, as in the case of the NESDP, the Urban Master Plan, and the Land-Use Plan. Thirdly, conflicts of interest amongst different stakeholders have to be addressed via the planning process (He et al. 2011).
2. *Leadership*: The Environmental Master Plan must take a leading role in the area of environmental protection. The principles of environmental protection, the basic requirements for achieving it, and the overall arrangement of efforts in this sector should be specified in the plan. Environmental Plan at lower level can be worked out on the basis of this plan, as can special plans for environmental protection like the Environmental Treatment Plan, the Facility Construction Plan, the Major Pollutants Emission Control Plan, etc.

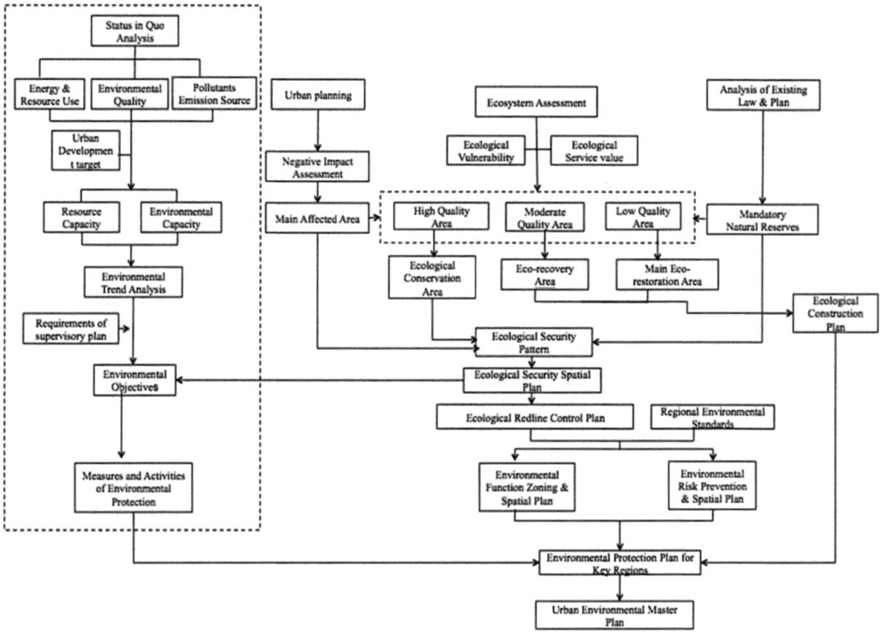

Fig. 3.6 The content and procedures of the Benxi Urban Environmental Master Plan

3. *Constraints*: The key function of the Urban Environmental Master Plan is to define the bottom line of environmental protection and to regulate economic growth and urban development accordingly. Here the MEP has laid down three red lines: ecological spatial controls, controls on the emission of major pollutants, and industrial environmental permission.

To fulfil these tasks, the content and procedures shown in Fig. 3.6 were included in the Benxi Urban Environmental Master Plan. Only the content and procedures within the dashed frame were traditionally included in the traditional Environmental Plan, with its focus on controlling and treating environmental pollution. The Urban Environmental Master Plan puts greater stress on spatial planning for achieving the integration of urban planning, land-use planning, and environmental planning as shown by the contents and procedures outside the dashed line areas in Fig. 3.6.

On the one hand, the interrelationship between environmental protection and urban planning is highlighted in the Urban Master Plan through the identification of negative impacts arising from urban planning and through the definition of the main affected area. On the other hand, the Benxi Master Plan puts great stress on the spatial analysis of ecological health and security patterns. Strategies and measures for spatial regulation in the process of urbanization have been systematically established, in the form of ecological redline controls and environmental function zoning.

The Benxi Master Plan includes promising elements that may help to overcome the traditional governance obstacles in the transition towards sustainability.

However, as the implementation of the plan has just started at the time of this writing, it is still too early to evaluate its success.

As mentioned above, urban planning takes place at multiple levels in China, in accordance with the country's administrative setup. From top to bottom, these levels are the state, the province, the city, and the county. The Benxi case can only demonstrate the integration of Urban Master Plan and Environmental Plan at the city level. It is evident that governments at all levels must make more attempts in this respect to find the best way to achieve multi-planning integration at each level of the planning system.

3.1.5 Is a Positive Reconstruction of the Environmental Plan Possible?

Could there be a positive reconstruction of the Environmental Plan, so as to make it more adaptive and effective within the context of multi-planning integration? And how could such a reconstruction be achieved? Will the new plans be strong enough to overcome the three governance challenges: departmentalism, self-reliance, and incrementalism?

Firstly, as compared with plans of longer standing (e.g. the NESDP, the Land-Use Plan, and the Urban Master Plan), the Environmental Plan is just in its initial stages. It is still defining its scope, building its theoretical foundations, establishing systematic methodologies, forming consulting markets, and engaging involved professionals. Moreover, departments of environmental protection have long been weaker than other planning authorities. Given these factors, it will be a challenge to ensure that the Environmental Plan is not weakened or even eliminated in the course of multi-planning integration.

Secondly, how can restrictions on development be implemented? What can prevent economic development from breaking through environmental bottom lines and ecological red lines? Environmental Plan must rise from a supplementary to a dominant position in relation to planning in the urban, economic, and industrial sectors. A mechanism can be thereby built up which will operate effectively and over the long term to protect natural resources and the environment.

Thirdly, we need to allow for different patterns of multi-planning integration, rather than applying a standardized format across the entire country. Consider the differences between, say, Shanghai and Xinjiang/Tibet. The former has vast economic resources but very little land; the latter have vast land areas but far more modest economic resources. In addition, multi-planning integration should be differentiated at the various administrative levels (e.g. province, city, county), in order to reflect the varied responsibilities of different levels of government. The central government, for example, must take responsibility for coordination at a macro level. Local governments such as urban governments should provide services, supervise markets, and furnish social governance and environmental protection on a local scale. Different local governments ought to be encouraged to seek out scenarios for multi-planning integration that best suit local conditions.

Fourthly, different patterns of multi-planning integration should be established for integrating the NESDP, the Urban Master Plan, the Land-Use Plan, and the Environmental Plan in diverse major function zones. The Environmental Impact Assessment process should be embedded in the NESDP, the Urban Master Plan, and the Land-Use Plan, thereby ensuring that imperatives of environmental and ecological protection are given sufficient consideration in other plans.

Fifthly, academic research on environmental planning is progressing too slowly, as compared with research on other types of planning (e.g. urban planning). With its existing theoretical foundations, environmental planning is thought incapable of meeting emerging demands for ecological protection or of adequately addressing recent challenges in connection with natural resources and the environment. It is urgent that both the central government and local governments establish workable mechanisms for environmental planning – mechanisms that give due consideration to the spatial and timeline characteristics of environmental issues and the complex interrelationship amongst environmental elements.

Furthermore, the various plans must all embody the values of the ecological civilization programme (cf. Introduction). The critical importance of environmental protection must permeate both their formulation and their implementation. Environmental Impact Assessment planning should be used as a tool to facilitate and coordinate planning in other sectors, thereby forming a common vision and a shared objective: to strive for a development model that furthers ecological conservation and is friendly to the environment.

Finally, it is essential to solve the problem of departmentalism for the success of multi-planning integration. The new plans aim to weaken departmentalism by providing a platform for communication and collaboration. This goal is supported by the current political emphasis on the construction of ecological civilization that creates conditions for substantial transformations, instead of incremental minor changes. The new model of planning also encourages departments to collaborate with other actors such as experts, companies, and communities, to diminish self-reliance. However, the deeply embedded hierarchies and power structures amongst departments (cf. Introduction; Delman, Chap. 8) and biased priorities and incentives of policy making in cities (Brehm and Scensson, Chap. 2) place tough challenges to the attainment of these goals.

Acknowledgements In writing this chapter, we have received support from the National Natural Science Foundation of China (Grant No. 41271508) and from the second batch of the Key Project of the National Social Science Foundation of China, 2013 (entitled "Public Involvement in City-level Development Planning").

References

Bao, C. K., He, J., Xu, M. L., et al. (2013). 我国环境规划体系框架设计(System design of environmental planning in China). *Urban and Rural Planning, 10*, 21–26.

Fang, C. L. (2007). 区域规划与空间管治论 (Theory of regional planning and spatial governance). Beijing: The Commercial Press.

Guo, L. Q. (2014). "一张蓝图多规融合" ("A blueprint for the integration of multi-planning"). *Intelligent Building and City Information, 211*(6), 24–25.

He, J., Bao, C. K., Zhou, L., et al. (2011). 规划环境影响评价中规划协调性分析实施框架 (The framework of coordination in planning of environmental impact assessment). *Tongji Academic Journal (Natural Science Edition), 39*(1), 94–99.

Hu, X. W. (2002). 我过区域规划的发展态势与面临问题 (Development trends and issues of regional planning in China). *Urban Planning, 26*(2), 23–26.

Hu, X. W. (2011). 着力健全规划协调机制 (Focus on the mechanism for the coordination of planning). *Urban Planning, 35*(1), 14–15.

Huang, Y. J. (2012). 体制改革与规划整合——对国内"三规合一"的观察与思考 (Institutional reform and integration of plans: Observations and thoughts of the "integration of the three plans"). *Modern Urban Research, 29*(2), 10–14.

Lü, H. D., Wan, J., Wang, C. X., et al. (2016). 环境规划参与"多规合一"多种模式的思考与建议(Thoughts and suggestions of multiple modes for environmental planning in multi-planning integration). *Emvironmental Protection Science, 42*(3), 24–27.

Smith, R., & Wiek, A. (2012). Achievements and opportunities in initiating governance for urban sustainability. *Environment and Planning C: Government and Policy, 30*, 429–447.

Tang, Y. Q., Liu, D. S., Li, J., et al. (2015). 关于环境规划在"多规合一"中定位的思考 (Reflections on the position of environmental planning in multi-planning integration). *Environmental Protection, 43*(7), 55–59.

Wang, W. S. (2009). "三规" 关系与城市总体规划技术重点的转移实践 (Relationship between the "three plans" and practices in shifting the technical emphasis in the urban master plan). *Urban Planning, 33*(5), 14–19.

Wang, J. Y. (2011). 分权下的"多规合一"——深圳新区发展历程与规划思考 ("Multi-planning integration" within the context of decentralization: A study on the development of Shenzhen new district). *Urban Studies, 28*(20), 23–29.

Wang, J. Y., Wu, D. Y., & Chang, X. (2008). 我国空间规划体系的时代困境与模式重构 (Limitations and reconstruction of the spatial planning system in China). *Urban Issues, 17*(4), 62–68.

Wang, J. N., Xu, K. P., Lu, J., et al. (2013). 国家环境功能区划制度的战略定位与体系框架 (Strategic role and framework of National Environmental Function Zoning). *Environmental Protection, 22*, 35–37.

Wei, G. J., Dong, W., & Sun, H. (2012). "多规整合"研究进展与评述 (Review of the progress in "multi-planning integration" in China). *Urban Planning Forum, 26*(1), 76–82.

Yao, K. (2010). "资源紧约束"条件下两规的有序衔接——基于上海"两规合一"工作的探索和实践 (Resource constraints and the integration of land-use planning and urban planning with Shanghai as a case). *Urban Planning*, *34*(3), 26–31.

Zhang, Q. (2014a). 基于多规合一的规划体制创新研究——以莆田实践为例 (Planning system innovation based on the "multi-planning integration with Fujian as a case"). *Fujian Construction, 193*(7), 7–10.

Zhang, W. W. (2014b). 中国政治制度的独特优势 (The unique advantage of China's political system). *Guangming Daily.* Accessed 25 Aug 2014.

Chapter 4
Environmental Policies Enter the Educational Sector: Different Shades of Green at District Level

Outi Luova

Abstract Environmental policies are cross-sectoral by nature, and during their implementation, they have to be adapted to the differing contexts of different sectors. Furthermore, there is variation in the implementation process between different geographical areas. In this chapter, Luova looks at how two inner urban districts in the city of Tianjin (one poor and one wealthy) implement their new environmental plans in the educational sector. Chinese urban districts have substantial freedom to adapt educational policies to fit local conditions, and schools themselves are encouraged to define and to emphasize their own specific priorities. Under these circumstances, how can environmental policy goals be reached in an effective way, asks Luova? Her findings show that, with regard to numerical targets, the poor district did better than the rich one. This was because a green profile was cheap to acquire, it opened the door to substantial external funding, and it made possible an improved bargaining position vis-à-vis the city government. However, the rich district was able to provide solid resources for its schools, enabling them to excel in terms of qualitative objectives: i.e. nurturing green lifestyles and promoting respect for the environment. The study highlights (1) specific features of environmental policy implementation in the educational sector, (2) various policy implementation processes in inner urban districts, and (3) the importance of clear qualitative targets in the definition of policy goals. In the conclusion, Luova proposes a model for examining governance dynamics in urban China.

Keywords Tianjin · Urban districts · Policy variation · Environmental education

Cities have a key role to play in addressing environmental issues, and urban governments are the decisive actors in planning and implementing any such policies. In order to get a comprehensive picture of what is actually going on in specific Chinese cities, we must look below the city level and study the sub-municipal areas. This is because, in Chinese cities, the effective power to determine policy outcomes rests with sub-municipal governments. Furthermore, environmental policies are cross-sectoral by

O. Luova (✉)
Centre for East Asian Studies, University of Turku, Turku, Etelä-Suomi, Finland
e-mail: outi.luova@utu.fi

J. Delman et al. (eds.), *Greening China's Urban Governance*, ARI - Springer Asia Series 7, https://doi.org/10.1007/978-981-13-0740-9_4

nature, and during their implementation, they have to be adapted to the differing contexts of different sectors. For example, different districts and even schools are encouraged to define and emphasize their own specific ways of organizing environmental education. Under these circumstances, how can environmental policy goals be reached in a city in an effective way? In this chapter, I compare two inner urban districts in Tianjin, one of China's eco-pilot cities, in an effort to throw light on the interplay between local features and the systems of planning and implementation. Based on this study, I propose a model for governance dynamics in urban China.

Sub-municipal units in China can be divided into five different categories, reflecting their differing administrative status: inner urban districts, outer urban districts, outer districts, rural counties, and new development areas. There is significant variety in the governance patterns of these different types of administrative area, so it is meaningful to focus on just one of them. The number of variables is also diminished thereby. I came by the idea of comparing Hedong and Heping districts in this chapter because of the intriguing fact that Hedong, one of the poorest inner urban districts in Tianjin, had the highest percentage of environmental "green" schools among its primary and middle schools in 2014. The richest inner urban district, Heping, came in third, after the wealthy outer urban district Jinnan. Common sense would suggest that rich districts are in a better position than poor ones to promote environmental education, which usually does not figure among the most important subjects in a school's teaching syllabus. An analysis of the two cases may be expected, therefore, to disclose some unexpected dynamics of environmental governance in inner urban districts.

Environmental awareness building is the focus of this chapter, for two reasons. Firstly, it is a neglected but central aspect of environmental policy. Secondly, unlike most other environmental policies – e.g. protection of the environment from harmful industrial discharges – environmental education has no negative impact on GDP growth. It does not necessarily cause extra costs; nor does it impose restrictions on production. It is thus easier to implement. However, all environmental policies in China – including those in the educational sector – are restricted in their implementation by the fact that they do not directly contribute to GDP growth. The latter objective, namely, still determines the priorities of local policy implementation in China. Most environmental efforts in that country belong to the area of so-called soft policy (Delman, Chap. 8).

The focus of my analysis is on the implementation of two Tianjin's policies: (1) its five-year plan for environmental education and communication (2011–2015) and (2) its regulations on environmental education. I have chosen the establishment of green schools as a specific case because it requires more effort and institutionalization than does conducting a few brief campaigns or arranging a certain number of environmental classes.

This study is based on official documents and media reports, as well as upon research mainly conducted during fieldwork in 2014 and 2015, during which time I interviewed staff members at various schools as well as at government departments at city and district levels. During these interviews, 2–7 persons took part in the discussion. I first provide an overview of educational policies and environmental policy frameworks at the municipal level in Tianjin. I then describe the implementation of educational policies in two of Tianjin's districts. In the conclusion, finally, I discuss my findings.

4.1 Chinese Urban Districts and Previous Studies on Diversity in Policy Implementation

Among the different types of sub-municipal units in China, inner urban districts are often referred to as "old urban districts". They form, namely, the historical centre of each respective city. Other districts may include large rural areas and may just recently have entered a period of rapid urbanization. Educational levels are usually higher among the residents of inner urban districts, and cultural offerings tend to be denser in such districts.

Recent studies have discovered divergences between these different types of sub-municipal areas in terms of urban development, administration, and community services. The different developmental paths reflect such features as history, geographical location, local modes of governance, traits of residents and leaders, and the economic basis of the district in question (Zhang 2005; Hsing 2010; Luova 2013).

Urban development projects in China today are generating significant new dynamics that tend to increase diversity among urban administrative areas. Different urban areas have differing potential for development, reflecting the strategic location of the district or county in question. The income which entrepreneurial local governments are able to generate from land speculation varies considerably. Inner urban districts tend to be much smaller in area than other districts, which means they have less potential for land speculation.

State entrepreneurialism is worth a closer look in a study of environmental governance too. The brilliant study by You-tien Hsing (2010) shows how land speculation-led state entrepreneurialism generates quite different social, political, and economic dynamics in inner-city districts, outer urban areas, and rural counties. The concept of state entrepreneurialism refers to the entrepreneurial management style of local-level government officials in their efforts to maximize their achievements. It is usually used in reference to the forceful drive of local officials to attract more inward capital for urban development in various public-private constellations (Duckett 1998; Chien and Wu 2011). It is equally helpful for describing the attempts of environmental administrators to gather resources in various creative ways. The entrepreneurial character of district-level leaders and school headmasters is an additional local feature to which this study pays attention.

As for the implementation of educational policy, recent studies on environmental education in China have mainly analysed its developmental trajectory and main challenges (Kwan and Lidstone 1998; Wu 2002; Lin and Ross 2005; Lee and Huang 2009;Tian and Wang 2016). Regional variation has been the topic of just a handful of works. McBeath with her colleagues (McBeath et al. 2015) as well as Zeng et al. (2009) have paid attention to divergences in green schools and in environmental education practices in China, but their analytical focus has been on macro-regional variation or on school-specific features. The divergences in question, they find, largely reflect differences between rural and urban areas, between the east and west of the country, and between different schools in terms of history, resources, and developmental strategies. These scholars have not paid attention, however, to the impact of differences on local modes of governance or at the sub-municipal level.

Thus, previous research on policy implementation in China has identified several different reasons for local variation. How then are we to place the different variables into a broader theoretical framework? Researchers tend to take the view that variation in policy implementation is natural and good, because the localities in which policies are to be implemented are usually highly divergent, whereas they regard variation in policy *outputs* as problematic. In that respect consistency should prevail. Since the 1970s, several waves of studies have tried to develop a comprehensive theory of policy implementation (for overviews, see deLeon and deLeon 2002; Matland 1995; Kohoutek 2013). The pioneers of this research strand developed the so-called top-down theory, which emphasizes the role of national-level decision-makers and planners in policy implementation. A second generation of scholars in this area called for a bottom-up approach instead. They criticized the top-down theory for neglecting the importance of local stakeholders and "street office" staff members, whom they saw as the actors that actually convert policy into action. Finally, a third wave of researchers has tried to find a golden mean between these two opposed approaches. As Jan Kohoutek (2013) concludes in his article on policy implementation in higher education, no theories have yet emerged which can satisfactorily address the complexity of the variables that explain variation in policy outcomes. Instead, research on implementation has tended to shift to alternative spheres, such as institutional, organizational, and governance theory.

In this chapter, I apply a bottom-up approach combined with theories of governance. My focus is on the meso level, because policies at both the city and the district level are important for policy outputs. I place this bottom-up approach within a governance-theory framework. Jan Kooiman (2007) maintains that governance issues arise in interactions between various political and social actors. He distinguishes, moreover, between *intentional* and *structural* levels of governance interaction. The intentional level refers to the goals, interests, and purposes of the interactions in question. At the intentional level, individuals who can commit an organization – in the case of this study department and school leaders – are the focus of analysis. The structural level concerns the circumstances that limit, broaden, or condition the intentional level. In my structural analysis in this chapter, I pay particular attention to incentive systems in the policy implementation process. Both material incentives (payments, material benefits) and nonfinancial ones (promotions, bonuses, rewards) have been found to motivate purposeful performance (Clark and Wilson 1961; Cai 2004), although sometimes in a perverse manner (Ran 2013). Kooiman contends that, due to the complexity of governance situations – as well as tensions within and between the roles and actors involved – governance unavoidably has unintended consequences too.

4.2 Environmental Education Policies in Tianjin

Tianjin is China's fourth largest city, with a permanent population of more than 15 million. It is located on the coast just southeast of Beijing. Among China's coastal cities, it has been one of the slower reformers, due its proximity to Beijing, the political (and conservative) centre of China (Hendrischke 1999). Since 2006, however, Tianjin has been the site of a state-initiated and state-supported development drive comparable to that seen in Guangzhou in the 1980s and Shanghai/Pudong in the 1990s. This is particularly true of the Binhai New Development Area within Tianjin.

Tianjin has also lagged behind other Chinese megacities in terms of environmental protection and awareness building. In 2002, however, an energetic new mayor by the name of Dai Xianglong took office and proceeded to put environmental issues on the agenda. In that year, under his leadership, the People's Government of Tianjin set out to become a model city of environmental protection (Tianjin EPB 2003). The district governments embraced this goal as well (Tianjin Government 2002). Tianjin's application for the status was successful, and the Ministry of Environmental Protection designated Tianjin as a model city of environmental protection in 2006. The following year the city government issued a master plan for the construction of an ecological city. This plan served as an overall local strategy for the city's ecological governance until the year 2015. Districts followed their own localized version of this plan. Environmental projects gained momentum, and in 2007 China and Singapore signed an agreement to build the Sino-Singapore Eco-city in Tianjin's Binhai New Area.[1] When Dai Xianglong left his post as Tianjin's mayor in 2007, he had created a regulatory basis for the improvement of environmental protection in Tianjin and started several flagship projects. Given Tianjin's weak starting point, the work was still at a very initial stage at the time of his departure. The city's plans and guidelines had not yet resulted in significant outcomes, and Tianjin still lagged behind other major cities in environmental protection and awareness building.

At the beginning of this decade, when the city's 12th five-year plan (FYP) – for the 2011–2015 period – was being prepared, Tianjin made a new push to improve its work on environmental protection. It seems there was now more pressure to do the work well. For example, in the outline for environmental communication and education work for the FYP12, it was frankly admitted that Tianjin was far behind other major cities, especially when it came to the funds provided for environmental education, the size of the staff involved, and the number of suitable environmental education units (Tianjin EPB 2011b). But there was a clear determination to reach national standards (at the very least) in all of the key areas.

In an indication of the city's new determination, one of the tasks scheduled for the 12th FYP was to prepare an evaluation system for the performance of civil servants on environmental issues (Tianjin EPB 2011a, b). This is the most effective way to get things done in China, because bonuses and opportunities for career advancement are the kind of incentives that call forth the most powerful motivation among civil servants and cadres to perform well. In 2012, accordingly, in an effort

[1] Such eco-city projects have been criticized for being land speculation-oriented local entrepreneurialism rather than sincere efforts to promote sustainable development (Chien 2013). Notwithstanding the criticism and reports on the challenges faced by eco-city projects, among China's larger eco-city projects, the Sino-Singapore Eco-city is regarded as a major effort to standardize eco-city criteria in China (Chang et al. 2016).

to achieve a substantial improvement in environmental education, the city issued its regulations on environmental education (Tianjin People's Congress 2012). Only the regulations themselves were binding; various related documents were not. The implementation of the latter had to be ensured in other ways. The annual work plans, for instance, included detailed targets, clear timetables for various tasks, descriptions of the division of labour to be applied between units, specific measures to safeguard the implementation of the most important tasks, and a clear accounting of which units would lead the work and which would participate as partners (Tianjin EPB 2013, 2014, 2015).

We must however remember that, despite the powerful new impetus to increase environmental education, the responsibilities mentioned in the new policy documents were just a minor part of all the tasks which fall on the shoulders of district governments. Still, the success of municipal leadership in China is not understood only in terms of economic development and social stability; it is also measured in terms of certain environmental objectives, such as the reduction of carbon emissions and PM2.5 and PM10 particles. Environmental education did not feature in the criteria for evaluating the performance of civil servants during the period of this study; hence it was of minor importance in the daily work of district governments. Since bad performance in the area of environmental education and communication was thought to taint Tianjin's national image, efforts in this area surely received specific attention during the 12th five-year plan.

4.3 New Plans for Environmental Education in Primary and Middle Schools

Environmental awareness building may be said to consist of three aspects: a knowledge of ecological and human systems, the use of investigation and evaluation to develop and apply skills and to heighten sensitivity to nature, and a sense of responsibility coupled with action (UNESCO-UNEP 1994, 1). The first two are important building blocks of environmental education. The third part, however, is crucial for furthering environmental awareness. Environmental education in China has developed within the framework of socialist moral education. In practice, environmental education in China has resembled that in many other countries in emphasizing environmental knowledge, while the nurturing of environmentally friendly values and lifestyles has been ignored (Lee and Williams 2009, 12).

In Tianjin, the ultimate objective of the educational plans was to cultivate environmentally aware residents. What kind of framework did the new policies provide for attaining this goal?

To start with, the conceptual framework for environmental awareness building was rather incoherent, with several parallel competing programmes that left many opportunities for flexible and selective local district-level approaches. In the spirit of the "green" 12th five-year plan, the city had prepared separate, parallel, and partly overlapping five-year plans for combatting climate change and promoting environmental protection, ecological urban construction, sustainable development, and environmental education. These plans used varying concepts and set out different targets – a pervasive problem of the Chinese planning system (Bao et al., Chap. 3).

In addition, a programme to modernize compulsory education in Tianjin (2008–2011) added to the diversity. It was decided that all schools and districts would emphasize their special characteristics.[2] This programme too paved the way for developing environmental education. It supplied additional funds for improving schools' facilities and promoting a green environment. These funds could be used to strengthen environmental education at the schools. The programme also provided resources for schools to strengthen their brand. A good brand is important for schools in China, and having a clear brand of its own also gives a school points in the evaluation process (Wang 2007; Shi 2015). A "green" brand was a popular choice, so schools used these funds to build a green profile. (Interview Hedong EB 2015).

The five-year plan for environmental education and communication complemented Tianjin's pilot programme of ecological urban construction. The FYP covered not just schools and kindergartens but also companies, media outlets, villages, residential communities, and administrative agencies. With regard to schools, the plan set out both general goals for all schools and specific targets for so-called green schools (also called environmentally friendly schools). Schools in general were encouraged to incorporate environmental issues into teaching in a wide variety of ways, such as through arts classes, essay-writing activities, and the reuse of old materials. Schools were also urged to hold environment-related competitions and to encourage pupils to participate in volunteer activities. All pupils at primary and middle schools would receive environmental education. The FYP further stipulated that, by the end of 2015, half of the primary and middle schools in the city would qualify for green school certificates, of which half would be city-level certificates with stricter criteria. At the city level, these targets were in fact already reached in 2012, which meant that the less enthusiastic districts did not need to make extra efforts to establish green schools. The regulations on environmental education stipulated among other things that districts must provide enough financial support for environmental education and that each district should establish a unit for environmental education. The main responsibility for arranging environmental education in schools was placed, naturally enough, on the Bureau of Education, but the Environmental Protection Bureau was tasked with providing expertise, materials, and other resources for schools.

Schools could apply for nomination as a district-level, city-level, national-level, or even international green school. Schools that won the nomination were reassessed every 4 years. The city-level evaluation criteria were divided into five groups, each worth 10–25 points, with 100 points in all (Table 4.1). To satisfy the city-level green school criteria, schools needed at least 90 points (see, e.g. Jinnan district 2014).

[2] The profiling of special characteristics (*tese* 特色) has been a nearly obligatory exercise in all development projects in China.

Table 4.1 Evaluation criteria for city-level environmentally friendly schools

Categories and points	Tasks	Evaluation method
School administration points: 5 + 5 + 5 + 5 + 5 (25)	(1) Establish a leading group and decide division of labour, (2) incorporate the idea of sustainable development into school administration, (3) responsible persons participate in training, (4) availability of study materials, (5) maintain good archive of relevant documents	Oral and written reports, inspection tours
In-class activities Points: 5 + 5 + 10 (20)	(1) Incorporate environmental issues into disciplines according to their specific features, (2) adopt new modes of teaching and lead pupils to study environment through research, (3) utilize new technology and multimedia	Review of schools' teaching plans and self-appraisal, pupils' research papers and other achievements, Internet databases, and multimedia materials
Participation in out-class activities Points: 5 + 10 + 5 + 5 (25)	(1) Arrange outreach activities for pupils; (2) include environment in science and arts activities, especially during specific environmental days; (3) utilize schools' information channels to increase awareness; (4) arrange education about environmental ethics and values as well as environmentally friendly lifestyles	Review reports of activities, check waste-sorting bins and recycled articles, check statistics for saving energy and water
Environment Points: 5 + 5 + 5 + 5 (20)	(1) Campus should be well-planned and with green areas, and 90% of pupils should participate in maintaining the green areas, (2) create a positive atmosphere for keeping the campus environment clean and tidy, (3) treat waste properly, (4) develop pupils' and teachers' routines for saving water and energy, reducing pollution, recycling waste, and reusing materials	Inspection tours
Achievements Points 5 + 3 + 2 (10)	Teachers or pupils receive (1) national-level awards (during 3 years), (2) city-level awards (during 2 years), (3) district-level awards (during 1 year)	Review award certificates and related publications

Source: Jinnan District (2014) Summary of the evaluation criteria for environmentally friendly schools (kindergarten)(*Tianjinshi huangjing youhaoxing xuexiao (youeryuan) pinggu biaozhun*天津市环境友好型学校(幼儿园)评估标准)

Among these key performance indicators, only a few of them related directly to awareness building. The criteria for in-class education concerned multidisciplinary, research-oriented teaching and the use of current e-learning technologies. Participation in out-class activities relating to environmental protection and the cultivation of environmental values were each worth five points in the assessment system. Promoting recycling, resource-saving habits, and the separation of waste was worth five points. A large proportion of the points could be collected by building infrastructure, improving the tidiness and greenness of the campus, and using modern educational techniques and methods. The methods of evaluation left opportunities for slack implementation: it is easy to write positive reports and to tidy up the campus before inspection visits. The criteria for green schools, with their emphasis on environmental knowledge, reflected the instilled old approach, while environmentally friendly values and lifestyles received less attention. Furthermore, as we shall see in the next section, environmental protection and sustainability could be defined in many different ways, making for vagueness in the definition of "greenness".

4.4 Divergent Adaptation and Implementation of Policies at District Level

The following analysis of interdistrict differences is based on a comparison of two inner urban districts. Heping and Hedong are both typical old urban districts with a small land area; however, Heping deviates from the general pattern of such districts in having a small population (Table 4.2). The residential population of Heping has fallen, due to the conversion of old residential areas to commercial and other non-residential use. This has generated substantial income for the district. Both districts devote an equal share of their public finances to education; however, given Hedong's smaller revenues and much larger population, its expenditures in this area only come to about one fourth of those in Heping.

However, these general features of inner urban districts do not explain why the poorer of the two districts, Hedong, has the highest percentage of green schools. Let us look at the district-specific features of these cases, in the hope of finding further explanations for the diversity. In what follows, accordingly, I conduct an overview of the early steps of environmental education, after which I examine how the new policies affected activities and policy outputs in the two districts.

Table 4.2 Basic indicators for Heping and Hedong

	Heping	Hedong
Area (km^2) 2003	10	40
Permanent population 2003	412,554	730,000
Permanent population 2013	364,200	966,800
GDP/permanent resident 2003	13,346	5452
GDP/permanent resident 2013	188,851	29,452
GDP 2013 (million CNY)	68,742	28,451
Public finance expenditure/permanent resident (RMB) 2013	16,323	4515
# operating expenditures for education of public finance expenditures in 2013	33%	32%

Sources: Tianjin Bureau of Statistics (2014) *Tianjin Statistical Yearbook 2014*; Tianjin People's Government (2004, 2014) Tianjin District and County Yearbook 2004 and Tianjin District and County Yearbook 2014

4.4.1 Heping District

Heping is the old central district of the city. It was also the administrative centre of Tianjin until the early 2010s, when the municipal administration moved to the more spacious Hexi District. Heping takes pride in "condensing the fine historical traditions of Tianjin, being the signal district of the city and being able to provide excellent education" (TCPBL 2012). As the traditional central district, Heping has been proud of being at the forefront in Tianjin in many policy areas, including environmental protection and education. In its official documents from various periods, the district government has stressed its aspiration to lead on environmental issues in Tianjin. In a document from circa 2007, for example, the Heping District Government stressed its leading role in promoting environmentally friendly activities in the city (Heping District n.d. ca 2007). Moreover, the district's environmental protection plan of 2013 specifically mentioned that, by actively arranging environmental education in schools and residential communities, Heping would be able to keep its lead in Tianjin in promoting environmental education (Heping District 2013a).

To excel in performance, the district leaders initiated various eco-pilot projects. In 2002, when Tianjin set out to become a model city of environmental protection, Heping was selected as one of the pilot districts (Heping District n.d. ca 2003). The district also incorporated environmental issues into its plan to maintain its nomination as a national-level civilized district.[3] The central theme in this plan was to save resources and to develop an environmentally friendly and livable Heping. One way to promote these goals was to establish more green schools (Heping District 2008). Hence, environmental education in Heping was mainly promoted within the framework of building a civilized district.

The first initiatives for environmental awareness building in Heping schools came in the late 1990s, from a teacher who was also the founding member of the Tianjin Friends of Green, a local NGO. This geography teacher integrated environmental protection into his classroom teaching at Middle School Number 90 and arranged related activities for his pupils as well. These activities were partly developed in cooperation with the Jane Goodall Institute. Self-study materials produced by the school gained attention throughout the city (Interview Heping Education Bureau (EB) 2015).

Middle School Number 90 was among the first in Tianjin to be nominated as a city-level green school (interview at the school). In the teaching activities at that school, the responsibility of individuals in protecting the environment was placed at centre from the beginning. This can be seen from one of the educational activities – known as "one plus six" – that was introduced there. The "one" in the name of the programme refers to one child who can pass on the knowledge of green habits to six adults, i.e. the child's own parents and grandparents. Later on, the district govern-

[3] The national civilized city (district) programme has a broad agenda that covers political, social, and cultural goals. A document from Heping district on this programme emphasized moral and ideological education, the spiritual and cultural needs of people, the conduct of civil servants and party officials, and the improvement of the ecological environment (Heping District 2008).

ment took responsibility for promoting environmental education in the whole district, including this particular programme. (ibid.)

In 2007, as a result of its continual efforts to support environmental education in schools, the Heping Environmental Protection Bureau was granted the title of national-level excellent unit of green school construction (Heping District n.d. ca 2008).

Compared to the schools in outer districts, those in inner-city districts had much smaller campuses, and their prospects for building greeneries or small gardens were limited. This disadvantage was countered through the establishment of an eco-environmental section at the district's Technology Education Centre. Pupils attending school classes at the centre could try out growing plants and conducting experiments in small indoor greenhouses (Interview Heping EB 2015).

By 2011, there were already two national-level green schools and 17 city-level green schools in the district, out of 49 primary and middle schools in all (Heping District 2011; Tianjin Education Commission 2012, 516, 524). So, when the five-year plan for environmental education and communication was issued in 2011, Heping was already doing well in terms of environmental awareness building, and after 2011, it continued its ambitious work and even improved its record. As the regulations from 2012 required, Heping started issuing annual plans for environmental education. In its first such plan, issued in 2013, the district decided on a division of labour among its different departments. With regard to schools, important tasks were given to the departments of finance, education, and environmental protection. Of these, the first mentioned was mandated to provide sufficient funds, while the second was urged to include environmental issues in curricula, to introduce environmental education into the schools' assessment system, and to encourage pupils to arrange many kinds of environment-related activities. The department of environmental protection, finally, was made responsible for arranging activities, conducting training, disseminating information, and carrying out overall coordination of environmental education in the district. Primary and middle schools were instructed to teach ecological life skills and to promote the spirit of "ecological civilization" (Heping District 2013b).

We can assume that, in accordance with its promises in the 2013 annual plan for environmental education, the district provided at least necessary if not generous financial and material support for schools. For example, the 2014 plan for environmental education explicitly stated that the district would provide free environmental education materials such as CD-ROMs to primary schools and kindergartens. The same document also stated, with a hint of self-praise, that due to their excellent resources, the different government departments would be promoting environmental education (Heping District 2014). In an interview, furthermore, the vice headmaster of Middle School Number 90 mentioned that the district's educational bureau had provided his school with large sums for environmental education.

To fulfil the new rules, the district established one unit for environmental education at the "Palace for Children and Youth". The Palace was elevated from a city-level to a national-level educational unit in December 2015 (MEP 2015). The work plan from 2014, moreover, stipulated that a similar unit – but now in accordance with district-level criteria – would be established as well; and it assigned the task of organizing environmental hobby clubs for school kids to the Palace for Children and Youth. In order to strengthen the institutionalization of environmental education, the 2014 plan also encouraged the establishment of additional green schools; however, it mentioned no numerical targets.

According to the vice director of the educational bureau, the main change brought by the 2012 regulations was to standardize activities. Schools were provided with teaching materials, and they were obliged to give classes and to arrange activities relating to environmental protection. All of them, moreover, could now use the facilities of the units for environmental education. All in all, the environmental education provided by Heping included – in addition to instruction about environmental protection and activities to keep campuses clean and green – a number of attempts to teach pupils to take action and to adopt green habits.

4.4.2 Hedong District

Hedong likes to boast about its long history: its Dazhigu area was the original centre of Tianjin in the seventeenth century. Over time, Hedong's position deteriorated, and today it is one of the poorest inner urban districts of Tianjin. In the current stage of Tianjin's urban development, Hedong's location is less advantageous. It is located on the inner urban fringe, between Heping and Dongli (the latter being a rich and fast-developing outer urban district). Hedong is known as a "workers' district", due to its industrial past; and although the factories have been moved and are now further away from the city centre, the residents of the district are still predominantly working class. Educational levels among them are lower than among their counterparts in Heping. Compared to Heping, moreover, Hedong has a much larger population and significantly lower revenues (see Table 4.2). Nevertheless, while it is less wealthy than most other inner urban districts, it puts a considerable emphasis on education. For example, it has sought to be best in the city in carrying out the modernization of compulsory education (Hedong District 2011). The high priority it gives to education can also be seen from the share of budgetary expenditures it assigns to that sector, which is as large as in Heping. However, the actual sums it allocates to education are much lower (see Table 4.2), which is probably why schools in Hedong are not considered to be among the best in Tianjin (unofficial discussions in Tianjin 2015).

Since 2002, when Tianjin set out to become a model city of environmental protection, Hedong for its part worked for the same goal. Indeed, it was nominated as an advanced district in this pursuit (Hedong District 2006). In its long-term eco-district plan for the 2006–2015 period, Hedong's central goal was to create a coherent system for saving natural resources and for optimizing the natural environment, the living environment, production, and culture. This would result, it was hoped, in the creation of an environmentally friendly urban ecological district (Hedong District 2008). So, in addition to emphasizing the need to protect the natural environment, Hedong stressed the importance of cultivating an environmentally friendly society. The development of green schools was an integral part of the eco-district plan (Hedong District SCPC 2010).

When Tianjin launched its green school programme in 1998, Liuwei Road Primary School in Hedong was among the first to accept the challenge. Environmental issues were emphasized in the school's curriculum, and its green activities were frequently publicized in the local media. The school also joined the "green future" activities of the Volkswagen Company (Liuwei Road Primary School 2010; Cuan and Guo 2013). It gained the title of both national-level environmental demonstration school and city-level green school.

The district's educational bureau allowed schools to develop their educational methods in a flexible way, but it wanted them to encourage pupils to study and observe their own environment and to find solutions to everyday problems. Hence, pupils were encouraged to take environmental action in the science field. In Middle School Number 102, for example, pupils were encouraged to use the school's excellent laboratory equipment to solve problems they detected in their everyday living environment and even to apply for patents on the basis of their research findings. As a matter of fact, pupils at this school did receive several patents for their innovations (Visit to the school 2015).

By the end of the year 2011, during which Tianjin tightened its requirements for environmental education, Hedong already had – out of 41 primary and middle schools – one international, three national-level, and 18 city-level green schools, as well as several district-level green schools. It had thus already reached its goal and so did not need to improve its performance (Hedong District 2012, 207, 320). The 2014 annual plan for environmental education explicitly mentioned that 2–3 new green schools and kindergartens would be striving for the title that year.[4] This goal was exceeded. By 2015, the number of national- and city-level green schools in Hedong had risen to 30 (Tianjin Daily 2016).

The plan also came with a list of different activities for schools to arrange in order to promote environmental awareness. But the 2014 work plan made no promises of financial support; nor did it mention establishing an educational unit. By the time 2015 came around, Hedong still had not been able to establish an educational unit. It can be assumed that the reason for this lay in a scarcity of funds. While the Heping plan had promised to include environmental education in schools' evaluation criteria, the Hedong document planned to utilize a stronger instrument: the responsibility system. According to the 2014 plan, specific targets for environmental education and communication (*Huanjing xuanchuan jiaoyu gongzuo mubiao zerenzhi* 环境宣传教育工作目标责任制) would be included in the responsibility system for civil servants. The head of the district's educational bureau described the use of administrative incentives as specific to Hedong. Moreover, in addition to using the responsibility system, the district urged schools to strive for green school nominations by including them in the schools' assessment system: the higher the

[4] Hedong's annual plan for environmental education is only available on the Internet for the year 2014. Nor was it possible to acquire the plans at the bureaus of the local government.

level of nomination (district-city-national-international), the better the assessment results (Interview Hedong EB 2015).

One of the reasons for the emphasis on environmental issues in Hedong was surely the fact that the head of the district's educational bureau was who he was. This man had previously been a headmaster of Middle School Number 102, known for its green activities. His ambitious work as headmaster had borne fruit, earning him promotion to the position of department head. In discussions I had with the head of the school district and with representatives from Middle School 102, a prominent topic was cooperation with companies, with institutions of higher learning, and with other actors. For example, Middle School 102 had received a 200,000 RMB donation from the municipal committee for science and technology, as well as financial support from Nankai University and Tianjin Medical University. The school had also cooperated with the British Embassy, as well as with NGOs such as Friends of Green (Interview at Hedong EB 2015). Other schools had received support for environmental awareness building from Shell, Samsung, and Amway (ibid.), not all of which are always associated with care for the environment! It would appear that, while they harboured a genuine interest in environmental education, schools in Hedong also found that the "environment" was a magic word that opened up new opportunities for external funding. Unlike their counterparts in the rich outer urban districts, headmasters in the poorer inner urban districts had to use much of their time collecting external funds (Interview at Dongli EB 2015); and promoting environmental education was apparently a good way of attracting such. This would seem to be a plausible explanation, inasmuch as Hongqiao, Tianjin's other poor blue-collar inner district, also had a high percentage of green schools. Among inner urban districts, Hongqiao ranked third, after Hedong and Heping (see Table 4.3). Hence, providing environmental education was among other things a way to get more funding and to improve conditions and performance at a school. Cooperating with international environmental programmes also brought prominence to a given school.

At the same time, the objectives and features of the various external funding schemes steered the development of environmental education in different directions – whether towards science, recycling, green areas, saving resources, or providing knowledge. Thus, in addition to increasing the disparity of resources, external funding led to greater diversity in terms of content.

A look at the other green schools in Hedong reveals this diversity of approach. Environmental education at Tianzhuang Middle School involved the integration of environmental issues into all classes, especially those in moral education, improvement of the environment and green areas on campus, and the training of pupils to recycle and to save water and energy (Tianzhuang Middle School 2016). On the other hand, Xiangshandao Primary School developed a brand of "green ecological culture", in which "green" referred to pupil's vitality and energy. Pupils were seen as green sprouts, for which the school was to provide a healthy, harmonious, and free environment for growth and development (see, e.g. Tianjin Daily n.d. ca 2013, China Education Newspaper 2012). A green and sustainable culture was defined

Table 4.3 Number and percentage of city-level green schools in Tianjin's inner urban districts (2014)

Inner urban district	Number of green schools	Percentage of green schools
Hedong	**22**	**(53.7%)**
Heping	**21**	**(48.8%)**
Hongqiao	17	(44.7%)
Hexi	23	(39%)
Hebei	17	(36.2%)
Nankai	14	(24.1%)

Source: Tianjin EP (2014) 天津市个区县市级环境友好型学校(绿色学校)创建进度统计表. (*Tianjinshi ge qu xian shi ji huangjing youhaoxing xuexiao (lüse xuexiao) chuangjian jindu tongjibiao*. Statistical table of the progress in the creation of environmentally friendly schools (green schools) in Tianjin's districts and counties)

here in terms of social and cultural well-being, with no specific reference to protection of the environment.

The uniting theme in the schools of Hedong District was the cultivation of pupils' moral and human qualities. Moral education had been a central feature of environmental education in China from the beginning. Nor did it require any additional funds, unlike improving the greeneries on campus or investing in laboratory equipment. For a poor district, therefore, it was a cost-effective way to improve performance.

In the main, then, the high percentage of city-level green schools in Hedong can be explained by (1) genuine interest on the part of the schools, (2) the presence of opportunities for them to show good performance and hence to improve their bargaining position with the municipal government, and (3) the improved fund-raising prospects that schools enjoyed – in a financially weak district – if they developed a green profile.

It was in order to create a green profile that Hedong's educational bureau sought to excel in environmental education. For the district's schools, moreover, few if any resources needed to be expended in order to qualify as green through the promotion of environmental moral education. Having a green brand improved the chances of a given school to develop cooperation with new partners and to attract external funding and other resources. And for the district, having a large number of green schools with improved resources enhanced its status and bargaining position vis-à-vis the city government, giving it greater muscle in negotiations for more resources and improving its leaders' chances for promotion. Clearly, the leaders of Hedong's educational sector and of some of its schools acted in an entrepreneurial manner. They used innovative measures to create a brand, to increase their resources, and to improve their negotiating position. The possibility of accruing "administrative credits" increased the district's chances for a favourable performance assessment, giving it an incentive to promote environmental education.

Hedong lacked the resources to build a strong basis for solid and coherent environmental education on its own. However, schools in the district with ambitious principals were able to open windows to external green funding, even if many schools did not avail themselves of the opportunity. "Greenness" was generally regarded as a good brand, and due to the priority the city gave to the construction of ecological districts, it was also a highly attractive and relevant tool for demonstrating performance.

4.5 Discussion

This chapter started with the initial question of why poor Hedong performed better in quantitative terms than did rich Heping with regard to green school nominations. The answer can be found in the different dynamics of interaction between plans, leaders, local features, and government incentive structures. My main finding is that the EBs of the two districts had partly differing goals in their pursuit of excellent performance. Leaders in both districts strove genuinely for improved environmental education. The difference lay in their other (perhaps primary) strategic goals. Heping aimed at a top-class performance in order to demonstrate a leading position in the city and to reinforce its brand as an ecological district; Hedong used greenness as a tool to acquire additional resources. These differing strategies and goals resulted mainly from different resource bases. The variation in resource bases, in turn, reflected differing location, history, economic structure, and population characteristics. Leaders in both districts could be described as entrepreneurial, but only in Hedong did the leaders of the educational bureau and school headmasters need to use their entrepreneurial skills in order to reach their strategic goals.

Both districts excelled in terms of their quantitative goals. In terms of the content of the environmental education provided, the goals were reached only unevenly in Hedong. Heping had enough assets to provide materials, expertise, and facilities for all schools. These good resources secured coherence in the district's efforts at environmental education. Hedong lacked the resources to build a strong basis for solid and coherent environmental education on its own. However, schools in the district with ambitious principals were able to open windows to external green funding, even if many schools did not avail themselves of the opportunity. Thus, two factors contributed to the emergence of variation in the content of the environmental education provided. Firstly, the assessment indicators for green schools allowed for wide variation in the definition of "green". In some cases the result was a quite pale green approach. In other cases a deep green school was the consequence, in which due heed was paid to all three aspects of environmental education: providing knowledge, developing a sensitivity to nature, and appreciating the need for responsibility and for action. Secondly, schools with active principals were able to create the conditions for excellent scientific inquiry into environmental issues and to offer exciting activities through various partnership programmes. Here too, wider variation was the result, due to the selection of different partners and funding organizations. The differing constellations of partners meant that different schools took diverse approaches in their efforts in this area, which again led to variation in terms of the emphasis different schools put on the three central dimensions of environmental education. If a combination of leadership and available resources appears to be central for environmental education, future studies should explore whether wealthy districts with leaders who perceive no utility in environmental education would show poor performance in their number of green schools.

In the case of environmental education, China's green governance system revealed not gaps in implementation but rather gaps in outputs, as well as distortions in incentives. Heping and Hedong both enthusiastically promoted the establishment of green schools, but green schools per se were not the main goal. Instead, the two districts pursued these policies in part because of other objectives. Greenness was instrumentalized in different manners and for different purposes. Furthermore, an excellent implementation of numerical targets did not ensure satisfactory results from a qualitative point of view.

Figure 4.1 illustrates governance interaction at district level and how it influences policy implementation in the case of environmental education. The party state produces plans which are transmuted during the process of stage-wise adoption from central to district level. At the crucial stage of implementation – in cities and districts – the policy-adoption process is influenced both by institutional factors and by a number of non-governmental actors. Borrowing Kooiman's distinction between structural and intentional levels of governance interaction, I show in the figure how policies are transformed when they encounter structural and intentional levels of governance and interact with non-governmental actors. Societal pressure is a relevant factor in environmental governance, but in the case of environmental educa-

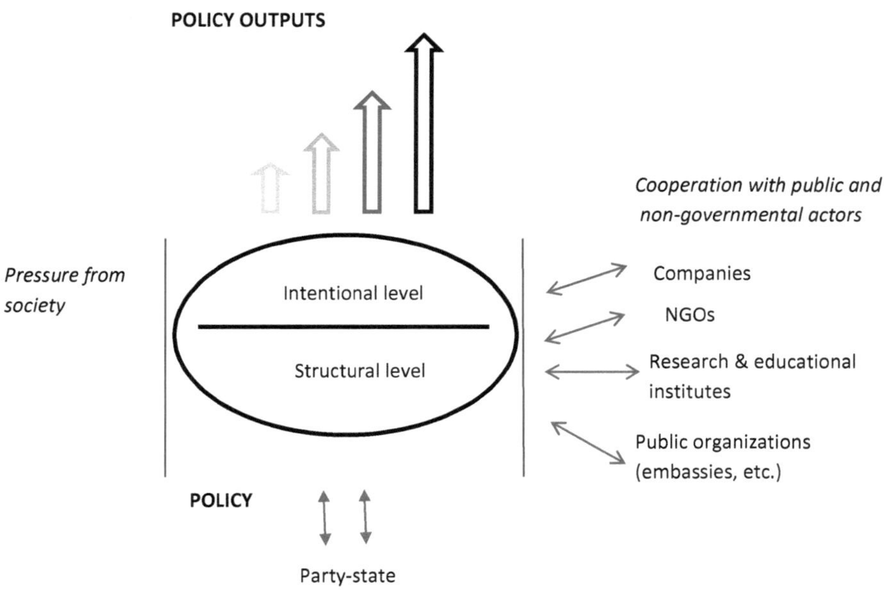

Fig. 4.1 Governance dynamics at district level

tion, it does not play a greater role. In this study, budget allocations to education, incentive systems, and cooperation partners emerged as significant factors at the structural level. These factors influenced the governance dynamics by limiting, broadening, steering, and conditioning the actions that the leaders took. Then, depending on the visions, goals, and capabilities of the district and school leaders (intentional level), the policies were implemented along diverse paths, resulting in a great variety in policy outputs, from pale and symbolic results to desired outputs.

If we relate these findings to the suggestions of Ren and Liu in Chap. 6, we find that current practices of environmental education in these two inner-city districts correspond to the needs of low-carbon development, that is, changes in behaviour among the wealthy and educated. The rich district, with its educated and wealthy residents, provided environmental education which aimed at building awareness and changing lifestyles – and not just among pupils but in their families too. In the absence of evidence, it is hard to say how effective this education has been in changing the behaviour of "high emitters".

Is it possible, then, to make any generalizations from this case about the implementation of environmental policies in inner urban districts? The extreme flexibility found in the implementation of educational policies is probably not typical for other sectors, which makes it difficult to draw general conclusions. However, it is possible to find aspects that reassert some recent trends. Firstly, greenness has become an asset in the performance-assessment system. Secondly, state entrepreneurialism has a wide foothold in the implementation of environmental policies, because of the availability of "green-tagged" resources through various domestic and international green, eco-, or low-carbon programmes and projects.

This study has confirmed the existence of different approaches to the implementation of environmental policy in inner urban districts. But more research is needed here. We need to clarify how typical these cases are and whether other types of approach are sometimes taken as well. Future studies should also analyse whether there are commonalities among inner- and outer-city districts or indeed whether it is at all possible to make any generalizations about policy implementation within different urban areas in contemporary China.

Acknowledgements In writing this chapter, I have received support from the Academy of Finland (grant no. 116755).

References

Cai, Y. S. (2004). Irresponsible state: Local cadres and image-building in China. *Journal of Communist Studies and Transition Politics, 20*(4), 20–41.

Chang, I.-C. C., Leitner, H., & Sheppard, E. (2016). A green leap forward? Eco-state restructuring and the Tianjin–Binhai Eco-City model. *Regional Studies, 50*(6), 929–943.

Chien, S.-S. (2013). Chinese eco-cities: A perspective of land-speculation-oriented local entrepreneurialism. *China Information, 27*(2), 173–196.

Chien, S.-S., & Wu, F. (2011). Transformation of China's urban entrepreneurialism- case study of the City of Kunshan. *Cross-Currents E-Journal, 1*. https://cross-currents.berkeley.edu/e-journal/inaugural-issue/transformation-chinas-urban-entrepreneurialism-case-study-city-kunshan. Accessed 4 Aug 2014.

China Education Newspaper. (2012). 绿色生态文化:让学校为师生的乐园 (Green ecological culture: make school into a paradise for teachers and pupils). 26 June 2012. http://paper.jyb.cn/zgjyb/page/1/2012-06/26/11/2012062611_pdf.pdf. Accessed 8 Feb 2016.

Clark, P. B., & Wilson, J. Q. (1961). Incentive systems: A theory of organizations. *Administrative Science Quarterly, 6*, 129–166.

Cuan, W., & Guo, D. (2013). 美丽的今天 绿色的未来——天津市河东区六纬路小学环境教育工作纪实 (Beautiful present, green future - report of environmental education work at the Liuwei Road primary school in Hedong, Tianjin). *Environmental Education, 4*, –103.

deLeon, P., & deLeon, L. (2002). What ever happened to policy implementation? An alternative approach. *Journal of Public Administration Research and Theory, 12*(4), 467–492.

Duckett, J. (1998). *The entrepreneurial state in China*. London: Routledge.

Hedong District. (2006) 天津市创建国家环境保护模范城市先进区. 关于表彰河东区创建国家环境保护模范城市先进单位、优秀组织工作者和先进个人的决定 (Tianjin establishes advanced districts as a part of the national environmental protection pilot city project. On awarding titles of advanced unit, excellent organization builders and individuals in Hedong district). http://zwgk.tjhd.gov.cn. Accessed 4 Mar 2013.

Hedong District. (2008). 关于编制《天津市河东区生态城区 建设规划》的说明(摘要) (Explanations of drawing up the "Programme for the establishment of Hedong ecological urban district in Tianjin". Summary). http://zwgk.tjhd.gov.cn. Accessed 4 Mar 2013.

Hedong District. (2011) 河东区国民经济和社会发展第十二个五年规划纲要的报告的通知 (Notice on the Report of the Outline for the Twelfth Five-Year Plan for National Economic and Social Development of Hedong District). Available via CNKI. http://www.cnki.net

Hedong District Standing Committee of the People's Congress (Hedong District SCPC). (2010). 关于河东区开展生态城区建设 工作情况的汇报 (Report on the work to develop an ecological urban district in Hedong). http://rd.tjhdq.gov.cn/news.aspx?subSystemNewsID=ca822438-4825-4603-af5e-79507a9d4784. Accessed 12 Mar 2013.

Hedong District. (2012). 河东年鉴2012 (Hedong Yearbook 2012). Tianjin: Tianjin People's Publishing House.

Hendrischke, H. (1999). Tianjin - quiet achiever? In H. J. Hendrischke & C. Feng (Eds.), *The political economy of China's provinces* (pp. 183–206). London: Competitive and Comparative Advantage. Routledge.

Heping District. (n.d. ca 2003) 环境保护 (Environmental protection) http://www.tjhp.gov.cn/servlet/ContentServer?cid=1115088864665&pagename=tjhp%2FPage%2Fsub-t46&lv=2&lv2=1100820937344&lv1=1087380370033&c=Page. Accessed 4 March 2013.

Heping District. (n.d. ca 2007). 加大城区环境保护力度努力营造可持续发展的生态环境 (Increase efforts to protect the urban environment, aim to promote sustainable development of the ecological environment). http://www.tjhp.gov.cn/futuretense_cs/wmb/sthj1.htm. Accessed 5 Mar 2013.

Heping District. (2008). 共创文明城区 共享文明成果——天津市和平区文明城区创建工作概述 (Create together and share the achievements of a civilized urban district – Summary of the work to create a civilized district in Heping, Tianjin). http://www.tjhp.gov.cn/futuretense_cs/wmb/cjgk.htm. Accessed 5 Mar 2013.

Heping District. (2011). 和平区绿色学校名录 (List of green schools in Heping district). http://www.tjhp.gov.cn/servlet/ContentServer?cid=1337048389498&pagename=tjhp%2FHPart%2Fxxhcont-1&c=HPart&p1=1336351147570. Accessed 5 Mar 2013.

Heping District. (2013a). 和平区2013年环境保护工作要点 (Key points on environmental protection work in Heping district in 2013). http://www.dowater.com/info/2013-04-25/141745.html. Accessed 3 Sept 2013.

Heping District. (2013b). 2013年和平区环境教育工作计划 (Work Plan for Environmental Education in 2013). http://221.239.83.12:8202/pub/root11/auto199/201508/t20150823_25154.html. Accessed 23 Aug 2017.

Heping District. (2014). 2014年和平区环境教育工作计划 (Work Plan for Environmental Education in 2014). http://221.239.83.12:8202/pub/root11/auto199/201508/t20150823_29710.html. Accessed 23 Aug 2017.
Hsing, Y.-T. (2010). *The great urban transformation: Politics of land & property in China*. New York: Oxford University Press.
Jinnan District. (2014). 津南区关于开展环境友好型学校(幼儿园)创建工作的通知 (Notice on the work to develop environmental friendly schools (kinder gartens) in Jinnan district). http://218.69.93.71/ConInfoParticular.jsp?id=8098; document 津南环教办发[2014]1号(新版). Accessed 12 July 2016.
Kohoutek, J. (2013). Three decades of implementation research in higher education: Limitations and prospects of theory development. *Higher Education Quarterly, 67*(1), 56–79.
Kooiman, J. (2007). *Governing as governance. First published 2003*. London: Sage.
Kwan, T. Y.-L., & Lidstone, J. (1998). Understanding environmental education in the People's Republic of China: A national policy, locally interpreted. *Environmental Education Research, 4*(1), 87–97.
Lee, J. C.-K., & Huang, Y. (2009). Education for sustainable development projects and curriculum reform in China: The EEI and the EPD. In J. Chi-Kin Lee & M. Williams (Eds.), *Schooling for sustainable development in Chinese communities: Experience with younger children*. Dordrecht: Springer.
Lee, J. C.-K., & Williams, M. (2009). Dimensions of environmental education and education for sustainable development. In J. C.-K. Lee & M. Williams (Eds.), *Schooling for sustainable development in Chinese communities: Experiences with younger children*. Dordrecht: Springer.
Lin, J., & Ross, H. (2005). Addressing urgent needs: The emergence of environmental education in China. *China Environment Series, 7*, 74–78.
Liuwei Road Primary School. (2010). 学校简介 (Introduction). http://www.tjlwlxx.com/News/ShowInfo.aspx?ID=56. Accessed 8 Feb 2016.
Luova, O. (2013). Divergent trajectories among Chinese community volunteer associations and urban governance: Comparing four districts in Tianjin. *Journal of Comparative Asian Development, 12*, 443–464.
Matland, R. E. (1995). Synthesizing the implementation literature: The Ambiguitu-conflict model of policy implementation. *Journal of Public Administration Research and Theory, 5*(2), 145–174.
McBeath, G. A., McBeath, J. H., Qing, T., et al. (2015). *Environmental education variation in China. Environmental Education in China*. Cheltenham: Edward Elgar Publishing.
Ministry of Environmental Protection (MEP). (2015). 环境保护部、教育部公布第二批全国中小学环境教育社会实践基地名单 (List of the second batch of national level environmental protection and social practice units for middle and primary schools). http://news.cenews.com.cn/html/2015-12/29/content_38072.htm. Accessed 9 Feb 2016.
Ran, R. (2013). Perverse incentive structure and policy implementation gap in China's local environmental politics. *Journal of Environmental Policy & Planning, 15*(1), 17–39.
Shi, L. (2015). 特色高中建设的困境与因应 (Creation of middle schools with special characteristics: Problems and responses). *Journal of Tianjin Academy of Educational Science, 6*, 76–77.
Tianjin Bureau of Statistics. (2014). 天津统计年鉴2014 (Tianjin Statistical Yearbook 2014). Beijing: China Statistics Press.
Tianjin Communist Party Branch Life (TCPBL). (2012). 和平区:兴民富商 力创"首善之区 (Heping district: Properous people, wealthy businessmen create the "best district"). Issue 11. http://www.jinyueya.com/magazine/19363/2012_11/. Accessed 3 Aug 2015.
Tianjin Daily. (2013). 天津市河东区香山道小学:以绿色生态文化育人 (Xiangshandao primary school in Hedong, Tianjin: using green ecological culture to educate people). Re-published on Tianjin Zhaoshangwang on 26 Nov 2013. http://tj.zhaoshang.net/2013-11-26/47846.html. Accessed 8 Feb 2016.
Tianjin Daily. (2016). 关于河东区"十二五"规划纲要实施情况的报告 (Report on the implementation of the outline for Hedong district's twelfth five-year plan). http://news.enorth.com.cn/system/2016/01/13/030753475.shtml. Accessed 11 Apr 2016.

Tianjin Education Commission. (2012). 天津教育年鉴2012 (Tianjin Education Yearbook 2012). Tianjin: Tianjin Academy of Social Sciences Press.

Tianjin People's Congress. (2012). 天津市环境教育条例 (Regulations on environmental education). http://www.tj.xinhuanet.com/gov/2012-10/08/c_113291448.htm. Accessed 5 Mar 2013.

Tianjin People's Government. (2004). 天津区县年鉴 2004 (Tianjin District and County Yearbook 2004). Beijing: China Statistics Press.

Tianjin People's Government. (2014). 天津区县年鉴 2014 (Tianjin District and County Yearbook 2014). Beijing: China Statistics Press.

Tianjin EP. (2014) 天津市个区县市级环境友好型学校(绿色学校) 创建进度统计表 (Statistics of the establishment of city-level environmental friendly (green) schools in Tianjin's each district and county). http://www.tianjinep.com/a/guanyuwomen/fazhanlicheng/20140618/1804.html. Accessed 21 Jan 2015.

Tianjin EPB. (2003) 2002年天津市环境状况公报 (Report on the condition of the environment in Tianjin, 2002).. http://www.tjhb.gov.cn/env/env_quality/the_state_of_the_environment_bulletin/tianjin/201411/t20141110_4955.html. Accessed 11 Aug 2015.

Tianjin EPB. (2011a) 天津市环境保护"十二五"规划 (12th Five-year plan for environmental protection in Tianjin) http://law.solidwaste.com.cn/upfile/2012/10/109_1350527205.pdf. Accessed 10 Sept 2013.

Tianjin EPB. (2011b). 关于印发《天津市"十二五"环境宣传教育工作纲要》的通知 (Notice on the outline for environmental communication and education work for the 12th five-year plan) http://www.tjzfxxgk.gov.cn/tjep/ConInfoParticular.jsp?id=26143. Accessed 10 Sept 2013.

Tianjin EPB. (2013). 天津市2013年环境教育工作计划 (Tianjin's Work Plan for Environmental Education in 2013). http://www.022net.com/2012/12-31/534452413327464.html. Accessed 10 Sept 2013.

Tianjin EPB. (2014). 天津市2014年环境教育工作计划 (Tianjin's work plan for environmental education in 2014). http://www.tianjinep.com/a/huanjingxuanchuan/lvsechuanmei/2013/1231/1411.html. Accessed 22 Jan 2015.

Tianjin EPB. (2015) 天津市2015年环境教育工作计划(Tianjin's work plan for environmental education in 2015). www.tianjinep.com/a/guanyuwomen/fazhanlicheng/20150104/2283.html. Accessed 22 Jan 2015.

Tianjin Government. (2002) 天津市人民政府关于开展创建国家环境保护模范城市活动的通知 (Tianjin People's Government's notice on the activities for the establishment of national level environmental protection pilot city). http://www.fawuzaixian.com/wenku/view/id/301035. Accessed 4 Aug 2014.

Tian, Y., & Wang, C. (2016). Environmental education in China: Development, difficulties and recommendations. *Journal of Social Science Studies, 3*(1), 31–41.

Tianzhuang Middle School. (2016). 天津市河东区田庄中学绿色学校领导小组构成、职责分工及工作情况 (The structure, division of responsibilities and working circumstances of the green school small leading group in the Tianzhuang Middle School) http://www.tjstzzx.com/showart.asp? id=2023. Accessed 8 Feb 2016.

UNESCO-UNEP. (1994). *A prototype environmental education curriculum for the Middle School*. http://www.unesco.org/education/pdf/333_49.pdf. Accessed 4 Aug 2015.

Wang, H. (2007). 学校品牌的经营与管理 (School brand management and administration). *Developing, 5*, 89–90.

Wu, Z. (2002). Green schools in China. *The Journal of Environmental Education., 34*(1), 21–25.

Zeng, H., Yang, G., & Lee, J. C.-K. (2009). Green schools in China. In J. Chi-Kin Lee & M. Williams (Eds.), *Schooling for sustainable development in Chinese communities: Experience with younger children*. Dordrecht: Springer.

Zhang, T. (2005). Uneven development among Shanghai's three urban districts. In L. J. C. Ma & F. Wu (Eds.), *Restructuring the Chinese city. Changing society, economy and space* (pp. 124–139). London: Routledge.

Interviews

Interview at Dongli Education Bureau, September 2015
Interview at Hedong Education Bureau, September 2015
Interview at Heping Education Bureau, September 2015

Chapter 5
Urban Water Management in Beijing and Copenhagen: Sustainability, Climate Resilience, and the Local Water Balance

Li Liu, Marina Bergen Jensen, and Xiaoxin Zhang

Abstract Cities play a critical role for the sustainable management of planetary freshwater. At the same time, cities need to adapt to climate change. This offers cities an opportunity to improve freshwater management. In this chapter, the authors describe the status of urban water management in Beijing and Copenhagen, from the standpoint of sustainable development and climate resilience. In particular, they look into the degree to which the local water balance has been displaced. They review key water challenges and instruments in both cities, as well as related governance aspects. They consider the potential impact of these instruments and the relevance of a green infrastructure (GI) approach, as well as the potential for improving urban sustainability governance. A special focus is on the role of GI for retracking the city towards sustainable urban water management and climate resilience.

Keywords Freshwater management · Sustainable city · Local water balance · Green infrastructure · Urban governance · Beijing · Copenhagen

5.1 Introduction

This chapter compares water management in Beijing and Copenhagen from a governance perspective. The two cities are compared because, firstly, both cities have huge water challenges, have a strong planning tradition, and are proactive with urban water management; secondly, the research was partially funded by Sino-Danish Center for Education and Research (SDC), through which the authors gained good channels for first-hand data in Beijing. We highlight governance approaches that entail a combination of traditional public management (TPM) with new public

L. Liu (✉) · M. B. Jensen
Department of Geosciences and Natural Resource Management, University of Copenhagen, Copenhagen, Denmark
e-mail: liu@ign.ku.dk

X. Zhang
Beijing Municipal Institute of City Planning and Design, Beijing, China

J. Delman et al. (eds.), *Greening China's Urban Governance*, ARI - Springer Asia Series 7, https://doi.org/10.1007/978-981-13-0740-9_5

management (NPM) and network governance (NG) tools to restore the local water balance and enhance urban sustainability and climate resilience.

Globally, freshwater is an essential and scarce resource which is under pressure from overexploitation, contamination, and climate change. If there is to be a transition to the sustainable use of freshwater resources, cities will have to play a significant role in it. Of all the water exploited, one third goes directly to households and industry, which are predominantly located in cities. The remaining two thirds, which go to agriculture, are affected indirectly by citizens' consumption of food and other agricultural products (Shiklomanov 1999).[1] Total water extraction increased from approximately 600 km^3/year in 1900 to about 3700 km^3/year in 2000, with most of the increase occurring after the 1950s. Modern pumps allow for ever deeper aquifers to be exploited, and huge piping systems with average leaching rates of 30% allow water to be transported to consumption sites further and further away (Ibid.). In combination with power dams, this heavy extraction of water causes a quantitative displacement of the local water balance in many places, resulting in severe ecological losses, increasing conflicts over diminishing wells, and abandoned farmland and villages (Pacific Institute 2015).

Cities also account for much of the displacement of the natural balance in terms of water quality. This takes place through stormwater runoff and the discharge of wastewater from households and industry into natural water bodies, with negative consequences resulting for the environment and communities downstream.

If sustainable management of freshwater is to be possible, the use of water will need to be improved in terms of both quantity and quality. While qualitative improvement must be achieved at a local level (every city has to ensure a good quality of discharged water to protect the local environment), it is less obvious how the quantitative aspect (overexploitation) is to be tackled, inasmuch as the pressure from cities is related both to the direct intake of water and to the virtual water. One approach is to focus on the amount of water drawn into the city for water supply from sources outside it. In this respect, a city has better quantitative water management if it takes in less water from outside the urbanized area (be it in the form of surface water or groundwater) and if its water supply originates in a higher proportion from the annually renewable water produced naturally within the local area by precipitation or made available by water reuse therein. In such a case, the wealth and population size of the city will correspond to the amount of water available and to the skills and technologies of the city in the reuse of water. A situation with a zero intake of water from outside the city is illustrated in Fig. 5.1. The situation illustrated is the goal of the fully urbanized island nation of Singapore, which is aiming at zero dependence on the import of water from Malaysia by 2060, and full reliance on wastewater cleansing and all-purpose reuse (PUB 2013). In this chapter, we refer to this environmentally ideal state, with its 'closed' urban water cycle, as a 'City with a Local Water Balance': i.e. the water

[1] This is so-called virtual water. The virtual-water content of a product (a commodity, good, or service) is the volume of freshwater used to produce the product, measured at the place where the product was actually produced.

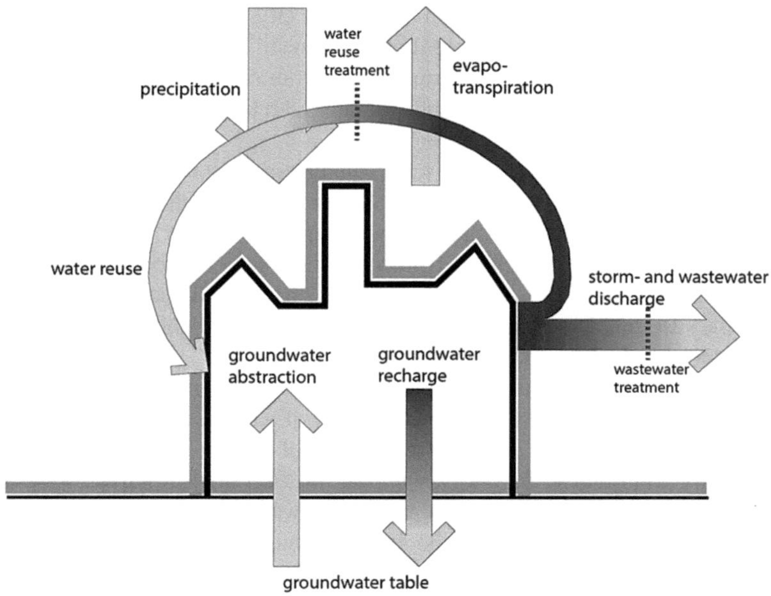

Fig. 5.1 City with a Local Water Balance. This city relies exclusively on precipitation and the reuse of wastewater. Its population size is matched to the extent of local water resources and to the ability of the society to reuse wastewater. The amount of groundwater extracted does not exceed recharging volume. The quality of water released from the city is as high as that of water entering it. The city maintains a significant green infrastructure for urban water management and multi-ecosystem services

supply is taken from renewable natural water resources within the city, in combination with water reuse; the city is resilient to pluvial flooding by means of retention (where stormwater runoff is permanently held back) and detention (where it is temporarily held back) within the urban landscape; the quality of water infiltrated or discharged is comparable to that of incoming natural water; and the city has a strong green infrastructure which is linked to urban water management and which provides multi-ecosystem services. We use this concept here to discuss the direct water supply to the city and the ways in which the water intake can be reduced, but not to calculate the city's local water balance in a strict sense. For the latter purpose, the imported virtual water (see note 1) of the city ought also to be considered, together with the freshwater needed for natural ecosystems outside the urbanized part of the catchment; but these are considerations which fall outside the scope of this chapter.

The degree to which the urban water cycle has been displaced quantitatively varies from place to place. In regions where freshwater resources are plentiful, and where much of the precipitation quickly discharges in any case into the ocean through rivers and groundwater flow, it would be a waste of resources for a city to strive for zero reliance on the import of water from outside the city. In such a case, namely, freshwater can be taken from outside the city without destroying surrounding ecosystems, and a fair local water balance can still be maintained. Another case where the import of physical and virtual water from outside city limits can be seen as sustainable is when the city is located on land with little agricultural or ecological value, such as wasteland or desert.

The need for sustainable urban water management governance approaches is accentuated today by climate change. This is forcing many cities to reconsider their drainage systems, to avoid frequent flooding due to heavier precipitation. If the concurrent investments to achieve climate resilience are linked with efforts to re-establish a local water balance in terms of both quality and quantity, then big steps towards the sustainable management of freshwater can probably be taken. This window of opportunity is the focus of this chapter.

We use the two capital cities—Beijing in China and Copenhagen in Denmark—as our cases., We first describe current practices of water exploitation and contaminant management, in order to assess the degree of local water balance in terms of both quantity and quality. Secondly, we look for discourses relating to the linkage between climate resilience and sustainable freshwater exploitation, with an eye to ascertaining the extent to which the two challenges are approached jointly by decision-makers in the two cities. Our chapter is based on a review of the literature, on an analysis of documents, and on interviews. The two cities have different social-economic contexts and face different challenges. Beijing Municipality, a megacity with strong institutions, is undergoing rapid economic development in a water-scarce region. Copenhagen, a relatively large city on a stable level of development, has strong institutions as well; it enjoys plentiful resources both economically and in terms of freshwater reserves; and it holds high ambitions when it comes to protecting the environment and mitigating and adapting to climate change. Despite the differences, both use a combined TPM-NPM governance model and institutional setup for urban water management, i.e. city administrations are responsible for water administration and planning, and public-owned water companies deliver the water supply and drainage services; being the capital cities, both cities have high ambition on climate resilience entailing growing NG for stormwater management. Therefore, through an investigation of the two cities' practice with some comparative perspectives, we hope to find more nuances in local governance practices on water management and enrich our understanding of governance approaches from different contexts.

We examine water management governance of Beijing and Copenhagen with reference to two other relevant aspects of water governance, i.e. urban climate resilience and urban green infrastructure.

Urban resilience relates to climate change: it is 'the degree to which cities are able to tolerate alteration before reorganizing around a new set of structures and processes' (Alberti et al. 2003). Urban climate resilience can be enhanced by two activities: mitigation and adaptation. Mitigation seeks to reduce greenhouse gas emissions to tackle climate change, while adaptation focuses on vulnerability to the unavoidable hazards and attempts to adjust to them (IPCC 2014; McEvoy et al. 2006). Spatial planning, through the arrangement of land use and development, is a powerful method of governance for both mitigation and adaptation. For at least two decades, an approach high on urban planning agendas globally has been to apply mitigation strategies, such as increasing the use of alternative energy sources and constructing more energy-efficient buildings and transport systems. Urban planning for climate adaptation arrived more recently, but it has become a major discourse of resilience—for instance, in connection with flood risk management (McEvoy et al. 2006; Stead 2014). This chapter focuses on the water-related aspects of urban climate resilience.

In an urban context, green infrastructure (GI) refers to green-blue (vegetated-aquatic) areas and some related open spaces of a city that provide social, biological, or environmental services. But the use of the term varies according to place and circumstance, reflecting the lack of any well-established common understanding. Mark A. Benedict and Edward T. McMahon (2006) refer to GI as 'an interconnected network of natural areas and other open spaces that conserves natural ecosystem values and functions...and provides a wide array of benefits for people and wildlife'. As suggested in Fig. 5.1, GI may offer unique options for establishing a good balance between a city (meaning both its physical structure and the social-economic activities carried on therein) and its local water reserves. Through the landscaping of urban surfaces, stormwater runoff can be managed in a way that mimics the natural water cycle, allowing for retention processes such as infiltration, groundwater recharge, and evaporation, as well as—in combination with substantial detention volumes—the reduction of flood risk (Austin 2014).

5.2 Management Systems in the Two Cities for Water Supply, Wastewater, and Stormwater

In 2012, Beijing Municipality had an area of 16,410.54 km^2 and a population of 20.7 million (Beijing Statistical Information Net 2013). It has a dry, temperate, and continental monsoon climate. Annual precipitation in the 1999–2009 period was 457 mm. About 3/4 of the precipitation falls in the months of June, July, and August. Potential annual evapotranspiration is 1200 mm (Meng 2014). In this chapter we focus on Beijing Central City, which roughly corresponds to the inner urban districts of Beijing Municipality. This definition, used by urban planners, excludes mountainous areas and some smaller towns from the central urban area.[2] In 2010, according to data from the Beijing Municipal Institute of City Planning and Design, Beijing Central City had an area of 1085 km^2 and a population of about 10 million (Zhang 2014). Copenhagen Municipality, on the other hand, had an area of 86.2 km^2 and a population of 566,906 in 2014 (Statistics Denmark 2016). It has a temperate

[2] Districts of Dongcheng, Xicheng, Chaoyang, Shijingshan, Fengtai (excluding the area west of the River Yongding), and Haidian (excluding the area north of the mountain with the four towns, i.e. Weiquan Town, Sujiatuo Town, Xibeiwang Town, and Shangzhuang Town), plus the area north of Beiyuan and Huilongguan in Changping District.

coastal climate. Average precipitation during the 1971–2000 period was 523 mm (Danish Meteorological Institute 2002). Precipitation is distributed relatively evenly over the 12 months of the year, with the highest occurring in July and August. Potential annual evapotranspiration is 600 mm. Beijing Central City and Copenhagen Municipality, henceforth referred to as Beijing and Copenhagen, are the focus of this chapter.

5.2.1 Water Systems in Beijing

Beijing is located in a region low in natural water reserves. Together with groundwater, Guanting and Miyun reservoirs have traditionally been important for the municipality's water supply; however, the inflow from the two reservoirs has dropped dramatically. Today they hold less than 10% of their original storage capacity. Guanting reservoir is no longer being used, due to pollution (Probe 2008). Rapid population growth and low water prices without any economic regulator have also given rise to overexploitation of the municipality's groundwater aquifers, especially in suburban areas. The water level has dropped from 3 m below surface in the 1960s to 24 m below surface in 2009—just 11 m above the dry bedrock (Liu et al. 2014; Probe 2008). For Beijing Municipality as a whole, about 6000 mill. m^3 of groundwater had been extracted above the safety limit as of 2008, and the lost groundwater resource capacity may never be replenished. To meet the demand for water, surface water has been transferred into the city and ever deeper groundwater extracted from neighbouring regions for several years, and restrictions have been placed on the use of surface water and groundwater in these regions (Probe 2008). It has been estimated that Beijing Central City can extract 600 mill. m^3 of groundwater annually in a sustainable way (Beijing Institute of Geology Survey 2003); the city's demand for water, however, is almost double that amount. In 2010, of the total 1126 mill. m^3 of tap water supplied to Beijing Central City, 531 mill. m^3 were from groundwater abstraction within the city; 232 mill. m^3 were from Miyun reservoir; 148 mill. m^3 were from emergency groundwater wells located in suburban areas; and 215 mill. m^3 were transferred from other regions. This water was used for households, for public buildings and industry (914 mill. m^3), and for maintaining urban landscapes and the like (49 mill. m^3); the remaining 163 mill. m^3 was lost due to leaching (Beijing Water Authority 2010) (see Fig. 5.2). 380 mill. m^3 of reclaimed water were also used for non-drinking purposes. The water supply network of Beijing Central City consists of nine waterworks, with a daily capacity of 2.95 mill. m^3, as well as a number of self-sustaining wells (*zibeijing* 自备井) (Meng 2014). According to a study done by Beijing Water Authority, the available stormwater-harvesting capacity in Beijing Central City is about 100 mill. m^3 annually (Zhang 2013).

Within Beijing Central City, ten wastewater-treatment plants (with a daily capacity of 2.62 mill. m^3) handle 95% of the wastewater produced within the area. A substantial amount of this wastewater is reclaimed and reused for the city's water supply (Meng 2014). Non-point pollution caused by runoff in the Central City is

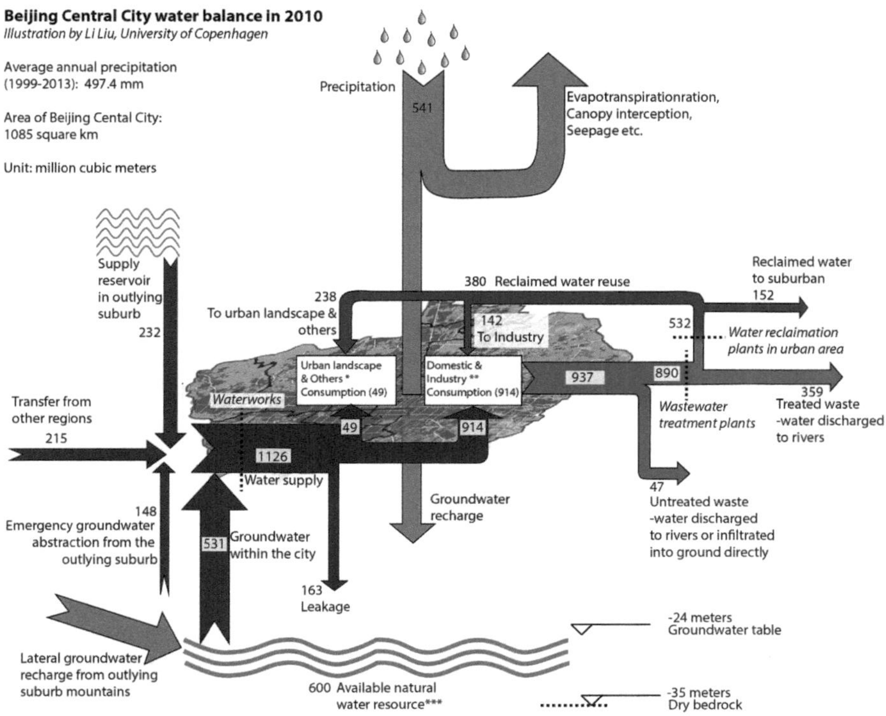

Fig. 5.2 Water balance of Beijing Central City in 2010. Unit: 10^6 m^3 (Beijing Water Authority 2010; Beijing Institute of Geology Survey 2003)
*Urban landscape and other consumption includes supplied water (excluding reclaimed water reuse) used for maintaining urban rivers and lakes, for watering urban green spaces, and for cleaning roads, fighting fires, etc.
**Domestic and industry consumption includes domestic water use (excluding reclaimed water reuse) for households and public buildings, together with industrial water consumption (excluding reclaimed water reuse).
***600 mill. m^3 available natural water resource in Beijing Central City is based on 'Capital Region Groundwater Resources and Environmental Assessment', by Beijing Institute of Geology Survey (2003).

considered more serious than point-source pollution. Of the five main rivers in Beijing Municipality, four fail to meet the national Environmental Quality Standards for Surface Water (GB 3838-2002) (the sole exception being the River Jumahe). The contamination is caused by combined sewer overflows (CSOs), poor sanitation and treatment, and direct discharge of sewage from industries. According to a field investigation carried out in Beijing Central City by Beijing Municipal Institute of City Planning and Design, there have been serious discharge problems in middle-sized rivers within that zone, including the River Qinghe, the River Liangshuihe, and (most polluted of all) the River Xiaotaihou (Zhang 2014).

Within its 2nd ring road, Beijing has a combined sewer system; from the 2nd ring road outward, separated sewer systems often prevail (Zhang 2013). Conventional rainwater systems in Beijing include drainage pipes, pumping stations, and urban rivers and lakes (Zhang 2013). Stormwater pipelines along the main roads are managed by the Beijing Drainage Group. The general service level[3] of the stormwater drainage systems is to manage three-year rain events—a level met by 85% of the system. The service level for future construction will be to manage five-year rain events. Waterlogging or pluvial flooding is increasingly prominent in Beijing, as a consequence of the combination of urbanization and climate change over the last decade (Zhang 2013; Zhao et al. 2014). The extreme flooding on 21 July 2012 involved a total precipitation of 164 mm, lasting 16 hours. Most local floods occurred on roads, under overpass bridges and intersections, and in areas with high pipe densities (Zhao et al. 2014). Frequent flooding and uncontrolled stormwater runoff do not just cause social-economic damage; they also aggravate problems of water pollution, through among other things combined sewer overflows (CSOs).

5.2.2 *Water Systems in Copenhagen*

Of the 61.5 mill. m^3 of annual precipitation that fell within Copenhagen in 2003, about one third (21.3 mill. m^3) was returned to the atmosphere through evapotranspiration. Another third (23 mill. m^3) was stormwater runoff, which was diverted through sewers to wastewater-treatment plants. Less than a tenth (4.9 mill. m^3) of the annual precipitation infiltrated the surface to charge the groundwater (Binning et al. 2006) (See Fig. 5.3). Almost all of the 32.8 mill. m^3 of drinking water consumed in Copenhagen that year was groundwater abstracted from well-fields distributed over a large part of Zealand, the island on which Copenhagen is located. Because of its relatively high quality, this water goes through a very simple treatment process after abstraction, with only aeration and filtration before being pumped out to consumers. Other water resources play a part as well, for secondary uses, but only on a very limited scale (2% at most): seawater is used for industrial process-cooling and stormwater harvested for laundry and toilet-flushing (Copenhagen Municipality 2012b). The water table in a large part of Zealand has fallen by up to 10 m. This increases the risk that saltwater will enter the aquifers (Binning et al. 2006). Climate change will likely bring changes in the groundwater level, which is expected to rise by 0.5 m along the coastline and to fall by 1 m in the rest of the municipality (Copenhagen Municipality 2011). This may pose a threat to the city's water supply. Moreover, due to contamination by pesticides and chlorinated solvents over recent decades, more than 10% of water supply wells around Copenhagen have been abandoned. Domestic water consumption in Copenhagen has fallen: from about 170 L per person per day in the 1980s to around 110 L per person today. The city achieved

[3] The service level describes the level of protection that the stormwater drainage system is designed to provide. With a service level of 3 years, the stormwater drainage system can be allowed to overflow every third year on average, but not more frequently than that. In other words, the stormwater drainage system is designed to handle rain up to the level of a 3-year rain event, which is the worst rain that occurs with a return period of 3 years.

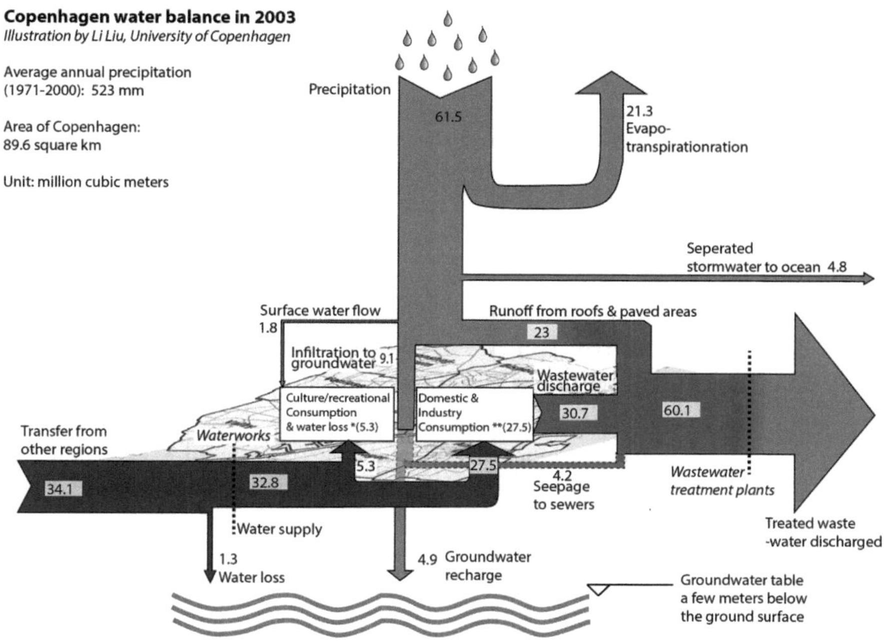

Fig. 5.3 Water balance of Copenhagen in 2003, (89.6 km^2). Unit: 10^6 m^3 (Binning et al. 2006)
*Culture/recreational consumption includes water used in urban ponds and fountains, for recreational and cultural uses, and the like; water loss includes water loss in pipelines, water use for firefighting and cleaning pipelines, etc.
**Domestic and industry consumption includes domestic water use for households and public buildings, together with industrial water consumption.

this by raising water prices, moving major industries out, stopping leakage in water pipes, conducting an intensive propaganda campaign involving school children and the public, and equipping every consumer tap with a water meter linking the bill to the level of consumption (Hansen 1996; Copenhagen Municipality 2012b).

Copenhagen's wastewater is treated in two wastewater-treatment plants, which take in all of the wastewater and most of the stormwater (Copenhagen Municipality and Copenhagen Energy 2008). In general, moreover, citizens in Copenhagen expect water in the harbour to be sufficiently clean for healthy bathing and other bodies of water in the city to be clean enough for recreational purposes. Their demands concerning water quality also seem to be rising. Unfortunately, however, combined sewer overflows bring contaminated water to the rivers and coasts in and around Copenhagen. More frequent and intensive rainfall brings more CSOs, burdening the ecosystem and undermining the well-being of Copenhageners.

Most of the discharge system of Copenhagen is a combined sewer system, with the oldest parts dating some 150 years back. It is sized for 10-year rain events (Copenhagen Municipality 2011; Copenhagen Municipality 2012a). Wastewater and stormwater are transported in the combined sewer to wastewater-treatment plants, from which the effluent is discharged into coastal waters. A minor fraction of stormwater runoff enters the stormwater pipes in a separate system, after which it flows into the ocean without any treatment. The current major challenge in Copenhagen is the increasing frequency and intensity of rainstorms. Meteorologists predict that, by 2100, the city will experience a 25–55% increase in precipitation during winter, up to a 40% reduction in precipitation during summer, and a 30–40% increase in thunderstorms, as compared with the situation at the beginning of the century (Copenhagen Municipality 2011).

5.3 Assessment of the Degree of Water Overexploitation and Contamination in the Two Cities

Major water flows in the two cities are shown in Figs. 5.2 and 5.3. Data are from 2010 for Beijing and from 2003 for Copenhagen. The figures refer to Beijing Central City (1085 km^2) and Copenhagen Municipality (89.6 km^2).

5.3.1 Overexploitation

According to the concept of a City with a Local Water Balance referred to above, a city's sustainable water sources consist of the annually renewable water produced naturally within its local area by precipitation, together with the water made available by reuse. Figure 5.2 shows the following for Beijing: of the 1506 mill. m^3 of water supplied in that city in 2010 (including reclaimed water), about 76% (1143 mill. m^3) came from sustainable water sources, i.e. available natural water resources (763 mill. m^3) in Beijing Central City and reclaimed water (380 mill. m^3), while 24% (363 mill. m^3) came from overexploitation of the city's own groundwater resources (148 mill. m^3) or other administrative regions' water resources (215 mill. m^3). In Copenhagen, as we see in Fig. 5.3, 100% of the 32.8 mill. m^3 of water supplied in 2003 was obtained from other regions' water resources. Beijing relies largely on groundwater abstraction within the city and on reclamation of wastewater. Copenhagen relies exclusively on distant water transfer (up to 50 kilometres), while local groundwater abstraction and reclamation of wastewater is almost non-existent. Stormwater harvesting in both cities is minimal.

Water consumption is correlated with total population. Table 5.1 shows that, in relation to its water consumption, Beijing's population density (9217) in 2010 was more than twice that (4007) which can be theoretically supported by water resources within the bounds of Beijing Central City. Copenhagen's population density (5599) in 2003 fell within the maximum density limit (6860). If per capita water consump-

Table 5.1 Maximum population density that natural water resources can support (Statistics Denmark 2016; Binning et al. 2006; Beijing Water Authority 2010; Beijing Institute of Geology Survey 2003; Zhang 2014)

		Beijing 2010	Copenhagen 2003	Beijing scenario
Population	Persons	10,000,000[a]	501,664	
Available local natural water resources[b]	m^3/year	600,000,000	40,200,000	600,000,000
Water consumption[c] (m^3/person/year)	m^3/person/year	138	65.4	65.4[d]
Theoretical population size supportable from locally available sources	Persons	4,347,826	614,678,899	9,174,312
City area (km^2)		1085	89.6	1085
Supportable population density (persons/ km^2)		4007	6860	8456
Actual population density (persons/km^2)		9217	5599	

[a]Data from 2010 for Beijing Central City, from Beijing Municipal Institute of City Planning & Design (Zhang 2014)

[b]Beijing 2010 data are based on 'Capital Region Groundwater Resources and Environmental Assessment', by Beijing Institute of Geology Survey (2003). Copenhagen 2003 data are based on Binning et al. (2006), as well as on the assumption that available local water resources are formed by precipitation (61.5 mill. m^3) in case area deducting evaporation (21.3 mill. m^3).

[c]Here water consumption is the total consumption per capita, including domestic consumption (household water use), industrial water consumption, and others.

[d]This is based on average annual water consumption per capita in Copenhagen in 2003 (Binning et al. 2006).

tion in Beijing had been reduced to the 2003 level in Copenhagen, Beijing's population density of 9217 in 2010 would nearly have been within the supportable by available local natural water resources (8456).

5.3.2 *Contamination*

In Beijing, of the 937 mill. m^3 of wastewater produced in 2010, more than half (532 mill. m^3) was reclaimed, while about another third (359 mill. m^3) was treated. This means that some 95% of wastewater was subject to quality control. The remaining 5% of wastewater (47 mill. m^3) was discharged directly into rivers or infiltrated into the ground, contributing to an imbalance in the city's local water cycle in terms of water quality. In Copenhagen in 2003, all 30.7 mill. m^3 of wastewater was subject to quality control, in wastewater-treatment plants. However, combined sewer overflows during heavy rain events put ecological pressure on local streams, contributing to an imbalance in the city's local water cycle in terms of water quality. Neither city had treatment procedures specifically designed for stormwater runoff during the years in question.

5.4 Water Management Instruments in the Two Cities

5.4.1 Water Management Instruments in Beijing

To secure Beijing's water supply and relieve pressure on its groundwater, the Chinese government has been carrying out a massive project: the South-to-North Water Diversion Project (SNWDP) with three routes (western, middle, and eastern routes). The middle route brings 1.2 bill. m^3 of water annually from the River Yangtze, in order to supply Beijing Municipality and its surrounding regions. The water is conveyed across 1267 km (Meng 2014). We have found no information on the possible ecological impact of the SNWDP in the River Yangtze catchment area. Besides the SNWDP, moreover, Beijing has a strategy for increasing alternative water sources, by such means as wastewater reclamation and stormwater harvesting. There have been attempts at reusing greywater, but the practice is not common in Beijing due to social concerns. The public, namely, distrusts greywater quality and fears risks to health, especially since the sudden and lethal outbreak of severe acute respiratory syndrome (SARS) in 2002–2003 (although this illness is not water-related) (Liu et al. 2014). The city is building new wastewater reclamation plants, improving water-use efficiency by adjusting its industrial structure, and exploring the desalination of sea water from Bohai Bay as a potential water source (Zhang 2014). In recent years, moreover, the Beijing Water Saving Office (a subdivision of the Beijing Water Authority) has issued guidelines and conducted water-saving campaigns, encouraging and requiring reductions in water consumption, mainly at the institutional level. Public education and propaganda for water-saving by citizens is increasing, but the effects are difficult to estimate (see also Luova, Chap. 4, in this volume). A three-tiered water pricing system was introduced on 1 May 2014, increasing the basic water price by 25% and imposing higher prices for additional use. This is expected to be effective at reducing water consumption (China News Net 2014).

Since 1990, Beijing has been working on improving its system for urban flood control and drainage. The idea is to develop retention and detention volumes in the western part of Beijing Central City and to increase drainage capacity in the eastern part (Meng 2014). The Flood Prevention and Waterlogging Reduction Plan for Beijing City Region being developed by the BWA proposes two deep tunnels in the eastern and western parts of the central city, respectively, in order to convey flood water to the two major rivers outside the 2nd ring road. Another Local Flooding Control Plan for Beijing Central City, under development by the Beijing Municipal Planning Commission, proposes not to dig deep tunnels but to establish a low-impact development (LID) greenbelt around the central city, based on existing and planned green areas, and to use this to manage stormwater by means of infiltration, retention, and detention (Liu and Jensen 2017). The city is also taking engineering measures, like upgrading pipes and pumping stations, in order to strengthen its urban drainage system (Meng 2014; Zhang 2013). In addition, the Beijing Municipal Planning Commission issued a new regulation in 2003, requiring all projects for new construction or reconstruction to apply rain-harvesting techniques (i.e. pervious pavements, sunken green spaces, and retention tanks). The Commission then followed with a detailed guideline in 2012, requiring every 10,000 m^2 of impervious pavement built to be equipped with a 500 m^3 rainwater detention facility (Meng 2014). The assumption in Beijing is that GI is only for managing 3–5-year rain events. Rivers and stormwater pipes are seen as the main methods for handling rain events more intense than the 50-year type, with GI only playing the role of delaying and reducing the peak flow. Beijing's strategy for handling stormwater thus integrates GI and grey infrastructure.

Beijing takes a highly top-down approach to urban water management. The BWA has charge of all water issues (water supply, wastewater management, and flood management) in the urban region. Its approach to governance is to apply traditional public management (TPM). This can be seen from its central control of waterworks, of the SNWDP, of the wastewater-treatment system, and of the city-wide pipe system for distributing reclaimed water. As state-owned companies, the Beijing Water Group and the Beijing Drainage Group administer water supply and wastewater management services, respectively. This institutional setup shows some characteristics of new public management (NPM), such as a strong focus on effectiveness in service delivery. The recent hike in water prices also shows a growing resort to NPM. However, a greater element of network governance (NG) can be expected in Beijing when it comes to managing stormwater and controlling floods by means of urban landscapes. In those areas a more decentralized approach can be expected, with multiple sectors aside from the BWA playing a role in administration. This can be seen in, among other places, the already highly privatized urban construction sector, although it is doubtful the general public will take part in the networks to any substantial extent.

5.4.2 *Water Management Instruments in Copenhagen*

Copenhagen faces problems of falling groundwater levels and reductions in water quality from aquifer oxidation (Binning et al. 2006). Greater Copenhagen Utility, HOFOR, has looked for ways, in collaboration with researchers, to short-circuit the supply and discharge systems and to use stormwater, greywater, and sewage directly from the wastewater stream for supply purposes. Thought has been given to replacing drinking water in certain places with other water sources for uses that do not require high water quality, in order to limit dependence on the extraction of drinking water in surrounding municipalities (Ibid.). The possibility of using local groundwater as part of the water supply has been investigated; the city's water supply plan, for instance, proposes to increase reliance on alternative water sources from the present 2% of total water supply to 4% by 2020 (Copenhagen Municipality 2012b).

To protect Copenhagen against extreme cloudbursts, two plans were implemented: the Copenhagen Climate Adaptation Plan (CAP) of 2011 and the Cloudburst Management Plan (CMP) of 2012. The objective of the CAP is to keep the 10-year return period service level, notwithstanding 30% heavier precipitation. The aim of the CMP is to avoid flood depths in excess of 10 cm for 100-year return period rain events. The CAP calls for a combination of green and grey infrastructures: diverting rainwater from sewers and managing it locally with green or low-tech solutions wherever feasible, enlarging sewers to reach the service level for 10-year rain events, and diverting floods to areas where they do the least damage (Copenhagen Municipality 2011). The assumption of the CMP is that expanding the ordinary sewer system and increasing retention-detention in green areas are not enough to solve flooding. Underground tunnels will be necessary, to divert flooding waters directly to the ocean. Streets and new cloudburst pipes will convey stormwater from events exceeding a 10-year return period to the tunnels (Copenhagen Municipality 2012b). Three types of stakeholders figure in the CMP: property owners, who are responsible for flood-proofing their properties; the utility company (HOFOR), which has to ensure that drainage systems meet service levels and that adaptive measures are implemented in accordance with the new risk dimensions; and the city administration, which must ensure that adaptive measures are incorporated into municipal master plans and are implemented. A detailed plan was also elaborated later, with more than 300 projects for implementing the CMP.

Copenhagen, with its centralized water infrastructure hardware system, has combined a top-down with a bottom-up approach to urban water management. The city administration is responsible for all water issues (water supply, wastewater management, and flood management), while HOFOR (Greater Copenhagen Utility) implements all water supply and wastewater-treatment services, using a combination of TPM and NPM. Within this general framework, particular attention has been paid to involving and mobilizing the public and the broader society, as reflected in the earlier water-saving campaigns, supplemented by an effort to reduce water consumption by raising water prices. When it comes to the management of floods and stormwater, a growing tendency towards decentralization and bottom-up administration can be observed in recent years. The city has already started to involve various actors from both public and private sectors, as well as property owners and citizens. This indicates a shift to NG for stormwater management.

5.5 Discussion

Water shortage is seen as an issue in Beijing, but not in Copenhagen. Beijing today has a huge system for reclaiming wastewater, which accounts for a significant fraction of the city's water supply. However, while the SNWDP ensures Beijing's water supply, the ecological impact of the project and its associated level of greenhouse gas emissions are unknown. Copenhagen relies very little for its water supply on secondary (alternative) water resources, such as reclaimed water.

Both cities focus strongly on flood control for climate resilience and have developed overall strategies and plans for this purpose. The two plans in Beijing stay at a more strategic level, and they are being developed by two different agencies with less consensus on the main emphasis, thus the conflict between the plans: the one focuses on digging deep tunnels and the other on establishing an LID greenbelt (see also He et al., Chap. 3). The importance of linking stormwater management to water supply, through harvesting or groundwater recharge, is given particular emphasis in the LID plan (although it figures as well in the tunnel plan). By the time of publication of this chapter, the plan for digging deep tunnels in Beijing has been held back in order to explore GI approach further. Copenhagen's plans have gone further, with implementation strategies more fully worked out and with better consensus among sectors and agencies. The goals, however, have changed from the CAP to the CMP, due to a perception of higher flood risk. The focus of the CMP is mainly on water discharge, with little coupling to water supply. In addition to the cloudburst management, the city does however has a plan to disconnect 30% of the city surfaces that today discharge to the sewer. This is to meet the expected 30% increase on annual precipitation without expanding the sewer. Green Infrastructure is among the solutions explored for meeting this goal. Additoinally, Copenhagen is exploring options for improving quality of stormwater runoff by use of filter soil, dual porosity filtration, and other nature-based solutions. Beijing uses top-down regulations and guidelines to implement its strategies, while Copenhagen seeks new ways in governance by involving public and private stakeholders in plan implementation, as well as to integrate the said plans with other urban projects. Both cities have made efforts to achieve clean water goals. Beijing puts the stress on technical solutions for treating water and protecting watersheds. Copenhagen emphasizes water catchment protection and quality control in water source regions. Both cities have a few examples of using green areas for controlling flood control and improving water quality.

To improve its local water balance, Beijing needs to continue reclaiming wastewater while at the same time exploring more innovative ways of using alternative water sources. As our analysis has shown, Beijing has a far higher population density than its natural water resources can sustain. If it retains this population density, it will become urgently necessary to reduce water consumption and to improve water-use efficiency. The effect of the recent hike in water prices remains to be seen, but a further step will likely be needed: the adoption of a full-cost pricing system

(Probe 2008). In the case of Copenhagen, the entire water supply comes from freshwater resources outside the city. While neither a shortage of water nor a danger to the ecosystem has been seen as a serious problem, problems and pressures may arise in the future. The discourse on improving the local water balance in Copenhagen is relatively weak. There is some discussion currently about combining stormwater harvesting with cloudburst management, and there was some discussion earlier about providing stormwater-harvesting tanks to private homeowners and about experimenting with groundwater abstraction wells in parks. In the present climate, however, the discourse in favour of flood control for climate resilience appears to be overwhelming. In neither city is stormwater used for the local water supply to any great extent. When the flood risk is high, priority is often given to securing the city from flooding by discharging flood water out of the city as fast as possible. Despite the risk they pose to cities, stormwater and flood water are part of the freshwater cycle; and they can be used as a local freshwater resource, thereby facilitating a local water balance. Energy can be saved by transporting less freshwater into cities from other regions and by transporting and treating less runoff as wastewater. Linking climate resilience to freshwater exploitation will reduce energy consumption and lessen the impact on the environment, thereby contributing to sustainable development.

Due to its close connection with urban landscapes and geo-biochemical processes, GI has the potential to combine solutions for flood control for climate resilience with solutions for sustainable freshwater exploitation, by such means as stormwater harvesting and infiltration for groundwater recharge. In addition, GI solutions have the potential to provide additional ecosystem services related to social and cultural values as well as biodiversity and crop production. GI solutions should be based on a city's local circumstances; they cannot be universal. A conventional mindset still rules in both Beijing and Copenhagen, with the alternative GI approach only being practiced on a small scale. Beijing puts the emphasis on infiltrating stormwater to its aquifers. It also uses green-blue infrastructure to receive lower-quality water. These practices help to reduce the imbalance of the urban water cycle. Copenhagen stresses the detention and discharge of stormwater and flood water, which is less helpful for reducing the imbalance of the urban water cycle. In Beijing, opportunities lie in solving some major challenges, i.e.. technical challenges posed by the concentrated rainy season, as well as operational challenges arising from distinct sectoral practices in water supply and in stormwater management (or flood control). The potential impact of projects like the SNWDP on climate and the environment may pose another difficulty. In Copenhagen, more awareness is needed of sustainable water supply practices and of the potential contribution of stormwater to this (although water supply pressure is not high yet). Both cities face time pressures in providing grey infrastructure for rain events more intense than the 50-year type, which GI has not proved able to manage. This is a strong barrier against making the necessary link between stormwater management and water supply through GI.

Our study has shown that Beijing mainly uses TPM in water management, together with a degree of NPM. The public has not yet been much involved. Copenhagen applies a combination of TPM, NPM, and NG, whereby the city seeks to involve property owners, citizens, and the private sector. The different choice of governance paradigms reflects differences in political systems, planning traditions, and land ownership patterns in the two national contexts. Water supply and wastewater management are basic and vital services which a city must provide. Neither Beijing nor Copenhagen can neglect to make some use of top-down approaches (TPM and NPM) with centralized hardware systems. However, there is a great potential to inject more in the way of bottom-up practices (NG) into the governance mix, thereby improving water supply and wastewater management in a sustainable way. Copenhagen's success in reducing water consumption over recent decades shows the results that can be achieved when citizens and the broader society are mobilized. (Then, if appropriate water pricing mechanisms are added to the mix besides, the results can be better still.) These successes may serve as to inspire Beijing. In the case of both cities, a greater reliance on NG would seem to be called for, in view of the increasing need for alternative approaches to managing floods and stormwater through urban landscapes and the associated need to engage a wider range of actors from both public and private sectors. The attraction of NG becomes yet more evident when we link issues of stormwater management to questions of water supply and wastewater management, given that decentralized stormwater facilities can provide water supply resources and alleviate pressures on municipal wastewater treatment.

References

Alberti, M., Marzluff, J. M., Shulenberger, E., et al. (2003). Integrating humans into ecology: opportunities and challenges for studying urban ecosystems. *BioScience, 53*, 1169–1179.

Austin, G. (2014). *Green infrastructure for landscape planning*. Glasgow: Routledge.

Beijing Institute of Geology Survey. (2003). 首都地区地下水资源和环境调查评价 (Capital Region Groundwater Resources and Environmental Assessment). Beijing: Internal report of Beijing Institute of Geology Survey

Beijing Statistical Information Net. (2013). 2012年分区县主要数据 (Districts and counties data in 2012). http://www.bjstats.gov.cn/sjfb/bssj/ndsjcs/201308/t20130828_255760.htm. Accessed 27 Mar 2014

Beijing Water Authority. (2010). 北京市水务统计年鉴2010 (Beijing Water Affairs Statistic Year Book 2010). Beijing: Internal report of Beijing Water Authority.

Benedict, M., & McMahon, E. (2006). *Green infrastructure: linking landscapes and communities*. London: Island Press.

Binning, P. J., Hauger, M. B., Rygaard, M., et al. (2006). Rethinking the urban water management of Copenhagen. *Water Practice & Technology, 1*(2). https://doi.org/10.2166/WPT.2006027.

China News Net. (2014). 北京阶梯水价5月1日实行 九成居民水价每方涨1元 (Beijing applies tiered water price from May 1, 90% residents with 1 yuan higher price per cubic water), 29 Apr 2014. http://finance.chinanews.com/ny/2014/04-29/6118058.shtml. Accessed 5 Apr 2016

Copenhagen Municipality. (2011). *Copenhagen climate adaptation plan*. http://subsite.kk.dk/sitecore/content/Subsites/CityOfCopenhagen/SubsiteFrontpage/LivingInCopenhagen/ClimateAndEnvironment/ClimateAdaptation/CopenhagenClimateAdaptionPlan.aspx. Accessed 28 Mar 2014

Copenhagen Municipality. (2012a). *Copenhagen cloudburst management plan*. http://www.deltacities.com/documents/WEB_UK_2013_skybrudsplan.pdf. Accessed 28 Mar 2014.

Copenhagen Municipality. (2012b). Vandforsyningsplan 2012 (Copenhagen water supply plan 2012). http://kk.sites.itera.dk/apps/kk_pub2/pdf/874_XFk5rWrVAG.pdf. Accessed 28 Mar 2013.

Copenhagen Municipality and Copenhagen Energy. (2008). Københavns Kommunes spildevandsplan 2008 (Copenhagen wastewater plan 2008). http://www.hofor.dk/wp-content/uploads/2012/06/spildevandsplanKE.pdf. Accessed 28 Mar 2014

Danish Meteorological Institute. (2002). *Danish climatological normals 1971–2000*. http://www.dmi.dk/fileadmin/user_upload/Rapporter/TR/2002/tr02-12.pdf. Accessed 5 Apr 2006

Hansen, A. (1996). *Copenhagen: water saving strategies*. http://infohouse.p2ric.org/ref/24/23427.htm. Accessed 5 Apr 2014

IPCC. (2014). *Climate change 2014: Synthesis report. Contribution of working groups I, II and III to the fifth assessment report of the intergovernmental panel on climate change* (Core Writing Team, R.K. Pachauri and L.A. Meyer, eds). Geneva: IPCC.

Liu, L., & Jensen, M. B. (2017). Climate resilience strategies of Beijing and Copenhagen and their links to sustainability. *Water Policy, 19*(6), 997–1013.

Liu, L., Jensen, M. B., & Meng, Q. Y. (2014). Potential contributions to Beijing's water supply from reuse of storm-and greywater. *Journal of Southeast University (English Edition), 30*(2), 150–157. https://doi.org/10.3969/j.issn.1003-7985.2014.02.003.

McEvoy, D., Lindley, S., & Handley, J. (2006). Adaptation and mitigation in urban areas: synergies and conflicts. *Municipal Eng., 159*, 185–191.

Meng, Q. Y. (2014). *Research and practices on water management in Beijing*. Presentation for 'Urban water management' course, Sino-Danish Center for Education and Research, Beijing, 16 Apr 2013

Pacific Institute. (2015). *Water conflict*. http://worldwater.org/water-conflict/. Accessed 3 Apr 2016.

Probe International Beijing Group. (2008). *Beijing's water crisis 1949–2008 Olympics*. http://www.chinaheritagequarterly.org/016/_docs/BeijingWaterCrisis1949-2008.pdf. Accessed 28 Mar 2014

Public Utility Burearu of Singapore (PUB). (2013). *Our water, our future*. http://www.pub.gov.sg/mpublications/OurWaterOurFuture/Documents/OurWaterOurFuture_2015.pdf. Accessed 12 Oct 2015

Shiklomanov, I. A. (1999). *World water resources: modern assessment and outlook for the 21st century*. Summary of world water resources at the beginning of the 21st century, prepared in the framework of the IHP UNESCO. Federal Service of Russia for Hydrometeorology & Environment Monitoring, State Hydrological Institute, St. Petersburg.

Statistics Denmark. (2016). http://statistikbanken.dk/statbank5a/default.asp?w=1680. Accessed 5 Apr 2016 (in Danish).

Stead, D. (2014). Urban planning, water management and climate change strategies: adaptation, mitigation and resilience narratives in the Netherlands. *International Journal of Sustainable Development & World Ecology, 21*(1), 15–27. https://doi.org/10.1080/13504509.2013.824928.

Zhang, S. H. (2013). *Practice of stormwater management in Beijing urban area – Current situation, problems and suggestions*. Presentation for 'Urban Water Management' course. Sino-Danish Center for Education and Research, Beijing, 23 Apr 2013

Zhang, X. X. (2014). *The water challenge in Beijing. Presentation for 'Urban Water Management' course*. Beijing: Sino-Danish Center for Education and Research 11 May 2014.

Zhao, J., Yu, K. J., & Li, D. H. (2014). Spatial characteristics of local floods in Beijing urban area. *Urban Water Journal, 11*(7), 557–572. https://doi.org/10.1080/1573062X.2013.833636.

Chapter 6
Direct Carbon Emissions by Urban Residents and Characteristics of High Emitters: The Case of Shanghai

Yuan Ren and Daisong Liu

Abstract The authors divide urban residents' direct carbon emissions (URDCEs) into two categories: residential energy consumption-related carbon emissions (RCEs) and transportation-related carbon emissions (TCEs). Based on a survey of urban residents' consumption, conducted in Shanghai in 2013, the authors analyse demographic and socioeconomic characteristics of high emitters and demonstrate the inequality of carbon emissions among different groups by income, consumption, and lifestyle. The chapter concludes that high emitters should take more social responsibility than others to help achieve a transition to a low-carbon society, and discusses how to encourage changes in their lifestyle so as to facilitate this transition.

Keywords Shanghai · Urban residents' direct carbon emissions (URDCEs) · Residential energy consumption-related carbon emissions (RCEs) · Transportation-related carbon emissions (TCEs) · Middle classes · High emitters

6.1 Domestic Consumption, Lifestyle, and Carbon Emissions

In an historic agreement, the Climate Change Conference in Paris in 2015 mobilized the common efforts of the world's nations in cutting global carbon and greenhouse gas emissions. The emphasis earlier has been on cutting carbon emissions

Y. Ren (✉) · D. Liu
School of Social Development and Public Policy, Fudan University, Shanghai, People's Republic of China
e-mail: yren@fudan.edu.cn

J. Delman et al. (eds.), *Greening China's Urban Governance*, ARI - Springer Asia Series 7, https://doi.org/10.1007/978-981-13-0740-9_6

from a production perspective. The implications include upgrading industrial structure and increasing efficiency with which energy and resources are used in production (Geyer and Jackson 2004; Sardianou 2005; Graedel et al. 1995; Guide and van Wassenhove 2004; Jackson 1996). However, as urbanization proceeds and service industries expand, consumption by urban residents is accounting for a greater and greater share of carbon emissions. People's lifestyle and their awareness and participation have strong influence on carbon emission, especially in rapid urbanizing China. The UN's world sustainable development goals also set sustainable production and consumption as important work agenda before 2030.

From a consumption perspective, carbon emissions include *urban residents' direct carbon emissions* (URDCEs), such as *residential energy consumption-related emissions* (RCEs) and *transportation/commuting-related carbon emissions* (TCEs), and also encompass the carbon emissions arising in connection with the consumption of commodities and services by consumers. In densely populated cities like New York, consumption may account for 70–80% of total emissions (New York Government, 2010). Schipper et al. (1989) conclude that consumption behaviour—domestic life, the consumption of services, the use of private cars, etc.—may account for 45–55% of total energy consumption. Bin and Dowlatabadi (2005) analyse the lifestyle, energy usage, and carbon emissions of Americans and discover that over 80% of energy usage and carbon emissions are caused by residents' direct and indirect consumption. Direct energy consumption (including domestic energy consumption and transportation) only takes up 4% of GDP in the USA, but it accounts for 28% of the country's energy consumption and generates 41% of its carbon emissions. Greening et al. (1998) finds, in a study of carbon emissions in ten OECD countries from 1970 to 1993, that a 30–70% drop took place in such emissions in these countries' industrial sectors over that period and a 1.4–15% drop in their transportation sectors; however, a 26–64% *rise* in such emissions took place in their residential sectors during the period, and the latter sectors accounted for 20–25% of total urban energy consumption in 1993.

The increased energy consumption and carbon emissions caused by additional consumption have attracted the attention of scholars. Many researchers stress the need to change lifestyles and patterns of behaviour in connection with consumption. Research into *sustainable livelihoods* and *sustainable patterns of consumption* is thus becoming an important focus for efforts to achieve sustainable development and a reduction in urban carbon emissions. As many studies have indicated, a change in patterns of cognition and of knowledge acquisition could raise energy-saving awareness among urban residents (Sardianou 2005; Parker 2005; Gyberg and Palm 2009; Ouyang and Hokao 2009; Carlsson-Kanyama and Eriksson 2005; Luova, Chapter 4). Frequently providing information about energy usage can impel residents to adopt more energy-saving behaviour (Brandon and Lewis 1999; McCalley and Midden 2002; Abrahamse et al. 2005; Egmond et al. 2005).

According to World Bank data, China had an annual average growth rate of 11.6% in residential electricity consumption during the 2005–2012 period and of 20.3% in residential natural gas consumption. Meanwhile, CO_2 emissions per capita increased from 4.89 tons in 2006 to 6.19 tons in 2010. The rate of increase in such emissions is much higher in China than in the major industrialized countries. Carbon emissions in China, a country in the middle stages of industrialization, originate above all in manufacturing and production processes. However, as countries urbanize and their predominant lifestyle gradually changes, carbon emissions associated with consumption tend to increase. To achieve its national target of starting to lower the overall carbon emission intensity by 2030, China will need more and more to contain carbon emissions caused by consumption. Meanwhile, current studies of consumption-related carbon emissions are usually based on macro-level data, which are used to do a decomposition analysis. These studies show how the scale of carbon emissions is affected by such factors as economic development, the demographic situation, the energy structure, the degree of efficiency in energy usage, and the like. However, carbon emissions have rarely been analysed at the micro level of the individual resident (Jin 2012; Zhu et al. 2009). In this paper, we compare levels of energy consumption-related carbon emissions by different groups of residents, with Shanghai as our case study. We also discuss challenges and pressures associated with the transition to a low-carbon society at national and city levels.

6.1.1 Energy Consumption-Related Carbon Emissions by Urban Residents

Scholars generally define the carbon emissions of urban residents as the greenhouse emissions produced by such residents' direct and indirect energy usage. Direct carbon emissions are often taken to refer to the CO_2 emissions generated by direct consumption of energy: coal, oil, natural gas, and electricity. Indirect carbon emissions are taken to refer to the CO_2 emissions generated during the production of commodities and before service processes (Vringer and Blok 1995; Golley et al. 2009). Some scholars categorize the CO_2 emissions produced by residential energy consumption and private transportation as direct carbon emissions and the emissions generated through consumption of various products and before services as indirect carbon emissions (Bin and Dowlatabadi 2005; Wei et al. 2007). The direct consumption of electricity, natural gas, and the like in daily activities—including cooking, hydraulic heating, lighting, and temperature regulation—is identified as residential energy consumption. Where transportation-related direct carbon emissions are concerned, some scholars only include carbon emissions caused by transportation, while others only include family gasoline usage (Vringer and Blok 1995). Still others include carbon emissions generated by the use of public transport (Bin and Dowlatabadi 2005; Wei et al. 2007). Direct carbon emissions can however be calculated using the emission coefficient provided in the *IPCC Guide for Greenhouse Gas Emissions List*. Indirect carbon emissions can be calculated by various methods, including an input-output method, a consumers' lifestyle method, and a life cycle evaluation method.

Studies show that residential energy consumption increases with rising household income, but at a slower rate than the latter. Residential energy consumption is inversely proportional to the price of energy, but it generally lacks price elasticity. Residential carbon emissions are thus mainly affected by other factors (Peter et al. 2004). However, residential carbon emissions by low-income households are sensitive to price fluctuations; those by high-income households, by contrast, are not (Nesbakken 1999; Anker-Nilssen 2003). Multiple studies have pointed to a significant correlation between energy consumption and carbon emissions, both of which are affected by customs, local climate, heating practices, temperature regulation, the type and size of rooms, and the number and energy efficiency of household appliances. A single household, for example, consumes more electricity per capita than a combined residence (Curtis et al. 1984; Black and Stern 1985; Huo et al. 2010; Sardianou 2005).

Transportation is another important source of direct carbon emissions. As a major energy consumer, transportation accounts for 20–25% of global carbon emissions, and this proportion is projected to rise to 30–50% by 2050 (Bin and Dowlatabadi 2005). Transportation, which keeps expanding in the wake of industrial production and the development of the service sector, is thought to account for a large part of urban carbon emissions and to be a central cause of the increase in global carbon emissions. According to a recent study, transportation-related carbon emissions account for about 40% of Beijing's emissions. Urban transportation includes the transportation of products and of residents. The transportation of residents shows an increasing trend, with a growing use of private cars and an increase in leisure tourism. Transportation-related carbon emissions related to family life account for an increasing proportion of direct carbon emissions by urban residents. Studies on carbon emissions by residents show that different population groups vary greatly in their transportation behaviour, according to age, income, social status, residence, and ownership/non-ownership of a private car. High emitters, who account for 10–20% of the population, produce 40–60% of the carbon emissions that derive from transportation (Brand and Boardman 2008; Brand 2009; Brand and Preston 2010; McNamara and Caulfield 2011).

6.2 Studying the Field

This study is based on a survey, named *Shanghai Residents' Carbon Consumption Household Survey*, conducted by the authors' research team in June 2013. Shanghai is China's important megacity and has a total population of 23 million in 2010 according to the census. Shanghai metropolitan area has totally 17 districts and 1 county, and totally it has 99 street office wards and 110 rural town level areas. Below the street office and town level, there are 3850 neighbourhood communities and 1632 village communities.

Spatially, Shanghai metropolitan area could be divided into three parts, central city area (including nine districts), inner suburb area (including four districts), and outer suburb area (including four districts and one county). In China's context, lifestyle in different location of a metropolitan area is very different. Comparing the three parts of Shanghai metropolitan area, central city is more urban; the outer suburb area is urban edge and still has large area of rural area, therefore, village is

an important community pattern in this area; and the inner suburban area is the mixed area of urban and rural life or a kind of urbanizing life pattern in a metropolitan area. We selected one district located in central city (Hongkou District), one district located in inner suburb district (Pudong New Area), and one district in outer suburb area (Jinshan District) as our survey areas. Figure 6.1 shows the distribution of these three districts.

Considering total population size and number of neighbourhood committee/village committee in each districts, we select 8 resident communities in two street offices in Hongkou district, 16 resident communities in two street offices and two towns in Pudong New Area District, and 12 resident communities in two street offices and two towns in Jinshan Districts as our survey locations. The total sampling sizes in these three districts are, respectively, 800, 800, and 600. We take an equal distance sampling method to select survey communities, that is, we make a list of all street office level administrative areas and resident committees sequencing with their total population and then we select surveyed location. In total we sampled 10 street offices/towns; below this street office/town administrative level, we then pick up 36 resident communities. The interviewed households are selected randomly with supports of resident committee, based on the household list they provided, in order to have an integral household pool to include all native population households and migrant population living in the areas.

Since the consumption of residential energy and of commodities looms large in overall household consumption, our study measures residential carbon emissions at a household level. We define 'household members' as those dwellers in a residence who eat breakfast and dinner together at least 15 days a month. When we encountered an interviewee with a mental impairment or functional incapacity, we substituted the next candidate for that person, in order to guarantee the effectiveness of the survey. We selected a total of 2200 sample households and were able to carry out 2118 questionnaire interviews. Meanwhile, to make the data more reliable, we discarded samples with incomplete data or with obvious logical mistakes. Our final database contained 1696 cases.

6.2.1 Calculation of Direct Carbon Emissions by Urban Residents

Direct carbon emissions by urban residents, as defined here, relate to direct energy consumption. They can be divided into *residential carbon emissions* (*RCEs*) and *transportation-related carbon emissions* (*TCEs*). The former are carbon emissions generated during the survey period by direct household energy consumption. The latter includes the use of electricity, natural gas, and coal for residential purposes: cooking, operating appliances, heating the living space, etc. We calculated RCEs based on carbon emission coefficients and the overall volume of energy consumed during the survey period. TCEs include commuting-related emissions, leisure-related emissions, and tourism-related emissions. We calculated

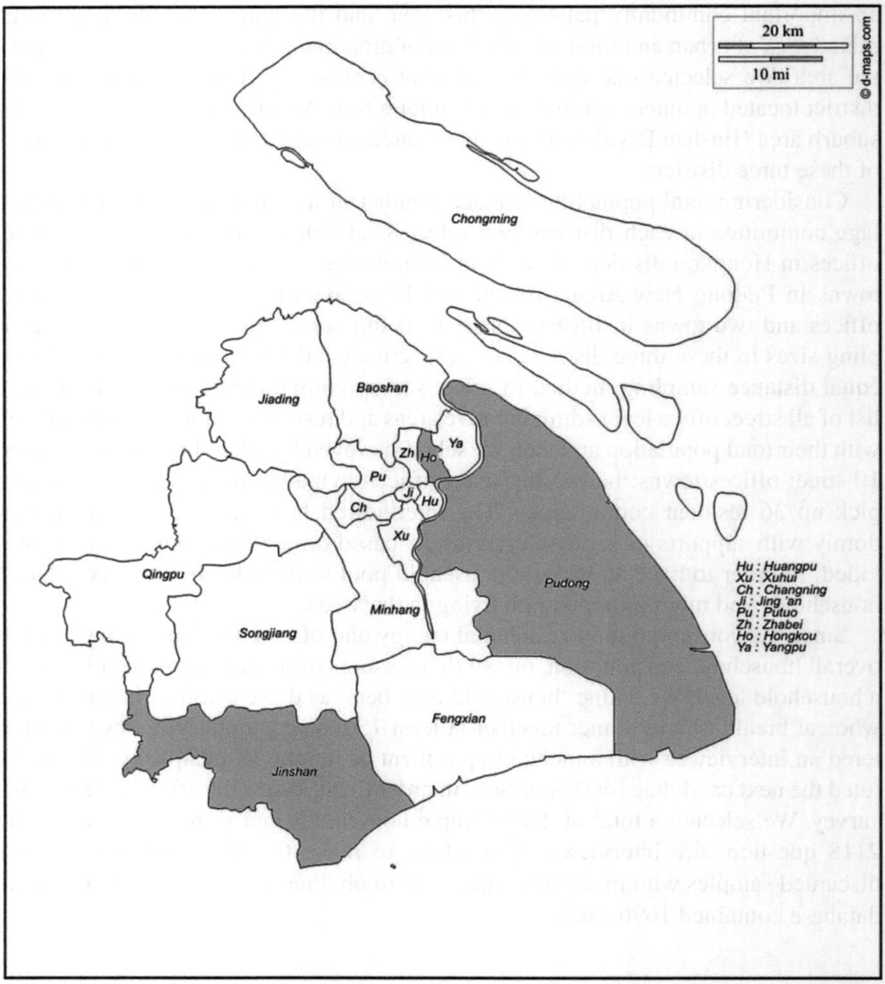

Fig. 6.1 Map of Shanghai and survey areas

commuting-related carbon emissions on the basis of the commuting distance and the carbon emission coefficient per kilometre of the specific commuting means adopted (car, bus, or metro). Residents' leisure-related carbon emissions are estimated based on the frequency of leisure travel, the mean travel distance of Shanghai residents (as acquired from other statistics), and a synthetic transportation-emission coefficient. Tourism-related carbon emissions are calculated based on the mean leisure travel distance (from residence to tourist destination) and the carbon emission coefficients of the main means of transport (plane, train, coach, and car). Specific formulas are as follows:

Household residential carbon emissions:

$$E_{\text{house}} = \Sigma\left(E_i \times F_i\right)$$

Commuting-related carbon emissions:

$$E_{\text{commuting}} = \Sigma\left(D_j \times C_j\right)$$

Leisure-related carbon emissions:

$$E_{\text{leisure}} = N \times D_0 \times C_0$$

Tourism-related carbon emissions:

$$E_{\text{tourism}} = \Sigma\left(D_k \times C_k\right)$$

where E_{house} means the household's residential direct carbon emissions, i means the type of energy directly consumed by the household (e.g. electricity, natural gas, coal), E_i refers to the direct consumption of domestic energy i, and F_i is the carbon emission factor; $E_{\text{commuting}}$ symbolizes transportation-related carbon emissions during workday commuting, j means the commuting method adopted, D_j refers to the driving distance of the method adopted, and C_j means its carbon emissions/km/person; E_{leisure} refers to transportation-related carbon emissions arising from leisure activities, N means the number of leisure trips during the statistic cycle, D_0 the average travel distance of Shanghai residents, and C_0 the full-transportation emission factor; and E_{tourism} means the transportation-related carbon emissions generated by family tourism, k means the number of trips in the past year, D_k the traffic distance for family tourism k, and C_k the unit passenger-turnover CO_2 emission coefficient for the transportation method of family travel k.

Since residential carbon emissions are calculated per household, while commuting-, leisure-, and tourism-related carbon emissions are calculated per capita, we define *urban residents' direct carbon emissions* as equal to residential carbon emissions per household member, plus the transportation-related carbon emissions that we have ascertained through interviews. This should facilitate a comparison between the different groups of emitters.

6.3 Direct Carbon Emissions: Inequality Among Urban Residents

6.3.1 Definition of High Emitters

Between May 2012 and June 2013, the mean value of residential carbon emissions per capita was 1093.2 kg. To divide the population into different emitter groups, we have taken 200% and 50% of that figure as our criteria for segmenting the surveyed

Table 6.1 Carbon emissions of various emitters

Average carbon emissions (kgCO2)	High emitters	Medium-high emitters	Medium-low emitters	Low emitters	Total
Classification standard	>2186.4	1093.2–2186.4	546.6–1093.2	<546.6	
Proportion	15.1%	34.9%	35.2%	14.8%	100%
Household per capita residential carbon emissions	1687.12	1147.90	741.40	395.90	974.99
Of which: Electricity-related carbon emissions per capita	1296.75	885.20	547.99	288.44	740.64
Fuel gas-related carbon emissions per capita	393.49	266.37	194.07	110.87	237.22
Transportation-related carbon emissions	1719.88	362.48	70.73	18.35	414.16
Of which: Commuting-related carbon emissions	1172.75	211.96	26.09	1.01	260.64
Leisure-related carbon emissions	44.67	25.85	12.71	7.46	21.35
Tourism-related emissions	502.46	124.67	31.93	9.88	132.18
Total	3407.00	1510.38	812.13	414.25	1389.15

population into four groups: (1) high emitters, (2) medium-high emitters, (3) medium-low emitters, and (4) low emitters. These constituted 15.1%, 34.9%, 35.2%, and 14.8% of the total, respectively; and they produced per capita carbon emissions of 3407 kg, 1510.4 kg, 812.1 kg, and 414.2 kg, respectively. High emitters thus accounted for 2.6 times the emissions of medium-high emitters, 4.2 times those of medium-low emitters, and 8.2 times those of low emitters (see Table 6.1).

6.3.2 *Structure of Urban Residents' Carbon Emissions*

Different population groups have different carbon emission patterns. In general, residential carbon emissions are greater than transportation-related carbon emissions. As carbon emissions increase, however, the proportion consisting of transportation-related emissions is expected to rise gradually and to surpass that consisting of residential carbon emissions. During the survey period, the per capita residential and transportation-related carbon emissions of high emitters were 4.26 times and 93.72 times as high as those of low emitters, respectively. Among low emitters, medium-low emitters, and high emitters, residential carbon emissions were 21 times, 10 times, and 0.98 1.02 times as high as transportation-related emissions, respectively. However, while residential carbon emissions thus accounted for most of residents' direct carbon emissions, transportation-related carbon emissions are expected to grow much faster than residential carbon emissions, and transportation-related carbon emissions already slightly exceeded residential carbon emissions among high emitters during the survey period.

Electricity and fuel gas account for about 75% and 25% of per capita residential carbon emissions, respectively, and they vary little across the different emitter groups. Transportation-related carbon emissions vary heavily in their structure among the different emitter groups. Commuting-related emissions account for about 60% on average and leisure- and tourism-related emissions for some 40%. However, the proportion of commuting-related carbon emissions among medium-low and low emitters is much lower than it is among medium-high and high emitters, mainly for two reasons: (1) a considerable proportion of the two former groups is made up of retirees, and (2) many persons in these groups use slow means of transportation, such as walking or biking. Figure 6.1 shows the structure of urban residents' direct carbon emissions across the four groups of emitters. Among low emitters, electricity accounts for most carbon emissions, followed by fuel gas; this group has few carbon emissions otherwise. Among higher emitters, the proportion of carbon emissions accounted for by fuel gas and electricity has declined, while the proportion accounted for by commuting and tourism has gotten higher and higher (Fig. 6.2).

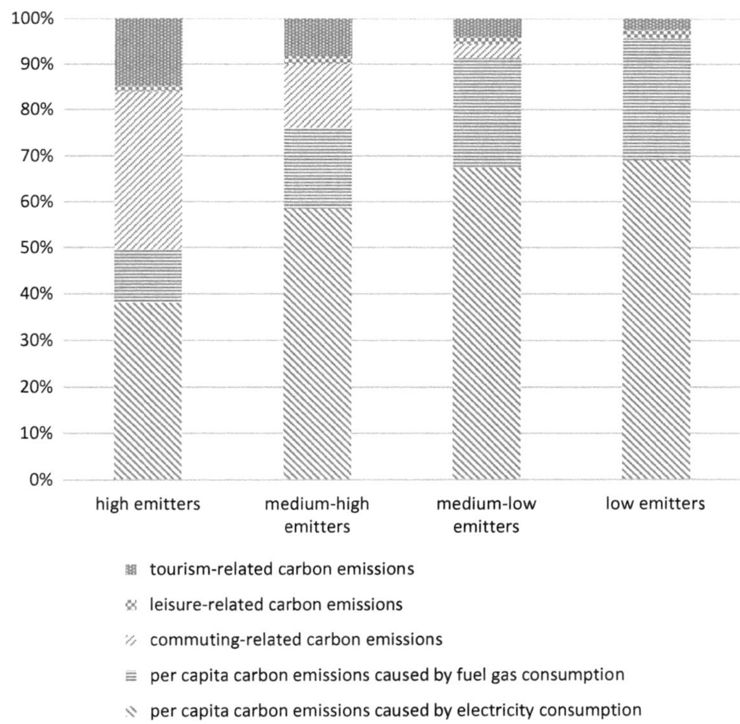

Fig. 6.2 Urban residents' direct carbon emissions: structure among different emitter groups

6.3.3 Carbon Emissions by Urban Residents: Inequality Among Different Groups

To illustrate the inequality among urban residents when it comes to carbon emissions, we calculate the carbon Gini coefficient for this inequality and draw a corresponding Lorenz curve to depict it. Based on their carbon emissions (from low to high), the households in our sample are divided into n groups, with an equal number of residents for each group. Wi is the proportion of cumulative carbon emissions of household 1 to household i in the total carbon emissions of residents.

This analysis shows that the carbon Gini efficient for per capita carbon emissions of residents is 0.37. The top 20% of emitters account for 44% of carbon emissions and the remaining 80% for 56%. This indicates a medium degree of inequality. The Gini coefficient for per capita residential carbon emissions stands at 0.30: the top 20% account for 38% of carbon emissions and the remaining 80% for 62%. In the case of transportation-related carbon emissions, however, the Gini coefficient reaches 0.78, with the top 20% accounting for 80.4% of carbon emissions and the remaining 80% for less than 20%—showing serious inequality in this area. Residential carbon emissions reflect the basic needs of residential living. They are characterized, therefore, by low elasticity and a generally equal distribution. Transportation-related carbon emissions are driven by residential location and by ownership/non-ownership of a private car; and they vary greatly according to the age, residence, and social status of emitters. Thus, since transportation-related carbon emissions are more elastic, they should be the focus of policy interventions for achieving a low-carbon urban society (Fig. 6.3).

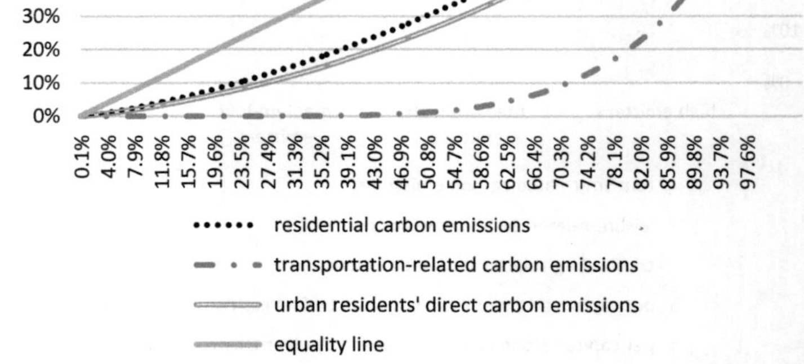

Fig. 6.3 Lorenz curve for residential and transportation-related carbon emissions

6.4 Demographic and Socioeconomic Characteristics of High Emitters

6.4.1 Demographic Characteristics

Household scale is inversely proportional to per capita direct carbon emissions by urban residents. The members of a household share residential energy usage, so larger households have economies of scale in regard to carbon emissions (see Table 6.2). As national census data shows, the average Chinese household had 3.96 members in 1990, but only 3.44 in 2000 and 3.10 in 2010. Smaller-scale households can be expected to result in a rise in per capita carbon emissions.

There is no significant difference in gender composition between the various groups of emitters. A slightly higher percentage of men (17.8%) than of women (13.0%) are high emitters. The opposite applies among medium-high emitters (33.9% of men and 35.7% of women belong to this group) and among medium-low emitters (33.1% of men, 36.9% of women). Where different age groups are concerned, middle-aged adults have higher carbon emissions on average than younger or older people. This is predicable enough, given that middle-aged adults are in the marriage and employment period in their life cycle: they need more housing space, and they must commute to work, which leads to higher direct carbon emissions.

6.4.2 Socioeconomic Characteristics of High Emitters

High carbon emissions correlate strongly with high income and high consumption. Thus, high emitters overlap to some extent with the middle classes. To study the relationship between occupation and carbon emissions, we have analysed the situation using the '10 Chinese classes' proposed by the Chinese Academy of Social Sciences (CASS). The different categories vary greatly in their occupational composition. For example, technicians and clerical staff constitute a combined 57% of

Table 6.2 Demographic characteristics of various emitter groups

Index	High emitters	Medium-high emitters	Medium-low emitters	Low emitters	In total
Household scale	2.65	2.78	3.03	3.62	2.97
Male	17.8%	33.9%	33.1%	15.2%	100%
Female	13.0%	35.7%	36.9%	14.5%	100%
Age > 25	14.1%	41.0%	35.9%	9.0%	100%
Age 25–50	19.9%	35.8%	30.7%	13.7%	100%
Age > 50	12.4%	33.9%	37.8%	16.0%	100%

Table 6.3 Carbon emissions among different occupations

	High emitters	Medium-high emitters	Medium-low emitters	Low emitters	
State and social administrative staff	19.0%	53.4%	20.7%	6.9%	100.0%
Private entrepreneurs	22.7%	50.0%	18.2%	9.1%	100.0%
Managers	29.8%	43.5%	22.4%	4.3%	100.0%
Technicians	17.1%	39.4%	32.0%	11.5%	100.0%
Clerical staff	16.7%	34.2%	35.4%	13.8%	100.0%
Individual business workers	16.7%	36.7%	30.0%	16.7%	100.0%
Commercial clerks	5.1%	38.5%	42.3%	14.1%	100.0%
Manufacturing workers	6.3%	26.0%	44.5%	23.2%	100.0%
Peasants	4.0%	9.3%	44.0%	42.7%	100.0%
White collar	18.0%	38.6%	31.9%	11.5%	100.0%
Blue collar	6.3%	26.0%	44.5%	23.2%	100.0%
Agricultural population	4.0%	9.3%	44.0%	42.7%	100.0%
Total	15.1%	34.9%	35.2%	14.8%	100.0%

total employed population, while state and social management staff, private entrepreneurs, and individual business workers account for less than 5%. This paper describes the relationship between occupation and carbon emissions via the distribution of different occupations in relation to carbon emissions of various types (see Table 6.3). We divide occupations into three groups: white collars, blue collars, and peasants. White collars are more likely to be high emitters and medium-high emitters. A manager has a 29.8% probability of being a high emitter (higher than for any of the other nine occupational groups), a 43.5% probability of being a medium-high emitter, and only a 4.3% probability of being a low emitter. Private business entrepreneurs, government administrators, and technical experts also have a higher than average likelihood of being high emitters. Conversely, peasants and manufacturing workers have the highest probability of being low emitters.

As shown in Table 6.4, persons with a higher level of education are more likely to be high emitters. Among those with postgraduate or higher degrees, 57.1% are high emitters, 38.1% are medium-high emitters, and only 4.8% are low emitters. With falling educational levels, the proportion of low emitters rises. Among the illiterate population, only 5.9% are high emitters, while 40% are low emitters and 38.8% are medium-low emitters.

6.4.3 *Lifestyle of High Emitters*

As mentioned earlier, the level of carbon emissions by residents increases with their income. The monthly per capita income of high emitters is 7222 yuan. This is 1.7 times the income of medium-high emitters, 2.5 times that of medium-low emitters,

Table 6.4 Carbon emissions among different educational groups

	High emitters	Medium-high emitters	Medium-low emitters	Low emitters	
Postgraduate	57.1%	38.1%	4.8%	0%	100.0%
Undergraduate	29.1%	45.3%	22.6%	3.0%	100.0%
Junior college	21.2%	41.7%	31.3%	5.9%	100.0%
Polytechnic schools	13.7%	36.6%	41.2%	8.5%	100.0%
High school	12.5%	37.5%	37.5%	12.5%	100.0%
Middle school	5.7%	29.0%	41.4%	23.8%	100.0%
Primary school	2.6%	19.7%	45.3%	32.5%	100.0%
Illiterate	5.9%	15.3%	38.8%	40.0%	100.0%
Higher education and above	27.2%	43.0%	25.5%	4.2%	100.0%
Secondary education	12.9%	37.2%	38.8%	11.1%	100.0%
Primary education and below	5.1%	25.4%	41.8%	27.7%	100.0%
Total	15.1%	34.9%	35.2%	14.8%	100.0%

and 3.5 times that of low emitters. The amount spent on housing by high emitters is 20% higher than that spent by medium-high emitters, 44% higher than that spent by medium-low emitters, and 109% higher than that spent by low emitters. In general, moreover, high emitters have more living space and a greater number of household appliances. However, the average family living space of low emitters is 119 square metres—higher than that of high emitters. That is because many low emitters are peasants and live in rural areas, where they have access to larger dwellings due to the lower price of housing in such areas. Families are also bigger in rural areas than in urban areas, on average. So, if we measure living space in per capita terms, we see that the living space of low emitters (at 35 square metres) is lower than that of high emitters (48 square metres), of medium-high emitters (39 square metres), and of medium-low emitters (35.5 square metres).

In addition, high emitters own a private car more often than others do, and they travel more frequently and across greater distances. They travel 1.69 times per year, and their average primary travel distance each time is 3965 kilometres. The corresponding figures for medium-high emitters are 1.56 times and 1702 kilometres; for medium-low emitters, they are 0.85 times and 956 kilometres; and for low emitters they are 0.52 times and 618 kilometres. The means of transport also varies: high and medium-high emitters are more likely to reach their tourist destination by plane, medium-low emitters by plane or bus, and low emitters by bus or private car.

Where residential location is concerned, persons living in inner suburb areas and in the central city are more likely to be high emitters, while residents of the outer urban edge have the lowest likelihood of being such. Residents of the central city tend to have a medium-high level of overall emissions, and they are the least likely to be low emitters (Table 6.5).

Table 6.5 Lifestyle of different emitter groups

Index	High emitters	Medium-high emitters	Medium-low emitters	Low emitters	In total
Interviewee's monthly income (CNY)	7221.89	4192.90	2860.21	2069.47	3869.70
Proportions in central city	17.6%	44.1%	30.3%	8.0%	100%
Proportions in inner suburbs	21.0%	34.9%	33.8%	10.3%	100%
Proportions in outer suburbs	9.6%	31.1%	42.1%	17.2%	100%
Housing price (ten thousand yuan/ square metre)	2.44	2.05	1.69	1.17	1.88
Family living space (square metre)	113.76	101.11	99.35	119.42	105.12
Number of household appliances (pcs.)	8.51	7.46	6.47	5.67	7.01
Time required from the residence to the bus station (minutes)	4.62	4.67	5.24	7.06	5.22
Private car ownership	56.35%	38.04%	24.23%	24.70%	33.97%
Number of times of family travel per year	1.69	1.56	0.85	0.52	1.18

For the most part, different population groups use different methods of commuting. High emitters mainly use private cars (or taxis or the like). Medium-high emitters rely largely on public transportation, such as the bus or the subway. Medium-low and low emitters mainly use slow means of transport: bicycles, motorbikes, walking, and so on. Higher emitters also have longer commutes, while lower emitters live much closer to their place of work (Table 6.6).

6.4.4 *Modelling High Emitters' Characteristics and Determinants*

At this point, our investigation should benefit from a closer look at the main demographic traits and socioeconomic characteristics of different emitter groups, as well as from a discussion of the main factors leading to higher emissions. To this end, we divide each of the four groups of carbon emitters into two levels (high and low), and we use a logistic model to portray the prime characteristics of high emitters. The stepwise modelling results can be seen in Table 6.7. It is clear, from models 1 through 7, that the models gradually fit better and better. Our findings point up some typical traits of high emitters:

1. The use of a private car for commuting, in preference to slow means of transport, is an obvious indicator of high emitters. Moreover, the use of public transportation points to low emitters in a much higher degree than the use of a private vehicle does, yet such emitters still produce significantly higher emissions than the reliance on slow means of transport indicates.

Table 6.6 Commute methods of different emitter groups

	High emitters	Medium-high emitters	Medium-low emitters	Low emitters	Total
By car: private car, taxi, reserved vehicles	56.5%	19.4%	5.4%		20.7%
Public transport: bus, subway, company shuttle	31.8%	48.5%	17.2%		29.6%
Slow transportation: motorbike, bicycle, walking	11.7%	32.1%	77.5%	100.0%	49.7%

2. Residential patterns are also revealing here: pricier housing is a significant indicator of high emitters.
3. The possession of a larger number of household electronic appliances is strongly correlated with higher carbon emissions.
4. Smaller household scale leads to higher carbon emissions, measured per capita.
5. Income is an important factor influencing people's consumption, and higher income is clearly correlated with heavier emissions.
6. Big houses usually mean high emissions.
7. According to the classical model of suburban gentrification in Western societies, those living on the urban edge are sometimes higher emitters because they commute across greater distances; in China, however, the central city and the suburbs typically have higher priced housing, and their population is richer and has higher social status. Residents of the central city and of the suburbs are therefore more likely to be high emitters.
8. Certain indicators, while not statistically significant, are nonetheless important in explaining high carbon emissions. The use of a private car usually means higher carbon emissions. Most middle-aged persons are married and have formed families, so members of this age group are higher emitters than either the young or the old (elders, e.g. tend to reduce their spending and thus their carbon emissions). Males are slightly higher emitters than females. The difference between the genders, while not very significant, mainly shows up in occupational and commuting patterns. Urban *hukou* residents and the local *hukou* population are higher emitters than non-*hukou* temporary migrants (see Introduction to this volume).

6.5 Conclusion and Discussion

Urbanization accompanies industrialization, and the two processes together are marked by increases in population, substantial shifts in social structure, and far-reaching changes in lifestyle—all of which lead to increasing consumption and rising carbon emissions in cities. We must therefore pay attention to carbon emissions caused by residents' consumption and work towards a transformation of lifestyles,

Table 6.7 Factors associated with residents' carbon emissions

	Model 1	Model 2	Model 3	Model 4	Model 5	Model 6	Model 7
Commuting method (ref: slow transportation)							
Private cars	2.880***	2.915***	2.746***	2.846***	2.493***	2.423***	2.233***
Public transport	1.341***	1.045***	1.034***	0.984***	0.895***	0.911***	0.878***
Housing price		0.476***	0.440***	0.423***	0.387***	0.386***	0.192
Household scale			0.178***	0.205***	0.190***	0.160***	0.130***
Electronic appliance				−0.507***	−0.522***	−0.566***	−0.619***
Household (ref: high income)							
Low-income					−1.597***	−1.623***	−1.433**
Medium-low-income					−1.006***	−1.011***	−0.714**
Medium-high-income					−0.666***	−0.660***	−0.469**
Housing space						0.006***	0.011***
Location (ref: suburban)							
Central city							0.094
Urban fringe							−0.868*
Has a private car? (ref: yes)							0.374
Age (ref: above 50)							
Below 25							−0.170
25–50							0.188
Gender (ref: female)							0.213
Education							0.036
Occupation (ref: blue collar)							−0.006
Shanghai citizenship							0.210

Urban *hukou*							0.876
Constant	−2.727	−3.559	−4.828	−3.597	−2.638	−2.871	−3.138
Chi-square value	220.393***	239.001***	266.768***	297.695***	322.073***	328.894***	305.828***
Cox and Snell R square	0.137	0.171	0.189	0.209	0.226	0.23	0.235
Nagelkerke R square	0.233	0.282	0.311	0.344	0.371	0.378	0.378
N	1666	1666	1666	1666	1666	1666	1666

Note: *** $p < 0.01$,** $p < 0.05$,* $p < 0.1$

in the direction of sustainability. In some developed countries, carbon emissions caused by consumption already make up the main source of total emissions. Carbon emissions in China are still mainly caused by industrial production, but the carbon emission pattern in developed countries suggests that China's cities will likely be following the same trend. The economy in large cities like Shanghai and Beijing is already being transformed into one dominated by consumption and services. It is in such cities that the emergence of consumption as the main source of overall urban carbon emissions will first become evident.

Our focus in this study is emphasizing consumption, and people's lifestyle has significant impacts to the increasing carbon emission in Chinese cities. We could draw the following main conclusions from our analysis:

1. Different population groups are unequal in their direct carbon emissions. High emitters constitute 15.1% of the total population, but their direct carbon emissions make up 37% of the total. Low emitters constitute some 14% of the total population (about the same as the proportion of high emitters), but they produce only 4.4% of total carbon emissions. Carbon emissions per capita by high emitters are 3.4 times as high as those by low emitters. The difference in this regard is relatively minor in the case of residential carbon emissions, but it is very significant in the case of transportation-related carbon emissions. Transportation-related carbon emissions by high emitters are 20 times as high as those by low emitters. This reflects not only the fact that high emitters mainly rely on a private car for commuting but also the fact that they produce higher emissions in connection with leisure travel and tourism. Transportation-related carbon emissions are more price-elastic, which means that transportation has an increasing impact on carbon emissions in cities.
2. High emitters usually live in smaller households, and men produce more emissions than women. In addition, adults with children produce higher emissions than unmarried youth or elders. From this point on, however, ageing may have decreasing effects on carbon emissions. Due to urbanization, more young people of family-forming age will congregate in the cities; thus, from an age-structure perspective as well, urbanization can be expected to result in higher carbon emissions.
3. The level of carbon emissions can be seen as an indicator of residents' socioeconomic status. White-collar persons produce higher carbon emissions than their blue-collar counterparts, and more educated people have higher carbon emissions than less educated people. Not surprisingly, these findings reflect a close correlation between income and consumption. In general, rich people produce higher emissions than poor people. Moreover, when we also take account of the indirect carbon emissions caused by consumption, the correlation between socioeconomic status and carbon emission levels becomes even more evident.
4. High emitters and low emitters have different lifestyles; in other words, the difference in people's lifestyles leads to different levels of carbon emissions. Residents of the central city and of suburban areas have higher emissions. The situation in Shanghai is very different from that in main European and North

American cities, since they might have an inner-city poverty model so people living in city centre produce less carbon, while residents of the urban edge are actually high emitters. The main characteristics of high emitters are their use of a private car and their residence in larger (and usually more expensive) housing units. Some low emitters, however, actually live in large houses. These are mainly rural dwellers in villages on the urban fringe. Nevertheless, spacious dwellings and private cars are typical indicators of high emitters, and this means that the middle class is the main component of the high-emitter population. High emitters are more likely to use a private car for commuting; low emitters tend to use slow means of transport.

5. The main characteristics of high emitters, then, are the receipt of high incomes, the use of a private car, the possession of numerous electric appliances, and residence in spacious and expensive dwellings. High emitters also tend to have fewer family members and to live in the central city or in suburbs, where housing prices are usually higher. Middle-class men are the typical high emitters. We can therefore expect the spread of middle-class urban lifestyles—and the associated spread of private cars and spacious dwellings—to result in a continuous increase in carbon emissions in cities. This poses a great challenge to the project of building a low-carbon society.

As the far-reaching inequality in energy-based carbon emissions between different social groups shows, high emitters exert a greater influence on levels of urban carbon emissions. It is therefore important to reduce the carbon emissions of high emitters in particular and to try to transform their lifestyles and consumption patterns in such a way as to make a transition to a low-carbon society possible.

The inequality in carbon emissions also indicates that high emitters must take greater responsibility for building a low-carbon urban society. Progressive pricing systems have already been established in China for water, electricity, and gas. We must now tackle the great inequality in transportation-related carbon emissions. High emitters are more likely to occupy a larger dwelling, to use a private car, and to live on the outskirts. It may thus be appropriate to introduce a progressive 'carbon tax' in housing and transportation, whereby high emitters pay a surcharge for their higher energy consumption. Such a policy might also promote the necessary changes in lifestyle. In practice, however, it is often the case that market prices for cars and spacious dwellings are lowered, in order to promote sales. The government's investments in road construction, housing construction, and land use actually function as a kind of stipend for high emitters. We must accordingly focus on private cars and on spacious dwellings, as the two key variables for residents' carbon emissions. If we are to succeed in reducing inequality in carbon emissions, and in achieving the necessary lifestyle changes, this will be critical.

In the wake of the social polarization between different social classes in China, changing patterns of social stratification in connection with consumption and carbon emissions have become evident. We would argue that there is now a 'carbon middle class', as measured by carbon emissions. Because carbon emissions are significantly correlated both with occupational status and with levels of income and education, this carbon middle class overlaps to some extent with ordinary middle-class population. However, government administrators and technicians have lower-carbon emissions than other middle-class groups. This means the carbon middle class also varies in relation to the ordinary middle classes, and education also has potential impacts to decrease people's carbon emission Nevertheless, higher occupational status and higher levels of income and education generally correlate with a higher-emission lifestyle. This lifestyle, which is spreading in step with advancing urbanization, will need to be changed. It will also be necessary to prevent medium and low emitters from becoming new high emitters. A new 'low-carbon middle class' is needed.

In China's context, urbanization is typically accompanying with, or actually expanding, the process of materialism and consumerism. Materialism and consumerism are typical in the early stages of modernization. The middle class is thus marked not only by its education, occupational status, and household income but also by its lifestyle. Private cars and spacious dwellings are widely regarded as key assets of the middle class. If, therefore, we are to transform ourselves from a high-emission society into a low-carbon one, we must achieve far-reaching changes in lifestyles. Large-scale efforts in environmental education will be needed (cf. Luova, Chap. 6). Such efforts must not only be geared at increasing citizens' knowledge about carbon emissions but also at enhancing their awareness of their social and environmental responsibilities. The purpose of such education must be to encourage citizens to adopt sustainable post-modern lifestyles. In this context, society must place a particular emphasis on the social responsibilities of high emitters. High emitters have higher levels of education, which makes them more likely to shape the idea of sustainable consumption and to shift to low-carbon patterns of behaviour and consumption. If put under pressure to show social responsibility and to follow the canons of environmental ethics, highly educated people can be persuaded to make a point of removing their 'high carbon' tag and to take the lead in changing behaviour. Members of this group may thus be induced to promote environmental protection, to engage in low-carbon consumption, to purchase low-carbon high-tech products, and to participate in environmental and social organizations. High emitters can presumably be persuaded to offset their air travel-related carbon emissions by investing in environmental and social development. If social pressures are put upon them to assume their responsibilities, high emitters will likely prove willing and able to develop new lifestyles and to take a leading role in the low-carbon transition. It ought to be possible to convince them to join a movement for a 'low-carbon middle class'—one whose lifestyle and patterns of consumption are adapted to the needs of a future low-carbon society.

Thus, in order to achieve a sustainable low-carbon society, there is a need to change the traditional materialistic and consumerist lifestyle during the current rapid urbanization of China and to develop a low-carbon pattern of daily life in the economic and social affairs of Chinese citizens. The approach to consumption must change in favour of a low-carbon orientation. When people have a stronger low-carbon awareness, when they engage in low-carbon consumption to a greater extent, and when they change their lifestyle and buy more low-carbon products—when they do these things—it will be easier to build a low-carbon society in connection with consumption. This may then lead to changing patterns of manufacturing and industrialization and be helpful in developing low-carbon production methods for satisfying people's low-carbon demands. In conclusion, the most crucial challenge in achieving a low-carbon society is to persuade high emitters and potential high emitters to adopt a low-carbon lifestyle as soon as possible.

Acknowledgements This study received support from a programme supported by the Fudan-Tyndall Center: *Exploration of the Role of High Emitters in Accelerating the Transition to a Low-carbon Society*. An earlier version of this paper was presented at the workshop *Governance of Sustainable Cities in China: Challenge and Practice*, held in 2015 by the Sino-Nordic Urban Governance for Sustainable Cities Network (UGN) at the University of Copenhagen, Denmark. The authors gratefully acknowledge the comments and suggestions of Stefan Brehm and Jesper Svensson. We would also like to acknowledge the editorial support of Hanzhang Ren and Peter Mayers.

References

Abrahamse, W., Steg, L., Vlek, C., & Rothengatter, T. (2005). A review of intervention studies aimed at household energy conservation. *Journal of Environmental Psychology, 25*(3), 273–291.

Anker-Nilssen, P. (2003). Household energy use and the environment: A conflicting issue. *Applied Energy, 76*(1), 189–196.

Bin, S., & Dowlatabadi, H. (2005). Consumer lifestyle approach to US energy use and the related CO2 emissions. *Energy Policy, 33*(2), 197–208.

Black, J., & Stern, P. (1985). Personal and contextual influences on household energy adaptations. *Journal of Applied Psychology*, 70(1), 3–21.

Brandon, G., & Lewis, A. (1999). Reducing household energy consumption: a qualitative and quantitative field study. *Journal of Environmental Psychology, 19*(1), 75–85.

Brand, C., & Preston, J. M. (2010). '60-20 emission'—The unequal distribution of greenhouse gas emissions from personal, non-business travel in the UK. *Transport Policy, 17*(1), 9–19.

Brand, C. (2009). Hockey sticks made of carbon: unequal distribution of greenhouse gas emissions from personal travel in the United Kingdom. *Transportation Research Record Journal of the Transportation Research Board.* paper #09-2729

Brand, C., & Boardman, B. (2008). Taming of the few—The unequal distribution of greenhouse gas emissions from personal travel in the UK. *Energy Policy, 36*(1), 224–238.

Carlsson-Kanyama, A., & Eriksson, B. (2005). Residential energy behaviour: Does generation matter? *International Journal of Consumer Studies, 29*(3), 239–253.

Curtis, F. A., Simpson-Housley, P., & Drevers, S. (1984). Household energy conservation. *Energy Policy., 12*(4), 452–456.

Egmond, C., Jonkers, R., & Kok, G. (2005). A strategy to encourage housing associations to invest in energy conservation. *Energy Policy, 33*(18), 2374–2384.

Geyer, R., & Jackson, T. (2004). Supply loops and their constraints: the industrial ecology of recycling and reuse. *California Management Review, 46*(2), 55–73.

Graedel, T. E., Allenby, B. R., & Comrie, P. R. (1995). Matrix approaches to abridged life cycle assessment. *Environmental Science & Technology, 29*(3), 134–139.

Greening, L., Davis, W., & Schipper, L. (1998). Decomposition of aggregate carbon intensity for the manufacturing sector: comparison of declining trends from 10 OECD countries for the period 1971–991. *Energy Economics, 20*(1), 43–65.

Guide D, van Wassenhove L (2004). Special issue on supply chain management. California Management Review 46(2):1-141

Golley, J., Meagher, D., & Meng, X. (2009). Chinese urban household energy demand and CO_2 emission. In C. Song & Y. T. Hu (Eds.), *China's policy choice: Economic growth, environment and climate change* (pp. 287–312). Beijing: Social Sciences Academic Press.

Gyberg, P., & Palm, J. (2009). Influencing households' energy behavior: How is this done and on what premises? *Energy Policy, 37*(7), 2807–2813.

Huo, Y., Zheng, S., & Yang, Z. (2010). 低碳生活的特征探索:基于2009年北京市家庭能源消耗与居住环境调查数据的分析 (The exploration of low carbon lifestyle features: an analysis based on the 2009 Beijing 'household energy consumption and living environment' survey). *Urban and Regional Planning Research,* 2, 56–72.

Jackson, T. (1996). *Material concerns: Pollution, profit and quality of life*. London: Routledge.

Jin, H. (2012). 我国城镇化进程中的低碳排放路径选择 (Low carbon route options in Chinese urbanization). *Macroeconomic Management, 4*, 48–50.

Mccalley, L. T., & Midden, C. J. H. (2002). Energy conservation through product-integrated feedback: The roles of goal-setting and social orientation. *Journal of Economic Psychology, 23*(5), 589–603.

McNamara, D., & Caulfield, B. (2011). Measuring the potential implications of introducing a cap and share scheme in Ireland to reduce green-house gas emissions. *Transport policy, 18*(4), 579–586.

New York Government. (2010). *Inventory of New York city greenhouse gas emissions*. Available via New York Government website. http://www.nyc.gov/html/om/pdf/2011/pr331-11_report.pdf. Accessed 20 Sept 2015

Nesbakken, R. (1999). Price sensitivity of residential energy consumption in Norway. *Energy Economics, 21*(6), 493–515.

Ouyang, J., & Hokao, K. (2009). Energy-saving potential by improving occupants' behavior in urban residential sector in hangzhou city, china. *Energy & Buildings, 41*(7), 711–720.

Parker, P. (2005). Who changes consumption following residential energy evalutions? Local programs need all income groups to achieve kyoto targets. *Local Environment, 10*(2), 173–187.

Peter, H., Ferrer-i-Carbonell, A., & Muskens, J. (2004). The ex post impact of an energy tax on household energy demand. *Energy Economics, 26*, 297–317.

Sardianou, E. (2005). *Household Energy Conservation Patterns: Evidence From Greece*. http://www.lse.ac.uk/europeanInstitute/research/hellenicObservatory/pdf/2nd_Symposium/Eleni_Sardianou_paper.pdf. Accessed 25 Aug 2016.

Schipper, L., Bartlett, S., Hawk, A., & Vine, E. (1989). Linking life-styles and energy use: A matter of time. *Annual Review of Energy, 14*(1), 273–320.

Vringer, K., & Blok, K. (1995). The direct and indirect energy requirements of households in the Netherlands. *Energy Policy, 23*(10), 893–910.

Wei, Y. M., Liu, L. C., Fan, Y., & Wu, G. (2007). The impact of lifestyle on energy use and co 2 emission: An empirical analysis of china's residents. *Energy Policy, 35*(1), 247–257.

Zhu, Q., Peng, X., Lu, Z., & Wu, K. (2009). 中国能源消费碳排放变化的因素分解及实证分析 (Factorial decomposition and empirical analysis of Chinese energy consumption variation). *Resource Science., 31*(12), 2072–2079.

Part III
The State's New Tools of Green Urban Governance

Chapter 7
Digital Environmental Monitoring in Urban China

Jesper Schlæger and Jingjing Zhou

Abstract In this chapter, the authors examine new technology as an integrative mechanism in environmental urban governance in China. The added value of this approach lies in its focus on how ICT (information and communication technology) affects administration from both a top-down and a bottom-up perspective. The main finding is that digital environmental monitoring serves to level out some of the information asymmetry between central government principals and local-level agents, thus paving the way for a more centralized model of environmental urban governance in China. Even so, many of the systems and software applications examined in this study are still in an early phase of adoption; therefore, the conclusions must not be seen as final, but rather as indicative of the direction of development.

Keywords Environmental monitoring · Internet · Air pollution · Water pollution · Solid waste · Fragmentation · Information disclosure · Microblogs · Software applications (apps)

7.1 Introduction

In this chapter, we explore how digital environmental monitoring technologies affect the attempts being made to achieve sustainable urban governance in China. With its track record of fast-paced urbanization, China has a depressing scorecard when it comes to environmental degradation, but unfortunately not a unique one among developing nations. The environment has emerged as a critical issue for the Communist Party of China, and in 2012 sustainability was written into the party constitution as a core value. Following this lead, the central government has placed tremendous weight on the application of information and communication technology

J. Schlæger (✉)
School of Public Administration, Sichuan University, Chengdu, People's Republic of China

J. Zhou
Sichuan University, Chengdu, People's Republic of China

J. Delman et al. (eds.), *Greening China's Urban Governance*, ARI - Springer Asia Series 7, https://doi.org/10.1007/978-981-13-0740-9_7

(ICT) to environmental governance. Premier Li Keqiang, who is responsible for government administrative reforms, adopted the term 'Internet Plus'. The notion encourages increased use of the Internet as an amplification of diverse areas of production to enhance economic growth and improve the ecological environment.

The insight regarding the Internet's potential to help reduce pollution has contributed to the shaping of environmental governance in China. For instance, the government has adopted 5-year plans for the informatization of environmental protection and embarked upon a project—the Golden Environment Project—aimed at linking the multiple platforms for environmental information with the nationwide e-government platform. The website of the Ministry of Environmental Protection had around 3 million visits per month in 2014 (Ministry of Environmental Protection 2015a). But citizens and businesses play a role too: for instance, the documentary film 'Under the Dome' by journalist Chai Jing went 'viral' on the Internet and was viewed more than 150 million times before being blocked by the authorities (Mufson 2015). There are at least a dozen websites and mobile software applications (apps) for checking air quality. Spending on equipment for environmental monitoring was close to RMB 5 billion in 2013, and that was just one part of the informatization of environmental governance (PRNewswire 2015). These are just a few examples, but they amply illustrate that ICT has become an integral part of environmental governance in China.

Digital monitoring technologies facilitate new and complementary modes of centralization and fragmentation in urban environmental governance. New technologies are often invented on the fringe of society but are then adapted by central state authorities at a later point in order to strengthen the existing power structure (Deibert 1997). This has been shown in research on the adoption of new technologies, both in the history of public administration generally (Danziger et al. 1982) and in the context of Chinese e-government more specifically (Schlæger 2013). It is thus unsurprising to find such a pattern in connection with efforts to develop urban green governance in China. Some additional reflection is needed, however, on what is going on in this particular context. Chinese governance is based on an intricate interplay between central strategy and local innovation, between top-down and bottom-up dynamics. As in many policy fields in China, the general direction of development in this area is spelled out at the central level. The differences between Chinese cities, nonetheless, are vast, so policy must be adapted to local conditions. The challenges posed to government by a more pluralistic society and by a growing economy with increasing market dynamics have led to the reinvention of development planning (Heilmann and Melton 2013). Experimentation and local adaptation are the core ingredients in the recipe for development which is now being pursued. Other chapters in this volume take up clear examples of increased fragmentation, as well as inclusion of a broader group of stakeholders, in the attempts being made to achieve urban green governance (e.g. Delman, Chap. 8). Nevertheless, once technological experiments have had a certain impact, they are shaped to serve the centralization of power. In this chapter, we analyse several new monitoring technologies and applications which have appeared in diverse fields of environmental governance, and we conclude that the central party state is able to leverage these new technologies in such a way as to strengthen central state power in China. As long as environmental sustainability remains a key policy goal of the central government, such ICT-based centralization bodes well for the environment in China's cities.

Theoretically, what is at stake in the implementation process is a principal—agent problem. Local governments slacken because the central government lacks information about the facts on the ground. The implementation chain breaks at two points, namely, (1) when local governments are faced with a trade-off between pursuing GDP growth and enforcing sustainability policy and (2) when local businesses perceive the risk associated with violating environmental rules to be low. Environmental goals already figure among the key performance indicators for local governments (cf. Delman, Chap. 8; Burrell and Almén, Chap. 10). Even so, pollution is much harder to measure than GDP, so it is difficult to find reliable measures to use in the yearly evaluations. However, ICT can change the cost-benefit calculation in both cases: if violations are easier and cheaper to identify, both the businesses that generate the pollution and the local governments which shield them will be at greater risk of suffering sanctions. In other words, there is less uncertainty to hide behind when the violations are presented in numbers. Of course, the risk of being discovered will only deter people from polluting if sanctions are in place and there is a credible commitment to their enforcement (Ma and Ortolano 2000). We do not address the question of enforcement in this paper; instead, we focus on the necessary groundwork for such an analysis, by pointing out the ways in which new monitoring technologies change the information asymmetry between central government principals and local-level agents.

Our theoretical contribution in this chapter, then, is to point out the mechanisms of ICT-based reinforcement of power that accompany and complement the tendencies towards fragmentation. It is arguably unsurprising to see an authoritarian state using new technologies to centralize power in its hands. Even so, many examples are cited in this book of how environmental governance is becoming increasingly fragmented. In that regard this chapter serves as a counterweight, illustrating some of the integrative mechanisms which are activated as a complement to fragmentation. On the one hand, Delman (Chap. 8) in this volume calls attention to the increased complexity in this area, as does J. Liu's (Chap. 10) contribution (albeit in more dystopic form). In this chapter, on the other hand, we argue that the dynamics of top-down and bottom-up governance are not mutually exclusive, but in fact complementary, and that centralization is the main tendency. In this way, our analysis dovetails with that of Almén and Burell, who show that central leaders allow protests because they find them useful—as a way of monitoring the behaviour of local governments.

In the first part of this chapter, we address attempts by the central state to tap into opportunities that new technologies afford it to gain control over policy implementation at the local level. Terms like centralization or command and control reflect such a top-down, traditional public administration (TPA) perspective (cf. Lund Chap. 1). The control aspect is important, and environmental management is relatively easy to carry out in this mode because of the existence of a distinct hierarchy with clear structures of authority. The main drawback of the approach lies in its inflexibility when faced with local conditions which may undermine the effectiveness and efficiency of implementation. Monitoring in the top-down perspective takes aim at the information asymmetry between central government principals and implementing agents at the local level. If monitoring mechanisms can be made both effective and centralized, such asymmetry will be eliminated.

In the second part of the chapter, we analyse the appearance of (apparently) bottom-up solutions to environmental problems in the form of online software applications. A normal first reaction would be to see these platforms as evidence of fragmentation and of citizen empowerment. So might they appear from a governance perspective, inasmuch as governance implies the involvement of multiple stakeholders in the making and implementation of environmental policy. The strength of such an approach lies in the adaptability which accompanies the devolution of discretion to lower levels. However, decentralizing authority adds to the complexity and fragmentation of the system. And it is very hard in a fragmented system to pinpoint who is accountable for the implementation of environmental policy. The risk of failure thus increases. Supervision from the bottom up could mean empowering citizens to hold the government accountable for failures in the implementation of environmental policy. In this paper, however, we suggest another mechanism, namely, that bottom-up monitoring is used as a fire alarm—which is a centralizing or integrative mechanism.

7.2 Methodology

To answer the question of how ICT affects urban environmental governance, we examine policies and software applications connected with digital environmental monitoring. The technologies examined in this paper include among other things the use of satellite geographical information systems, as well as the automated real-time monitoring of waste water and air quality. We have collected data from government websites, news websites, corporate websites, and online software application stores, supplemented by industry reports and articles in peer-reviewed journals. We have paid particular attention to Chinese-language sources—usually inaccessible to an international audience—to give a state-of-the-art picture of the situation in digital environmental monitoring. Our reliance on secondary data sources, however, makes for some limitations. As we were unable to collect any observational, interview, or survey data, we can only infer the actual use to which the different software applications and computer systems were put, on the basis of news accounts, downloaded statistics, and academic work on the topic. However, while the scarcity of first-hand sources leaves substantial room for further research, there are still enough data to yield a useful overall map of current developments in environmental monitoring.

7.3 Top-Down: Centralized Control over Efforts to Achieve Sustainable Urban Governance

Before bureaucratic sanctions can be enforced, the administration must first establish that a violation of existing regulations has occurred and then identify who is responsible. Identifying implies monitoring—understood as the repeated testing of ongoing activities for purposes of supervision. Enforcing environmental regulations requires strong evidence, in the form of test results proving that certain limits have been exceeded. Technologies for testing the level of pollutants and identifying their source consist mostly of well-known types of chemical test, airborne particulate measuring, etc. However, such tests are often expensive and labour-intensive. Monitoring processes are an important and often overlooked aspect of attempts to achieve environmental sustainability, and they can be transformed by new technologies. Digital environmental monitoring has attracted only limited attention among researchers (yet see Deng et al. forthcoming; Mai and Francesch-Huidobro 2015; Ma and Ortolano 2000). Practitioners, however, have perceived it as important, as can be seen from the size of the fiscal expenditures devoted to it.

7.3.1 Government Expenditures on Environmental Monitoring

To get an idea of the magnitude of China's investment in environmental monitoring, we can take some figures from the budget of the Ministry of Environmental Protection. In 2014, the Ministry's expenditures on emission reduction came to RMB 827 million (Ministry of Environmental Protection 2014a). This figure includes RMB 736 million for environmental monitoring and information. A comparison with certain other numbers contained within the budget of 827 million may be revealing here: expenditures for preventing water and air pollution, respectively, came to RMB 59 million and RMB 3.8 million. To be sure, budgetary allocations for the latter items at the Ministry do not reflect those at the local level, where pollution prevention and environmental enforcement take up a larger share of the budget. Moreover, the central government has in fact budgeted RMB 10 billion for the 2016–2020 period for projects aimed at reducing air pollution (Ministry of Environmental Protection 2014a). Even so, these figures give a first impression of the great importance ascribed to environmental monitoring.

Expenditures on environmental monitoring equipment have surged from roughly RMB 1 billion in 2006 to RMB 5 billion in 2013. The suppliers are both international and Chinese companies, of which a handful have turnovers of several hundred million RMB per year (PRNewswire 2015). So far the equipment has mainly been purchased from abroad, but this is changing. There are now strict requirements in a large number of procurement-tendering processes that the equipment be purchased from Chinese producers (Chinese Government Procurement Network 2014).

As an example of the costs connected with environmental service platforms on the local level, Jinwan District can be considered. Jinwan, a district marked by urban sprawl, lies on the southwestern outskirts of Zhuhai Municipality in South China. In 2015, the tender for an ICT platform in the district was won by a local company. It included a process information system priced at RMB 450,000, a laboratory information system priced at 504,000, 2 servers at RMB 60,000, 12 mobile units at 42,000, and network access cards (including 20 GB of prepaid fees) at 18,000—making for a total cost of RMB 1,074,000 (Jinwan District Bureau of Environmental Protection 2015).

Moreover, environmental protection in China has entered the era of the Internet of things (Wang, 2015). In Jiaozhou City near Qingdao in East China, a small city of mixed urban and suburban composition, RMB 8.9 million were allocated in 2015 to construct a cloud computing system for handling trade in pollution emission rights. The system comprises pollution source online monitoring, environmental quality online monitoring, motor vehicle emissions online monitoring, disassembling of 'yellow license plate' (*huangbiao*黄标) vehicles, control of total amounts and emission rights trading, and environmental crisis management (Ministry of Environmental Protection, 2014b). Initial investments at the time of writing were RMB 3.48 million. This money has been spent on the following items: 13 online continuous emission monitoring systems (CEMSs) for thermal plant air pollution, 14 online craft manufacturing air pollution systems, 15 waste water emission monitoring systems, 6 surface water monitoring systems, 2 regional high-altitude video systems, 1 central environmental information control centre, and 5 branches of the control centre. To this can be added 50 hand-held monitoring devices, 8 'toolboxes' for environmental monitoring, and 1 mobile monitoring car (Ministry of Environmental Protection 2014b).

The added value of these investments is that the top-down system becomes cheaper to manage, because transferring information to the centre can be done at a lower cost. It is cheaper, faster, and easier to share information on an electronic network than to do so with paper documents that require physical transportation. In particular, the centralization of networks (i.e. the development of a national-level network) becomes feasible.

7.3.2 Macro-policies of Environmental Monitoring

The rules and regulations that guide environmental monitoring in China give an idea of the government's intentions. We can start with a look at the overarching policies. Since the inauguration of the 11th Environmental Five-Year Plan (Ministry of Environmental Protection 2008), the creation of automated online monitoring techniques has been a key project, under the heading of the Golden Environment Project. This project, however, is rarely mentioned in the academic literature. Instead, frequent mention is made of the digital environment in recent academic work (*shuzi huanbao* 数字环保). In spite of challenges (immature technologies, a low uptake of public platforms, an official environmental database of dubious quality), present policy gives many indications that the current massive growth in environmental monitoring will continue for years to come. A 2015 national action plan on big data suggests as much (State Council 2015). This action plan has implications for digital environmental monitoring. For example, it stipulates that information resources be joined up in order to avoid isolated information islands and that the information be open to the public. Where data-sharing is concerned, there are a number of standards for information systems relating to environmental governance in China. These standards are important for ensuring that the different sensors are able to communicate and to form networks. Openness paves the way for the 'sharing economy', a point also stressed by Chinese Premier Li Keqiang. This entails a renewed emphasis on the pursuit of the Golden Environment Project, with its aim of central control.

7.3.3 Disclosure of Environmental Information

The main area of e-monitoring which is currently relevant is that of information dissemination. Yet, even information disclosure is difficult to ensure, because of local interests. Success has so far been limited when it comes to the publicizing of environmental information. An excerpt from an official newspaper, the *China Daily*, explains:

> Regarding what has been described as 'chaotic environmental information' released from different channels and by different standards, the government said it would establish a national environment monitoring network by 2020 for the unified release of environmental data. The Ministry of Environmental Protection will look into local governments' involvement in environment monitoring to reduce interference. Some local governments have manipulated environmental data as pollution levels are linked to performance reviews. (Xinhua 2015)

In order to discourage local governments from neglecting their duty to publicize environmental information, the Ministry of Environmental Protection has devised a Pollution Information Transparency Index (PITI) and published a map on its website of which cities and provinces deliver the required environmental data in a timely manner (MEP Environment Supervision Centre n.d).

In regard to air pollution, the central government is taking measures to ensure compliance by local authorities. An increased use of satellites in national-level monitoring of air quality was planned in the 2013 Air Pollution Preventive Governance Action Plan. These satellites are a complement to monitoring stations (described below) under the direct control of the central government, as well as to direct feeds from key industries, which are required to install real-time monitoring sensors. According to official figures, the incidence of stubble burning fell by 36.9% during the 2013–2014 period. In 2013, namely, the satellites picked up 7983 incidents of this kind; in 2014 they only detected 5034 (State Council 2013). There are probably multiple reasons for this decrease, inasmuch as plant residue is

Table 7.1 Standard for construction of air quality monitoring stations

Urban population	Area in square kilometres	Smallest number of monitoring stations
< 250,000	<20	1
250–500,000	20–50	2
500,000–1,000,000	50–100	4
1,000,000–2,000,000	100–200	6
2,000,000–3,000,000	200–400	8
>3,000,000	>400	Construct one monitoring station for every 50–60 km^2 area, and not fewer than ten stations

Source: Ministry of Environmental Protection (2013), authors' translation

increasingly used for other purposes and thus is worth money for farmers; even so, such monitoring from above changes the incentive structure for local governments. Satellite monitoring reveals quite clearly where fires are being caused by illegal burning, making it possible to hold local authorities to account for violating rules or for failing to enforce them. Local governments in turn have a stronger incentive to enforce the rules, in order to avoid confrontation with higher levels of government.

As mentioned, the central government has built an independent network of air pollution monitoring stations. There is a national policy for how many monitoring stations there should be in each city. These are presented in Table 7.1 below (Ministry of Environmental Protection 2013).

As can be seen from Table 7.1, the Ministry of Environmental Protection has its own monitoring system in place in the various provinces (Chinese Government Procurement Network 2014). This system is controlled and accessed directly from the Ministry. Daily operation and maintenance have been outsourced to a number of private companies. The number of monitoring points in each province varies, from a high of 23 in Guangzhou to a low of 2 in Qinghai and Tibet. Running costs for a single monitoring station come to about RMB 200,000 per year. With around 200 stations nationwide, the total yearly cost is close to RMB 40 million. These stations furnish the Ministry with its own reliable data, which makes it much harder for the local governments to hide falsifications of environmental monitoring data and statistics. This is a good example of how something that may appear to be a purely technical system has important consequences for the degree of centralized control.

Upon being assembled the data are digitized, collected into databases, and (to some extent) published online. Among the most important online databases are those found on the website of the Ministry of Environmental Protection Data Center which provides a broad range of data. The used data for the website of the Ministry of Environmental Protection, presented in Table 7.2, show an increasing trend in visits and page views per month. In 2014, there were 3 million visits per month on average. There is thus a steady base of users with an interest in using the data available on this website.

Table 7.2 User statistics for website of the Ministry of Environmental Protection

	Page views/month	Visits/month
2008	14,720,000	1,060,000
2009	17,000,000	1,300,000
2010	12,000,000	1,600,000
2011	12,000,000	2,100,000
2012	20,000,000	1,600,000
2013	20,000,000	1,600,000
2014	26,000,000	3,000,000

Source: authors' compilation, based on Ministry of Environmental Protection Government Openness Reports (various years); available on www.mep.gov.cn

One of the core problems for environmental agencies is their inability to enforce hard sanctions. Data from environmental monitoring systems can identify companies that pollute the environment, but some companies simply refuse to pay the fines imposed on them and continue to pollute. The 1831 system in Jiangsu addresses this problem by linking environmental information to companies' access to bank loans (Ministry of Environmental Protection 2015b). If a company fails to provide the government with the required environmental data, banks are not allowed to lend money to the company. This approach can be implemented in practice, because the banks are under government control.

The overall impression given by the top-down strategy of environmental monitoring is that the systems currently adopted, in combination with those under construction, will level out information asymmetries between the central government and local governments. An important element in this respect is the ability of the central government to monitor key environmental variables independently of its local counterparts. In addition, data provided by the central government serve as the basis for a broad range of software applications that enable citizens to engage in environmental activism and to take part in the production of public services. In the next section, we look at some examples of such bottom-up environmental monitoring.

7.4 Bottom Up: Empowering or Engaging Citizens?

Citizens in China have been expected to be passive receivers in the process of social management. Even so, facing the increasingly intractable issues which economic reforms have brought to the fore, they have shown signs of undertaking a new role—that of active or even co-producing citizens. Industrial production in the urban environment leads to greater economic welfare, but it also leads to worsened pollution and to heightened social tension. Dissatisfaction among citizens is visible in contestation both online and offline, and a large number of environmental protests have been held in China (J. Liu, Chap. 9). Examples include the demonstrations in Xiamen against a proposed petrochemical factory (2007), the expressions of public outrage about dead pigs floating in the Huangpu River (2013), and the online response to the environmental documentary movie 'Under the Dome' mentioned above. And while this activism targets public authorities, it helps the central government to implement its policies. Engaging social forces in implementation forms a key part, namely, of the central government's strategy for developing a new green governance approach to reduce pollution. Members of the public can further the implementation process by complementing top-down monitoring by environmental agencies with their own efforts at bottom-up monitoring. Having acquired the ability to access data in real time, moreover, entrepreneurs can now create software applications to suit local (or national) needs for information, such as about levels of air pollution. In this way, cross pressures can be put on local governments: key performance indicators are gathered from above, while a public pillory emerges as a network governance (NG) instrument (cf. Lund, Chap. 1) from below in the shape of pollution figures posted online.

Nevertheless, the benefits of the Internet to civil society in times of social tension are ambiguous, because online content is censored. The government can prevent the spread or continuation of discussions that it deems undesirable. There is also a bias when it comes to the availability of online information, because it only reaches those citizens who can get online and who are able to find the relevant information (cf. Walravens 2015, p. 220). Yet, this is changing. In recent years the threshold has fallen dramatically, with rapid growth in rates of Internet penetration and popularization of online environmental discourse. By popularization, we mean that environmental information is conveyed in everyday language rather than in technical terms. The information disseminated in such venues as bulletin boards, microblogs, and instant-messaging systems can take the form of short status updates or of longer accounts or reports. Links to original research are also provided therein in many cases.

The following analysis of a range of online applications illustrates how ICT tends to reinforce existing power structures, at the same time that it can empower citizens to press local governments in the direction of more sustainable urban governance.

7.4.1 Environmental Information Disclosure and Empowerment

The central government provides information on its own communication platforms (Ministry of Environmental Protection 2014c; School of Public Affairs New Media 2015). Among the most important of these are the 'double micro' (*shuangwei* 双微) ones, i.e. microblogs (*weibo* 微博) and WeChat (*weixin* 微信). These platforms are managed together as part of the government's social media strategy. By March

Table 7.3 Official microblogs of various environmental protection agencies

Agency	Following	Fans	Posts
Ministry of Environmental Protection	194	5380	1364
Sichuan environmental protection agency	72	35,507	3669
Chengdu environmental protection agency	258	87,724	4677
Tianjin environmental protection agency	180	256,148	4172

Source: authors' compilation based on Sina Weibo 15 September 2015 http://www.weibo.com/p/1001603819219119685614 www.weibo.com/u/3732167165 www.weibo.com/u/3225377512; www.weibo.com/u/3250739191.

Note: Chengdu represents major cities outside the capital region; Tianjin is part of the capital region.

2015, about half of the provinces had opened weibo accounts. The provinces which had not yet done so were mainly the less economically developed ones, with the notable exception of Guangdong.

As the numbers in Table 7.3 show, the Ministry of Environmental Protection has just 5380 fans, while the local environmental protection agency in Tianjin (representing the capital region) has more than a quarter of a million. The local level, then, seems to hold the highest interest for users. The numbers for Tianjin also doubtlessly reflect massive local interest (cf. Luova, Chap. 4) after the explosions at a chemical storage facility on 12 August 2015. Nevertheless, the popularity of local-level channels should not be taken as an indication that the system is undergoing uncontrolled fragmentation or even that it is just becoming less centralized. The most relevant and targeted information can be delivered by local governments, so information furnished by the latter has greater value as a public service than do environmental updates from all over the country. An urbanite in Tianjin may not be interested in updates on the air quality in Chengdu (a large city in Western China). Thus, it is in the interest of the central government that local news be provided. Actually, providing such a service is a new task for local governments. It also helps empower citizens, who gain a platform for interacting with government.

The websites and social media interfaces provided by the environmental protection agencies in various provinces are a recent addition to their communication tools (Ministry of Environmental Protection 2014c). The information portals in the province-level megacity Chongqing and in Shaanxi province, for instance, were opened in 2014. Yet the microblogs are already well-institutionalized, even though they are new; this can be seen from the fact that benchmarking is taking place (People's Daily 2015). It is thus possible to find a list with a ranking of the top 50 official environmental microblogs. So, while government environmental microblogging is a new area, it is developing rapidly. The monitoring information that forms the basis for some of the content on these microblogs and websites comes from a variety of sources, and it is used in a range of ways. We discuss this question further below.

An automated water monitoring system is under construction. Water management is the most expensive element in Chinese environmental policy (cf. Liu et al., Chap.

5), as is typical in developing countries (Liu and Wang 2008). The treatment of waste water and the purification of drinking water are particularly important for urban residents. In 2014, 202 cities of prefecture level and above were participating in groundwater testing, and there were 4896 monitoring points in all (Ministry of Environmental Protection 2015c). Only 38% of the water rated as good to superior; 62% was either poor or very poor. The automated monitoring system consists of monitoring points that extract samples and analyse them automatically. The digitized results are then sent via the Internet to an analysis centre that connects several points. The quality of the water in question is ranked according to the standards of Surface Water Environmental Quality Standard GB3838–2002 (*dibiao shuihuanjing zhiliang biaozhun*地表水环境质量标准). The measurements are usually done in 2 or 4 h intervals; however, the water quality can be measured continuously in case of sudden pollution events. Like the necessary weekly recalibration of the system, such continuous monitoring requires manual recoding of the system (China Environmental Monitoring Center 2008).

The provision of data by the government and their distribution through software applications can serve to raise awareness of water quality issues among citizens. There are only a few water quality applications, however. One such is A-La Drinking Water from 2013, which allows users to check the quality of drinking water at the district level (Eastday 2014). This application covers Shanghai. The data are based on four different standard levels of water quality. All water plants in Shanghai are required by regulation (the Management Rules for Shanghai Public Water Supply Water Quality Information Openness) to publish data on the quality of their water on a daily basis. Another example is the Hangzhou River water quality application (Pan 2015), which collects information at the municipal and county levels about water quality in the rivers of Hangzhou. It features maps setting out the administrative responsibility of different parties involved in governing the quality of the water in question. Using a colour scheme with six grades, this app monitors the different rivers on a monthly basis. More advanced and user-friendly examples of information provision are featured in the next section, on air pollution monitoring.

Air pollution apps are available from about a dozen fairly large providers. The start-ups dating from the early 2010s, moreover, have recently been joined by giants like Microsoft (*xiaoyu tianqi*小鱼天气). The different apps vary in their operating logic. Some are based solely on official Chinese data; others include the US consulate data mentioned earlier; still others (like that offered by Microsoft) rely on simulation models based on official data, but paired with a big data approach including meteorological, topographical, and traffic data. There are also experimental apps based on the logic of crowdsourcing, such as clarity, which uses wearable equipment to test air quality and then links the data thus collected to a cloud service (Alba 2014).

The Close Urban Manager (*tiexin chengguan* 贴心城管) from Hangzhou (Hangzhou Municipal Government, 2014) has been downloaded more than 250,000 times. It integrates information provided by environmental protection agencies at both national and local levels to provide granulated air pollution information for

367 cities. The main function of this application is to inform citizens. As in the case of the water quality apps reviewed above, its users are consumers of information, rather than participants in the provision of a public service. By contrast, in the examples considered below—which concern garbage management—the role envisioned for users is an active one.

7.4.2 The Citizen as Co-provider of a Public Service

Waste Collector (*shou feipin* 收废品) is a local application developed by students from the Zhejiang Institute of Technology (XinhuaNet 2014). The functionality covers a single residential area. Prizes are given to people who sort their garbage correctly and dispose of it in containers in the area. When recycling the material, the user scans its barcode, and the system can assist with the sorting. In Beijing, the Help to Home (*bangaojia* 帮到家) application provides a platform for the residents of 50 communities to link up with front-line waste collectors (ChinaNet 2015). This increases the number of 'customers' for the collectors, and it makes life more convenient for the residents, who can order a pickup instead of waiting for the weekly round. The material promoting the application refers to it as an 'Internet + recycling' model, alluding to the Internet Plus phrase mentioned earlier (ChinaNet 2015). In Guangzhou, the 92 Recycling (*92 huishou* 92回收) app applies the same logic (Baidu Baike n.d). Its focus to begin with has been on the collection of high-value glass and wood, rather than of other types of waste. Users also get additional points for sorting household waste. The quality of their sorting is reviewed by inspectors who visit the community on two regular daily stops.

Through an application, a WeChat group, and an Internet website, Recycling Brother (*huishouge* 回收哥) lets urban residents in Wuhan announce their demand for recycling services (Huishouge n.d). Garbage collectors can then compete for the contract (*qiangdan*抢单) to pick up the waste. This application, developed by a private firm, is sponsored by the provincial and municipal governments. The waste to be recycled includes paper, plastic, batteries, large appliances, small appliances, lamps, and scrap metal. The website provides educational materials on the harmful effects of electronic waste.

In addition to the garbage sorting apps, there are a large number of urban management applications that link directly into a city's digital urban management system. In many cases, they make it possible to take a photo of garbage which has been thrown out on to the streets and to send the picture in to the urban management control centre for further action. The relevant renovation teams will then be sent to clean it up. Examples of such applications are Xuancheng Shimintong (Xuancheng Municipal Government n.d), Changsha Shimintong (Changsha Wanbao, 2014), Chongqing Chengguantong (Zhongguo Jianshe Bao 2015), Suzhou City All People Urban Manager (Yao 2014), Guiyang Baixingpai, and Hangzhou Tiexin Chengguan.

Many applications are specifically targeted on one kind of pollution or one aspect of public service provision; however, these are being supplemented by an increasing number of software programmes which integrate data from multiple sources, with a focus on furnishing a basis for citizen activism. We turn to these in our final section.

7.4.3 Integrated Platforms as a Fire Alarm

I Love Beijing is a '*chengguantong*' (城管通) application for urban residents (Beijing Municipal Government n.d). Launched in July 2012, it has had 288,648 downloads of its android version at the time of writing (15 September 2015). It integrates a wide range of different facilities, e.g. maps, an information service, a system for lodging grievances and complaints, a hotline to the Bureau of Urban Management, and local information about grocery markets. The app has the slogan: 'My city, I decide'. Yet, notwithstanding the empowerment suggested in the slogan, the main impact of the application is to provide information and an access point to government.

On the national level, the Weilan maps are an example of the integration of different types of pollution information source into a single platform (Pollution Maps n.d). They offer air quality measurements, an air quality map, a water quality map, and a pollution source map. The motivation for publishing these maps online is to increase public awareness of the existence of such data (Xie 2015). Public pressure can then be generated for more data and for more reliable data. The availability of maps that pinpoint heavy polluters makes it easier to take action. For example, public media campaigns can be waged against companies that fail to live up to environmental standards, and questions can be put to local environmental protection agencies as to why rules are not being enforced. In this way, local 'policy entrepreneurs' gain an additional instrument for voicing their concerns (Mertha 2009).

Another national-level application is Environment on the Beat (*huanbao shuishou pai* 环保随手拍), supported by the Ministry of Environmental Protection (HBSSP 2014). The Environmental News Workers' Association encourages the use of this app, as does the Committee of Environmental News, Culture, and Arts. Users of the app can take photos and record videos with their mobile phones, use location services to identify cases of environmental pollution, and then send in the evidence to a back-office platform where environmental specialists assess the evidence. After assessment by these specialists, those users who have given valuable information can win honourable mention or gifts as a reward for their efforts. The system requires real-name registration, in order to prevent abuse. Verified complaints are forwarded to the news media and to relevant environmental enforcement agencies. The logic behind this application may be described as that of a fire alarm, since the app encourages citizens to take part in a bottom-up process of environmental monitoring. The same is true of the final application, reviewed below.

Green Hero (*lüxia* 绿侠) is a mobile software application that links users directly to environmental enforcement agencies (Ministry of Environmental Protection 2014d). The app requires real-name registration, and users who have made false accusations are blacklisted. In a way, this application provides citizens with the same functionality as that enjoyed by environmental enforcement officers, albeit without the power actually to enforce rules.

As we can see from the national-level applications, there are fora where citizens can vent their grievances to central organs if and when local enforcement agencies have been remiss in carrying out their duties. So, even though most environmental enforcement takes place on the local level, there are ways to tackle issues of protectionism or corruption at that level through platforms that allow direct communication with central organs and the national media.

To sum up our main point in this section, the Chinese central government has been successful in fostering local innovations that can induce the public to work as a fire alarm. But there are no clear indications of empowerment in the applications analysed here. That is to say, such apps do not offer urban residents any mechanisms by which they can actually hold local governments to account for failing to implement or to enforce environmental regulations. Our findings tend, therefore, to support the interpretation that ICT reinforces the existing power structure and contributes to the centralizing tendency pointed out earlier. On a more upbeat note, however, we may note that ICT has facilitated a shift on to a more sustainable path of development, by making information about the ills of environmental degradation available to a broader group of citizens. Furthermore, citizens enjoy increasing opportunities to participate in the co-production of basic public services.

It is not essential to our argument whether these Internet apps are used or not. As J. Liu shows in Chap. 9 in this volume, platforms in many cases are not used. The main virtue of the environmental microblogs, from the standpoint of the central government, lies in their function as a fire alarm—through the opportunity they afford citizens to react to pollution in their local area. The objective is not to enhance civic participation for its own sake. In fact, the platforms play their part in the overall information ecology aimed at monitoring the environment. The capacity of the central government to implement its environmental policies is strengthened thereby.

7.5 Conclusion

In this chapter, we have first described the top-down strategies of environmental monitoring pursued in China in connection with urban governance. The Ministry of Environmental Protection has initiated a series of policies for building up a comprehensive network of environmental monitoring, the contours of which are becoming clear. A major feature of these policies is the ability they afford the Ministry to monitor key environmental variables independently of environmental agencies at the provincial and municipal levels.

In the second part of this paper, we have explored the bottom-up processes which are found in this area. Numerous software applications invite citizen participation by providing information about air quality, garbage sorting, and to a lesser extent water quality. In some cases, real-time monitoring (CEMS) of the emission levels of polluting companies is linked to strong incentives: a company's prospects for obtaining a bank loan, for example, can be affected thereby.

We have then asked how the new technologies which have appeared in this area affect the attempts being made to achieve sustainable urban governance. A core finding of ours—and one that in important ways cuts against the thesis of fragmentation—is that municipal governments are more and more likely to be caught if they shirk their duty to enforce environmental policies. The availability of open data results in unified or joined-up management, which serves to counter the information asymmetry that otherwise undermines the implementation of environmental policy. ICT in the Chinese context is therefore best understood as an integrative mechanism. It functions as a fire alarm, helping to uncover big scandals and to publicize critical events that sometimes become tipping points in environmental policy. As a countervailing tendency to fragmentation, we have identified an incremental—but momentous and forceful—push towards centralizing the everyday efforts made to achieve green urban governance. Top-down strategies of supervision are changed by the adoption of new technologies, as are bottom-up processes of co-production of public services. The technologically mediated centralization plays out in less than spectacular ways; nor does it draw many headlines in the newspapers. Yet it counters the tendency towards bureaucratic fragmentation and promotes movement in the opposite direction, thereby marking a tectonic shift in patterns of governance.

In sum, the effect on environmental governance is to make local governments more open to control, both from above (through systems of centralized control) and from below (through citizen-activated fire alarms). Citizens also have greater opportunities to participate in the co-production of public services, such as garbage sorting. The remaining question is whether the different bottom-up models examined in this paper can be used in environmental governance more generally.

This chapter has been in the nature of an exploratory overview, which naturally limits our ability to delve into details. Future research may see more fine-grained analyses, such as case studies or surveys that focus on the actual use of systems and software applications. Nevertheless, we believe we have furnished a useful map for future research and provided practitioners with some indications of how digital technologies can be employed in efforts to achieve sustainable urban development. The challenges faced by Chinese cities are by no means exceptional, and understanding how they have been handled may be beneficial for other countries as well.

Acknowledgement The research for this chapter was supported by a Basic Research Grant for the Central Universities, Sichuan University (skyb201201).

Bibliography

Alba, D. (2014). This wearable detects pollution to build air quality maps in real time | WIRED [WWW document]. http://www.wired.com/2014/11/clarity-wearable/. Accessed on 20 Feb 2016.

Baike, B. (n.d.). 92回收 [WWW document]. http://baike.baidu.com/item/92回收?fr=aladdin. Accessed 15 Feb 2016.

Beijing Municipal Government. (n.d.). 我爱北京市民城管通 [WWW Document]. http://map.bjcg.gov.cn/m/. Accessed 15 Feb 2016.

Changsha Wanbao. (2014). 长沙推数字化城市管理模式 市民可点APP反映问题-中新网 [WWW document]. http://www.chinanews.com/sh/2014/05-16/6177772.shtml. Accessed 18 Feb 2016.

China Environmental Monitoring Center. (2008). Surface water quality monitoring network 地表水水质自动监测网 [WWW document]. http://www.cnemc.cn/publish/totalWebSite/news/news_5029.html. Accessed 15 Feb 2016.

ChinaNet. (2015). '帮到家'手机APP [WWW Document]. http://hn.cnr.cn/zyjjq/jjqzsyz/20150716/t20150716_519230379.shtml. Accessed 15 Feb 2016.

Chinese Government Procurement Network. (2014). Notification on China environmental monitoring base national air quality testing directly controlled stations operational managment project procurement results中国环境监测总站国家城市空气质量监测直管站运行管理项目中标公告 [WWW document]. www.ccgp.gov.cn/cggg/zygg/zbgg/201411/t20141117_4737649.htm. Accessed 15 Feb 2016.

Danziger, J. N., Dutton, W. H., Kling, R., & Kraemer, K. L. (1982). *Computers and politics: High technology in American local governments*. New York: Colombia University Press.

Deibert, R. (1997). *Parchment, printing, and hypermedia: Communication and world order transformation*. New York: Columbia University Press.

Deng, Y., Brombal, D., Farah, P., Moriggi, A., Critto, A., Zhou, Y., & Marcomini, A. (forthcoming). China's water environmental management towards institutional integration: A review of current progress and constraints vis-à-vis the European experience. *Journal of Cleaner Production*. https://doi.org/10.1016/j.jclepro.2015.08.022.

Eastday. (2014). A-La drinking water [WWW document]. http://citynews.eastday.com/csdb/html/2014-08/14/content_129514.htm. Accessed 15 Feb 2016.

Hangzhou Municipal Government. (2014). Hangzhou municipal government E-government performance evaluation indicator system 2014, Appendix 2 [WWW document]. http://www.hangzhou.gov.cn/art/2014/9/9/art_811152_78684.html. Accessed 15 July 2015.

HBSSP. (2014). 环保随手拍 [WWW document]. www.hbssp.org/china/. Accessed 15 Sept 2015.

Heilmann, S., & Melton, O. (2013). The reinvention of development planning in China, 19932012. Modern China 0097700413497551.

Huishouge. (n.d.). 回收哥首页 [WWW document]. http://www.huishouge.cn/. Accessed 18 Feb 2016.

Jinwan District Bureau of Environmental Protection. (2015). 珠海市金湾区环保综合服务平台采购项目(项目编号0851-1561ZH00CL97).

Liu, D., & Wang, Y. (2008). China: Water monitoring technology and instruments [WWW document]. egs.apec.org/uploads/docs/China_WaterMonitoringTechnology.pdf. Accessed 15 Feb 2016.

Ma, X., & Ortolano, L. (2000). *Environmental regulation in China: Institutions, enforcement, and compliance*. Lanham: Rowman & Littlefield Publishers.

Mai, Q., & Francesch-Huidobro, M. (2015). *Climate change governance in Chinese cities*. London: Routledge.

MEP Environment Supervision Centre. (n.d.). 全国数据传输有效率公示 [WWW Document]. http://www.envsc.cn/ValidTransferRatePublicityPlatform/Module/Main/Main.html. Accessed 15 Feb 2016.

Mertha, A. (2009). 'Fragmented Authoritarianism 2.0': Political pluralization in the Chinese policy process. *The China Quarterly, 200*, 995. https://doi.org/10.1017/S0305741009990592.

Ministry of Environmental Protection. (2008). The National Eleventh Five-year Plan for Environmental Protection (2006-2010) [WWW document]. http://english.mep.gov.cn/Plans_Reports/11th_five_year_plan/200803/t20080305_119001_4.htm. Accessed 15 Feb 2016.

Ministry of Environmental Protection. (2013). Technical regulation for selection of ambient air quality monitoring stations (on trial) 环境空气质量监测点位布设技术规范(试行) [WWW document]. http://kjs.mep.gov.cn/hjbhbz/bzwb/dqhjbh/jcgfffbz/201309/t20130925_260810.htm. Accessed 15 Feb 2016.

Ministry of Environmental Protection. (2014a). Ministry of Environmental Protection Budget 2015环境保护部部门预算2015 [WWW document]. http://gcs.mep.gov.cn/zhxx/201504/P020150417699313492424.pdf. Accessed 15 Feb 2016.

Ministry of Environmental Protection. (2014b). 监察'掌中宝' 排污随时看 胶州市环保物联网系统实现重点污染源在线监控 [WWW document]. www.mep.gov.cn/ztbd/szhb/201408/t20140804_287310.htm. Accessed 15 Feb 2016.

Ministry of Environmental Protection. (2014c). Chongqing environmental protection public affairs WeChat New 'Menu'重庆环保政务微信新增'菜单' [WWW document]. http://www.mep.gov.cn/ztbd/szhb/201408/t20140825_288173.htm. Accessed 17 Feb 2016.

Ministry of Environmental Protection. (2014d). 绿侠 [WWW document]. www.zhb.gov.cn/zhxx/hjyw/201406/t20140623_277270.htm

Ministry of Environmental Protection. (2015a). 环境保护部政府信息公开工作2014年度报告 [WWW document]. http://www.mep.gov.cn/gkml/hbb/bgg/201503/W020150320561295777882.pdf. Accessed 15 Feb 2016.

Ministry of Environmental Protection. (2015b). 监测数据要成为证据? 绍兴构建公众、企业、政府共监管模式 [WWW document]. http://www.mep.gov.cn/ztbd/szhb/201509/t20150907_309419.htm. Accessed 19 Feb 2016.

Ministry of Environmental Protection. (2015c). Water environment 淡水环境 [WWW document]. http://jcs.mep.gov.cn/hjzl/zkgb/2014zkgb/201506/t20150605_303011.htm. Accessed 15 Feb 2016.

Mufson, S. (2015). This documentary went viral in China. Then it was censored. It won't be forgotten [WWW document]. Washington Post. https://www.washingtonpost.com/news/energy-environment/wp/2015/03/16/this-documentary-went-viral-in-china-then-it-was-censored-it-wont-be-forgotten/. Accessed 18 Feb 2016.

Pan, J., 2015. 按个键就能找河长投诉 杭州发布'河道水质'APP [WWW document]. http://zjnews.zjol.com.cn/system/2015/06/05/020685810.shtml. Accessed 15 Feb 2016.

People's Daily. (2015). 环保系统微信排行榜2015年6月28日-7月4日 [WWW document]. URL. Accessed 15 Feb 2016.

Pollution Maps. (n.d.). 蔚蓝地图2.0 [WWW document]. https://itunes.apple.com/cn/app/wu-rande-tu/id838211758?mt=8. Accessed 15 Feb 2016.

PRNewswire. (2015). China environmental monitoring instrument industry report, 2013-2015 [WWW document]. http://www.prnewswire.com/news-releases/china-environmental-monitoring-instrument-industry-report-2013-2015-300086758.html. Accessed 15 Feb 2016.

Schlæger, J. (2013). *E-government in China: Technology, power and local government reform*. London: Routledge.

School of Public Affairs New Media. (2015). 全国环保政务微博盘点半数省级环保机构尚未开通官微.

State Council, 2013. 国务院关于印发大气污染防治行动计划的通知 [WWW Document]. http://www.gov.cn/zwgk/2013-09/12/content_2486773.htm. Accessed 18 Feb 2016.

State Council. (2015). State council on promulgating speed up big data development action plan. 国务院关于印发促进大数据发展行动纲要的通知 [WWW Document]. http://www.scemc.cn/tzgg/9541.htm. Accessed 15 Feb 2016.

Walravens, N. (2015). Qualitative indicators for smart city business models: The case of mobile services and applications. *Telecommunications Policy, 39*, 218–240. https://doi.org/10.1016/j.telpol.2014.12.011.

Wang, Q. (2015). The internet of things and government management innovation [wulianwang yu zhengfu guanli chuangxin]. Chengdu: Sichuan University Press.

Xie, W. (2015). 蔚蓝地图:一幅民间绘制的蓝天路线图_检察日报社多媒体数字报刊平台 [WWW document]. http://newspaper.jcrb.com/html/2015-03/30/content_182613.htm. Accessed 18 Feb 2016.

Xinhua. (2015). China to regulate environment monitoring – China – Chinadaily.com.cn [WWW document]. http://www.chinadaily.com.cn/china/2015-08/12/content_21581187.htm. Accessed 19 Feb 2016.

XinhuaNet. (2014). 杭州推出手机APP'收废品' 鼓励垃圾分类回收 [WWW document]. http://www.zj.xinhuanet.com/newscenter/sociology/2014-06/23/c_1111265820.htm. Accessed 17 Feb 2016.

Xuancheng Municipal Government. (n.d.). 宣城市城市管理监督指挥中心 [WWW document]. http://www.xc12319.gov.cn/. Accessed 18 Feb 2016.

Yao, Q. (2014). 下个APP你也是城管 | 《江淮晨报》多媒体数字报平台 [WWW document]. http://epaper.hf365.com/jhcb/html/2014-04/02/content_786452.htm. Accessed 18 Feb 2016.

Zhongguo Jianshe Bao. (2015). 重庆市民城管通APP客户端上线 [WWW document]. http://www.chinajsb.cn/bz/content/2015-03/26/content_154129.htm. Accessed 18 Feb 2016.

Kirby W. (2015) [illegible] [WWW document]. http://news.[illegible].com/html/2015/[illegible]. Accessed 15 Feb 2016.

Xinhua (2015) China to regulate environmental monitoring [illegible] [WWW document]. http://www.chinadaily.com.cn/china/2015-0[illegible]. Accessed 17 Feb 2016.

XinhuaNet (2014) [illegible] [WWW document]. http://www.xinhuanet.com/newscenter/society/2014-06/03/c_1[illegible]. Accessed [illegible] Feb 2016.

Xuancheng Municipal Government (n.d.) [illegible] [WWW document]. http://www.xc[illegible].gov.cn/. Accessed 18 Feb 2016.

Yao O (2014) [illegible] [WWW document]. http://epaper.[illegible].com/html/2014-[illegible]. Accessed [illegible] Feb 2016.

Zhongguo Minsheng Bao (2015) [illegible] APP [illegible] [WWW document]. http://www.chinagsn.cn/[illegible]/2015-0[illegible]_154129.html. Accessed [illegible].

Chapter 8
Performance Reviews, Public Accountability, and Green Governance in Hangzhou

Jørgen Delman

Abstract The study asks whether Hangzhou's system of public sector performance reviews promotes sustainability governance. The author begins by examining how ideas and concepts relating to sustainable urban development are incorporated into urban development programmes in Hangzhou. The chapter then turns to analyse the design of the system for performance reviews of party-state organizations and their leaders. Performance reviews comprise various measures for performance assessment and performance management. In addition, however, they have come increasingly to involve social participation, in order to allow for the public's voice and public participation in governance. The analysis demonstrates how the public performance review system in Hangzhou has potential to incentivize major government departments working with sustainability-related issues to toe the green Communist Party line, and how innovative provisions to enhance the public accountability of the local administration may help promote local environmental governance. This may pave the way for narrowing the well-known implementation gap in connection with implementation of urban environmental policies.

Keywords Hangzhou · Eco-civilization · Environmental policy implementation gap · Performance reviews and results management · Public accountability · Green urban governance

8.1 Introduction

Most of China's provinces have eco-footprints that significantly exceed their biological carrying capacity (Xie et al. 2012, 21), and China's major cities have per capita carbon footprints that are larger than those of many other world cities (Baeumler et al. 2012, xlii). While China's environmental policies are increasingly aimed at dealing with these challenges, some influential studies (Kostka 2014; Van

J. Delman (✉)
University of Copenhagen, Copenhagen, Denmark
e-mail: jorgen.delman@hum.ku.dk

J. Delman et al. (eds.), *Greening China's Urban Governance*, ARI - Springer Asia Series 7, https://doi.org/10.1007/978-981-13-0740-9_8

Rooij 2006; Economy 2004) have argued that there is an implementation gap between national policies and a preference amongst local governments for economic growth and social stability. The consequence is, they note, that local governments in China are often characterized by bureaucratic fragmentation, a weak motivation for green change, and a lack of environmental awareness – making the gap vis-à-vis national policies all the greater.

To address the challenges, new governance approaches and instruments are therefore required. The party-state and stakeholders from society and business have to be mobilized to find common solutions (Jabareen 2008; Lund, Chap. 1) with the party-state (see Introduction) as the lead. It must monitor developments, negotiate responsibilities, engage stakeholders to take part, mobilize and distribute resources, reward compliers, punish trespassers, and guide and regulate the activities of the various actors (Mai and Francesch-Huidobro 2015; Kostka 2014).

In this case study, I analyse how challenges to sustainable development are addressed in Hangzhou, the capital city of Zhejiang Province. The authorities in that city combine extensive green programming with innovative performance reviews, the aim of which is to incentivize government agencies to address sustainable development more seriously than in the past. My focus is primarily on how sustainability goals are taken into account in these performance reviews, which are important instruments for controlling and guiding local governments and leaders. I also consider how social participation and public accountability can help narrow the implementation gap and circumvent the existing restrictions on social participation discussed by Kostka (2014, 29–33).

I use "sustainability" generically, as a concept describing interlinked approaches and processes that primarily address environmental and climate change issues while also considering economic and social sustainability. The focus is on processes that can enhance local capacity for urban green governance while narrowing the implementation gap. I first examine Hangzhou's green turn and its associated sustainability policies, and I look at how these are reflected in the local performance review system. I then analyse the system of performance reviews in relation to the development of sustainability governance, considering not least the role of social assessments and public accountability reviews. Finally, I discuss the extent to which the integration of green policies and sustainability dimensions in these performance reviews supports the argument that the approach chosen in Hangzhou can promote green governance and narrow the implementation gap.

8.2 Hangzhou's Green Turn

For some years now, Hangzhou has been seeking to become a "liveable city" (*shenghuopinzhizhicheng* 生活品质之城) (Delman 2014). At the same time, as elsewhere in China, there has been mounting social pressure on the city's government to provide better public environmental goods and services (Wu 2014). This has prompted the city government to embark on major green reforms and reconstruction to clean up its environment. A series of policies, programmes, and plans in connection with the environment, climate change, and clean and renewable energy have been promulgated. While these policies have been aligned with those pursued on the national level, Hangzhou has sought to become a frontrunner in this respect (Delman 2014). Furthermore, the city's leadership has systematically developed and implemented a programme of governance innovations to improve the efficiency and accountability of its government services, including services relating to green and sustainable development (Wu 2014; Wu and Chen 2014).

A major step was taken in 2009, when the municipal leadership declared Hangzhou as one of the first low-carbon cities in China under a programme pursued under the aegis of the National Development and Reform Commission (NDRC) (Zhonggong 2009). This programme capitalized on prior urban environmental programmes to turn cities into "garden", "liveable", or "environmental" cities (Delman 2014). Hangzhou's overall climate change target was set at a reduction of CO2 emissions per GDP (gross domestic product) unit by 50% by 2020 (Zhonggong 2009; Yin and Yan 2011; Table 8.2). Hangzhou's leadership was tasked with preparing and implementing comprehensive plans to accelerate the transition to an urban system with lower carbon emissions and with low-carbon production, consumption, and lifestyles. These initiatives would be supported by a consolidated system for collecting and managing data on greenhouse gas emissions (Zhou et al. 2012, 7).

The NDRC did not provide a uniform model for the low-carbon cities, and it is still unclear what defines a low-carbon eco-city, or which indicators should be used for making decisions and assessing progress and performance, especially in connection with effectiveness and efficiency (Zhou et al. 2012, 12 and 34). Hangzhou therefore had to define its own roadmap, called "6 dimensions, 1 system" (*liuweiyiti* 六位一体), the main priorities of which were to shift the economic structure away from low-end, carbon-intensive manufacturing towards high-end, low-carbon manufacturing and more generally to effect a transition to low-carbon construction, life, environment, and society. The 2009 decision included specific sub-targets and measures to attain the overall target for CO2 reduction. These were subsequently incorporated into sub-plans (e.g. for the different sectors), with associated budgets, project lists, and measures for implementation (Wang 2010; see also He et al., Chap. 3).

Furthermore, since 2008 the concept of a "low-carbon eco-city" (*ditan shengtai chengshi* 低碳生态城市) has been used as a more comprehensive green development concept (Zhou et al. 2012). In 2011, Hangzhou's government responded with an indicative "eco-civilization plan" (*shengtai wenming guihua* 生态文明规划) (Hangzhou Eco 2011), which largely incorporated the objectives and targets of the other plans relating to sustainable urban development. In 2014, however, Hangzhou's leadership found – after some years of implementation – that the city was still facing challenges to its green transition. The nationally imposed requirements for reduction of emissions were more demanding than elsewhere in China. Hangzhou was suffering from resource constraints; and with regard to energy, the city is still largely dependent on imported "dirty" energy, which has contributed to make Hangzhou exceed its eco-capacity. In order to maintain its competitive position, finally, Hangzhou is therefore forced not only to save on energy but also to turn to new and cleaner types of production. While the city leadership describes these challenges as critical, it also sees them as offering opportunities for the city to become a frontrunner in energy saving and in new energy technologies (Hangzhou CC Plan 2014).

8.3 Incentivizing Local Departments

Whereas China's low-carbon eco-city programmes reflect a new developmental approach that works with the generally accepted sustainability dimensions – i.e. economic, social, and environmental sustainability (Jiang and Zhang 2012) – there are many indications that policy compliance remains a challenge. To stimulate vertical policy compliance, China's central leadership has devised various means to incentivize and control local governments and their leaders. Performance reviews (*zhengji kaohe*政绩考核) have become important in this context (Ran 2013; Liang bumen 2011). Such reviews were launched in the 1980s to improve public sector performance. They normally rely on a responsibility system that employs "management by objectives". In principle, this involves cascading objectives and targets down the administrative levels to the individual departments and posts, through result contracts and annual performance reviews that monitor implementation (Burns and Zhou 2010).

These techniques have been inspired by Western methods, especially new public management (NPM) (Burns 2014). NPM focuses on the need to move towards more explicit and measurable standards of public sector performance, with output controls and material or other incentives designed to make organizations, their leaders, and their staff responsive to preset output measures (Hood 1995). Theoretically, NPM engages the public in various ways in enhancing the performance and accountability of the public sector, and the effectiveness of the approach depends on the influence of the stakeholders concerned on the monitoring and/or incentive systems of service providers (Paul 1992:1047).

In the Chinese context, the performance reviews have three types of key performance indicators (KPI, *mubiao kaohe*目标考核): (1) "guidance indicators" (*zhidaoxing* 指导性) or "general indicators" (*yiban zhibiao*一般指标), also called "soft" indicators; (2) "hard indicators" (*ying zhibiao* 硬指标); and (3) "indicators with veto power" (*yipiao foujue mubiao* 一票否决目标). "Veto" indicators are critical: failure here leads automatically to sanctions (Wang 2013).

The Chinese leadership now considers performance reviews essential for stimulating the implementation of green policies (Kostka 2014; Gao 2009). In 2006, the 11th national 5-year plan (2006–2010) defined 22 key national environmental indicators, eight of which were described as "binding" (*yuesuxing* 约束性) for the first time. They comprised targets to reduce energy intensity, sulphur dioxide emissions, and chemical oxygen demand (COD) pollution. Responsibility for reaching these targets was put on the shoulders of leading cadres at all levels of the party-state (Wang 2013). In 2011, the same was done in the area of climate change policy (Liang bumen 2011). As for energy conservation, performance reviews with KPIs were already incorporated into the Energy Conservation Law promulgated in 2007 (Energy Conservation Law 2007).

In connection with the decision in 2009 to make Hangzhou a low-carbon city (Zhonggong 2009), the city's party secretary, Wang Guoping, declared that: "We must establish an assessment system for measuring actual performance of leading cadres with regard to protection and development of the eco-environment, and make the completion of the targets and tasks associated with low carbon development an important component in the performance reviews of leading cadres and officials at all levels" [author's translation] (Wang 2010, 11). The goal was to comply with central requirements so that "organizations would comply, leaders would comply, and that the work would be done" (Zhonggong 2009, 22).

While such declarations by local leaders are not uncommon in China, Ran (6) finds that local leaders frequently treat environmental objectives as secondary to economic and social priorities and that the mechanisms designed to monitor local implementation and to sanction non-compliance have focused on hard or "veto" targets like growth and social stability, rather than on softer targets like the environment. Thus, there is a significant implementation gap between the aspirations of the central government and the environmental policies pursued at local level, mainly because of the biased nature of the incentives designed to guide local policy implementation. Ran writes: "…[the incentives are] perversely structured, meaning that the central government provides more incentives for local government's non-implementation or poor implementation of its environmental policies than for full implementation" (Ran 2013, 34).

Similarly, Kostka argues that reliance on the target-based implementation system as the main instrument for environmental policy management has yielded mixed results. While environmental issues have quickly become priorities on the policy agenda of local governments, the skewed focus of the performance review procedures has produced unanticipated and undesirable results, because procedures are aligned neither with central political priorities in the environmental field nor with local realities (Kostka 2014, 43). The Performance Review Office (*Kaopingban* 考评办) in Hangzhou, which was established under the party leadership and the government in 2006 (I-1016),[1] has made similar observations in relation to the city's eco-civilization plan. The incentives employed do not necessarily promote long-term environmental change, and bureaucratic fragmentation often becomes a barrier to cross-departmental coordination (Wu and Chen 2014, 290).

However, Kostka and Ran's focus on the implementation gap may not fully capture the scope of the efforts made to promote sustainable urban development in a city like Hangzhou. In 2013, the city's leadership made it clear that the focus on

[1] Full name: 杭州综合考评委员会办公室. Website: http://kpb.hz.gov.cn/showpage.aspx?id=181. The *Kaopingban* refers to a 考评委员会 [Performance Review Committee] (I-1016) that comprises leaders from the relevant party and state organs as well as from the People's Congress and the Political Consultative Conference (http://www.hzkpb.gov.cn/showpage.aspx?id=204, accessed 22.9.2015) . In 2012, *Kaopingban* became responsible for performance management as well and added a new signboard: 杭州市绩效管理委员会办公室(http://kpb.hz.gov.cn/showpage.aspx?id=206, accessed 22.9.2015). It is now a full-fledged department (正局级) (I-1016).

economic growth would need to be balanced from then on against eco-construction and social well-being and that the city's performance review system would have to reflect this change (Zhonggong, 2013). There is thus evidence that the city's leadership has recognized the implementation gap and that – as we shall see below – it has also started to address it. The question is: how is it doing this, and how is it trying to mitigate the challenges to compliance discussed above?

8.4 Hangzhou as a Case

I have chosen Hangzhou as a case because it has been at the forefront of a variety of national environmental programmes since the 1990s and since it was one of the first cities to engage with eco-city and low-carbon development. Furthermore, Hangzhou was the second city after Harbin to be selected by the central government for pilot implementation of new types of procedures for performance reviews and management in 2011 (I-1016; I-1019). Since then, the city has experimented extensively with refining its performance review system for government organizations and officials (cf. Almén 2016).

The performance review process in Hangzhou has been documented extensively by the local authorities (e.g. Wu 2012b), but it has received only limited attention in the scholarly literature (Almén 2016; Dai and Teng 2008). Almén (2016) anchors his analysis of this process within debates about policy entrepreneurship, governance innovations, and democratization. My focus here is somewhat different: I ask how performance reviews can make environmental policy implementation more compliant?

I use an exploratory case-study approach (Streb 2010; Yin 2009) to examine the relevant developments in Hangzhou as phenomena that occur in a contemporary political and social context. I trace political initiatives over time that can document whether sustainability dimensions have gained prominence in Hangzhou's approach to performance reviews and how procedures and tools for performance reviews and management are redesigned to ensure policy compliance.

While almost all city departments and subgovernments deal with sustainability-related issues, the departments examined here are those most closely involved, i.e. the Development and Reform Commission (DRC: overall planning, climate change-related planning, resource use), the Economic and Information Commission (EIC: industry, energy efficiency, renewable energy), and the Environmental Protection Bureau (EPB: air and water pollution, general environment). This paper builds on documentary evidence and official statistics, as well as on data from fieldwork and interviews conducted in Hangzhou between 2013 and 2015.[2] It bears stressing, finally, that it is still too early to conclude whether the system works effectively for its purpose.

[2] The interviewees were primarily government officials engaged in environmental and climate politics in Hangzhou. Partners at Zhejiang University helped organize the interviews. The interviews were recorded and subsequently transcribed by research assistants.

8.5 Performance Reviews and Sustainability Indicators in Hangzhou

8.5.1 Development of a Performance Review System

Hangzhou has conducted regular performance reviews since 1992 (Wu, 2012b, 31, 56–57, 251–89; Dai and Teng 2008). The designers of the system from the *Kaopingban* have drawn on NPM and social capital theory, as well as on national and international experiences, e.g. the UK experience with social "commitments" or "promises" (*chengnuo* 承诺). Eventually, Hangzhou's leadership opted for a bidirectional top-down/bottom-up approach with a strong focus on social assessment and social participation (Wu, 2012b, 1–18). This approach is called "three dimensions in one system" (*sanwei yiti* 三位一体). The three dimensions are target performance assessment (*mubiao kaohe* 目标考核), leadership assessment (*lingdao kaoping* 领导考评), and social assessment (*shehuipingjia*社会评价). A bonus dimension known as "innovation" (*chuangxin chuangyou* 创新创优) comes on top of these three (Wu, 2012b, 87; Almén 2016).

Reviews of performance on KPIs and public satisfaction surveys were originally at the centre of the performance reviews, but a more sophisticated system was developed between 2005 and 2012. It incorporates a set of procedures based on both peer and top-down reviews of party and government departments, combined with citizen satisfaction surveys and other instruments for measuring public accountability (Wu 2012b; Dai and Teng 2008). This system has been retained since 2012, with only small adjustments since then (I-1016). In 2015, for example, the social assessment based on visits to interview households was combined with a mobile phone-based survey (I-1019).

The three core dimensions of the system add up to 100 points, with 45, 50, and 5 points for each assessment category, respectively (see Table 8.1). At the level of subgovernments (city districts and counties), the weights are different, viz. 30, 65, and 5, respectively.

The additional innovation and excellence dimension was introduced after initial experiments to incentivize government organizations and officials to further improve public services and make them more effective. This dimension can add 3 more percentage points at city level and 5 points at subgovernment level. In the opinion of one of the chief designers, it provides a direct link between performance reviews and governance innovation (I-1016; see also Almén 2016; Wu and Chen 2014).

Essentially, the system operates with two types of basic indicators: "results indicators" (绩效指标) which are "binding" (*yuesuxing* 约束性) and "work targets" (工作目标) which are "indicative/expected indicators" (*yuqixing zhibiao* 预期性指标). While the nomenclature is different from Wang's (6) taxonomy, the effect of the indicators should be the same. If a binding indicator is not fulfilled, the organization will not receive any points in relation to this indicator. In the case of the indicative indicators, a reduction is made in proportion to the degree of target fulfilment (I-1016). Different types of assessors who represent different interests are involved with different aspects of the system. The various types of assessment are integrated into one system that ranks the party and state organizations on the basis of their performance (more below). At the level of implementation, the system is associated with performance management contracts, known as "responsibility contracts" (*zerenshu* 责任书), which are signed by the municipal leadership and by the relevant subordinate department, e.g. the environmental bureau or a district government (I-2019). The contracts specify the tasks and targets to be achieved by the designated contractor and how performance is to be measured. Many of these contracts can be found in the public domain, which adds to the transparency of the system and its public accountability.

In June 2015, the public accountability approach was reinforced in relation to the eco-civilization programme. The City Government's Office and its Eco-civilization Office (*shengtaiban*生态办) published a bundle of responsibility contracts with departments and subgovernments. These outlined tasks, work targets, and performance procedures relating to the eco-civilization programme and the construction of a "Beautiful Hangzhou", a new development concept aligned with Xi Jinping's idea about a "Beautiful China" (Hangzhou Shengtai 2015). The performance assessment system was also written into local law in 2015 (I-1019; Hangzhou Shi Jixiao Guanli 2015).

8.5.2 Sustainability Indicators of City Departments

The first dimension of Hangzhou's performance review system focuses on attainment of KPIs (cf. Table 8.1). The main sustainability-related indicators were already specified in the 2009 decision to make Hangzhou a low-carbon city (Zhonggong 2009); however, in 2011 the implementation guideline for the city's energy-efficiency and climate change programme specifically instructed the city's departments to institute a rigid system of performance review, with key indicators pertaining to these two areas and with provisions for punishing departments at sub-district level, in case of non-compliance, by means of the city government's "veto power" (*yi piao foujue* 一票否决) (Hangzhou Shi 2011a). The city's climate change plan from 2014 provided for a set of "measures to safeguard implementation" (*baozhang cuoshi* 保障措施), including better leadership performance and – again – the establishment of a specific low-carbon performance review system (Hangzhou CC Plan 2014). A DRC leader saw leadership as the most critical dimension, since a lack of committed leadership may shift the department's focus towards other priorities (I-1015).

In the sections below, I briefly examine the 2015 targets for the three departments studied here. My purpose is to ascertain the extent to which sustainability indicators have become part of their procedures for performance review and management.

Table 8.1 Performance review procedures in Hangzhou

Category	Format	Weight in % (city level)	Weight in % (subgovernment level)
Assessment of KPIs (目标考核)	Assessment by the Performance Review Office (*Kaopingban*) and relevant technical departments of organizations directly under the city government: fulfilment of specific targets, work indicators, management of target fulfilment, and other types of progress assessments. The indicators are divided into "development indicators" (economic construction, social management and public service, and development potential) and "work targets" relating to specific types of activities	45	30
Social assessment (社会评价)	Based on about 12,000 assessment forms from representative samples of the following nine social groups: (1) citizens; (2) enterprises; (3) city-level CPC representatives; (4) city-level members of People's Congress; (5) members of the city-level Political Consultative Conference; (6) provincial organization reps, retired cadres, academics, and assessors specially appointed by departments (e.g. from social organizations); (7) members of leadership groups at city and subgovernment level; (8) responsible people within delegated offices at subgovernment and community level; (9) local community and village leaders	50	65
	The survey measures level of satisfaction and ranking of specific daily challenges relating to public services and asks for suggestions for improvements		
Performance assessment by leadership (领导考评)	Comprehensive assessment by four leading groups 四套班子 (party committee, People's Congress, city government, Political Consultative Conference), the chief justice, and the chief procurator of comprehensive progress of all organizations under the city government	5	5
Total		**100**	**100**
Assessment of innovation and excellence (创新创优)	This assessment is based on applications from organizations under the city's party-state organizations. The *Kaopingban* and a panel of experts will make the assessment	+ max 3 points	+ max 5 points

Sources: 考评体系 [The performance assessment system]. 杭州考评网 http://kpb.hz.gov.cn/showpage.aspx?id=223. Accessed 26.10.2013; Wu 2012b, 90-3; I-1019

Table 8.2 Hangzhou's key energy-related low-carbon development targets, 2005–2020

		2005	2010	2015	2020
1	Reduction of CO2/GDP per unit (t./10.000 CNY)	–	–	[−40%]	[−50%]
2	Use of coal in primary energy consumption	–	–	35%	30%
3	Energy use/GDP per unit (t./10.000 CNY)	0.87	0.68	0.55	–
4	Share of non-fossil fuel in total primary energy consumption (%)	3.7	5.8 (year: 2009)	10	15
5	Share of cars using new energy or saving energy (%)	5	9	15	20

Sources: Hangzhou Shi 2011b; Hangzhou CC Plan 2014

- The Development and Reform Commission (DRC)

As the host of the secretariat of the city's Leading Group for Climate Change and Energy Savings (I-1007; I-1015; Delman 2016), the DRC has formulated a number of climate change policies and plans that stipulate key targets for climate change mitigation and adaptation (cf. Table 8.2). Until now, the two main documents have been the 12th 5-year plan for low-carbon city development (Hangzhou Shi 2011b) and the climate change plan covering the 2013–2020 period (Hangzhou CC Plan 2014). As for 2015, the main indicators relating to climate change under the DRC's authority concern the completion of the 13th 5-year plan (2016–2020) (2015 niandu (DRC) 2015). The 13th 5-year plan comprised 25 key projects, including projects relating to resource and environmental management and 30 lesser environmental projects on topics like garbage handling and new energy vehicles. The responsible departments will subsequently elaborate detailed sector plans. There would be 84 plans altogether (Liu 2015), and many of these were to cover sustainability-related tasks and projects designed to help attain the specified targets. In addition, the DRC was expected to ensure that data from government agencies is collated systematically (2015 niandu (DRC) 2015). A new climate change centre would eventually take responsibility for this work (I-1020).

The performance review system for the DRC appears to be shifting gradually away from a one-sided focus on economic growth towards an approach that incorporates sustainability-related indicators. This is reflected in the DRC performance contract for 2015 under the eco-civilization programme (Hangzhou Shengtai 2015). The main assessors of performance are so-called third-party organizations, which are recruited through bidding but which remain subordinate to the DRC (I-1015). The independence of such organizations may thus be called into question.

- The Economy and Information Commission[3]

Energy, energy efficiency, renewable energy, and circular economy fall under the responsibility of the Economy and Information Commission (EIC), which deals with industrial sectors. Energy-savings targets are set in specific plans and handled through energy contracts with enterprises (I-1015). The two most important energy

[3] This section draws on Guan and Delman (2017).

efficiency indicators are energy intensity and total energy consumption.[4] In the 11th 5-year plan, Hangzhou's overriding goal for energy efficiency was to reduce energy intensity by 20% between 2005 and 2010 (Hangzhou Government 2005). A further 15–18% reduction was introduced in the 12th 5-year plan (2010–2015) (Hangzhou Government 2010). Finally, the climate change plan for 2013–2020 polished the target, so that the 2015 goal for energy-intensity reduction would be attained ahead of time – in 2014 – while the medium-term goal for 2020 was maintained. The overall goals for the 2010–2020 period are shown in Table 8.2.

The only "binding" target for the EIC in 2015 was the reduction in energy use per unit of GDP (cf. Table 8.2). The EIC expected to surpass the target, by reaching a reduction of 23.7%. Other targets for energy savings or green energy were "indicative only" and comprised closure or retrofitting of high-energy-use enterprises and introduction of new energy vehicles (Shi Jingxinwei 2015).

Hangzhou's local government has also put a mandatory cap on energy use (Guan and Delman 2017). This is disaggregated into specific KPIs in the responsibility contracts of local governments and enterprises. These contracts appear to be compulsory rather than voluntary, although the city's government is unable to control the enterprises effectively, especially the private ones (I-1015).

These traditional top-down instruments are crucial for incentivizing officials and enterprise managers. In order to increase the pressure for better performance, the targets for energy savings and reduction of CO2 emissions have now been separated. In the past they were combined as a single "veto" indicator, with each target counting for half. Now they count as one full "veto" indicator each (I-1019). In 2013, 9 out of 13 districts and counties under Hangzhou city failed to meet their targets for energy saving, and they were all sanctioned accordingly, i.e. they got 4–5 points less in their overall score, indicating the relative importance of this key indicator in the overall assessment (I-1016). A DRC representative found that the targets were ambitious and could generally be difficult to attain. But she said the bar had to be set high, since Hangzhou is a low-carbon pilot city (I-1007).

- The Environmental Protection Bureau (EPB)

The work of the EPB is guided by a series of critical sustainability indicators. In 2015, however, the Bureau was subject to only one "binding indicator", which concerned the reduction of emissions from a number of key chemical pollutants. This indicator is set nationally (cf. above), and it has been "binding" for some years. Other "key work targets" are only "indicative", e.g. clean water supplies, environmental monitoring, eco-civilization construction, year-on-year reduction of air pollution, curtailment of vehicles with high particle emissions, inspection and

[4] Energy intensity, i.e. energy consumption per unit of GDP, has been set as a national indicator for energy conservation since 2006. Additionally, total energy consumption was set as a second national indicator for energy conservation in late January 2013, when China's State Council approved an "energy consumption control target" below the equivalent of 4 bill. t. of coal per year by 2015 (Guan and Delman 2017).

implementation of relevant regulation, and finally implementation of investment projects worth about 3 bill. Yuan (Huanbaoju kaohe 2014).

The work of the Xiacheng District EPB in central Hangzhou encompasses many of the targets set by the DRC and the EIC. Polluting factories are being closed or forced to move away, while remaining enterprises have to use natural gas or electricity to the extent possible. The EPB also participates in a programme to curtail high-emission vehicles. In general, many departments are required to collaborate at the operational level under the leadership of the district government to implement policies and attain the targets set (I-1018).

In June 2015, a bundle of responsibility contracts under the city's eco-civilization programme, including one for Xiacheng District, was issued to all departments and subgovernments (Hangzhou Shengtai 2015). Xiagenh District's contract contained a series of tasks relating to performance management, especially in connection with past nonperformance, such as failure to reduce emissions sufficiently. It also set out current key policy targets for the improvement of water quality and the reduction of air and water pollution.

The Xiacheng EPB is now expected to establish its own performance review system. This in itself will earn the Bureau up to 130 points, testifying to the importance of the system for performance review and management at the level of grassroots operational units. The technical tasks are assigned a maximum number of points individually, adding up to a total of 1000 points. There are also indicators that may add extra points, e.g. if the EPB receives prizes or if it proactively institutes civil litigation on behalf of the public interest. But penalties do not just follow from failure to meet targets: they are also imposed if the Bureau shows itself unable to complete projects assigned by the province or higher-level agencies, or if unsolved problems are identified by higher-level inspections, or if environmental problems are disclosed by the media and create a bad image. However, if the EPB mobilizes the media itself to address environmental issues, sanctions are not applied. Finally, the occurrences of mass incidents in the district that relate to local environmental issues are punished, depending on the consequences; but major mass incidents that do not relate to the mandate of the local EPB are not punished (Hangzhou Shengtai 2015, 4–5; I; I-1019).[5]

The Xiacheng EPB is thus under considerable pressure to perform. It must become more professional and more proactive, in order to avoid "incidents". The fact that the EPB can use the media itself is an innovation and shows how public accountability can be enhanced when social actors are engaged. Furthermore, local residents may use knowledge about the performance contract to institute protests in order to force their agenda through. However, popular satisfaction with the EPB only needs to reach a minimum of 65% to achieve a maximum of 100 points. This may seem to contradict other efforts, described below, to afford the general public greater influence through the performance reviews.

[5]The Hangzhou city government's EPB has a different type of performance contract. It specifies tasks and a few targets but does not provide for specific assessment criteria (Hangzhou Shengtai 2015, 78–81).

8.5.3 Leadership Assessments

Leaders at a range of administrative levels are supposed to assess their subordinate organizations and their leaders, using a 5-point scale. These leadership assessments are implemented through a cascading top-down approach but account for only 5% in the overall review (cf. Table 8.1). To receive 5 points from its leadership, a subordinate organization must be able to do the following: "In splendid fashion to complete normal work goals, as well as special tasks assigned by the leaders of the city party committee and the city government, and to demonstrate outstanding results in innovation and improved solutions. Social results must be good, and the organization must be at an advanced level compared with other similar organizations around the country. Finally, the contributions of the organization to the city must be outstanding". The minimum, 1 point, is allocated to an organization that "cannot complete normal work goals, as well as special tasks assigned by the city party committee and the city leadership, thereby influencing work [negatively]. Compared with other similar organizations around the country, the organization is falling behind. Finally, the organization's contributions to the city are very small" [author's translation] (Wu 2012b, 123–4). The points in between are associated with different criteria in between the two extremes. The system at subgovernment level is more differentiated with regard to points, but it follows the same principles (Wu 2012b, 124–5).

8.5.4 Public Accountability and Sustainability Governance

The importance accorded the social assessment – weighing 50% at city level and 65% at subgovernment level (cf. Table 8.1) – is remarkable. The current configuration of the system is the result of a long process of development (Wu 2012b, 90–106), and Hangzhou is one of the first cities in China to have taken such a step. Technically, nine different groups of assessors are included in the social assessment (Table 8.3), which comprises a sample survey of ordinary citizens, combined with comments and suggestions from citizens and from a variety of cohorts, the relative shares and weights of which are shown in Table 8.3. Government websites are also used to elicit comments and suggestions from the public. The information harvested is used as part of the performance review.

Responses to the citizen surveys are processed by the local statistical bureau (I-1019), and the comments and suggestions made by respondents are forwarded to the responsible departments by the *Kaopingban* (I-1016). The use of the statistical bureau as an external partner and the gradual empowerment of the *Kaopingban* indicate that the city leadership is committed to making the review as unbiased as possible.

The first comprehensive survey was conducted in 2011. It generated 10,702 opinions and suggestions from the public, to which the city leadership responded by instituting a number of administrative adjustments (I1016; Hangzhou Kaopingban

Table 8.3 Composition of sampled cohorts of social survey

	Cohort	Share/ weight
1	(1) Citizens (with and without *hukou*)	40%
2	(2) Enterprises	15%
3	(3) City-level CPC representatives; (4) city-level members of People's Congress; (5) members of the city-level CPPCC; (6) provincial organization representatives, retired cadres, academics, and assessors specially appointed by departments (e.g. from social organizations); (7) members of four leading groups (*sitaobanzi* 四套班子) (party committee, People's Congress, government, Political Consultative Conference) at the level of city district and county/county town	25%
4	(8) Responsible people within delegated offices at subgovernment and community level	10%
5	(9) Local community and village leaders	10%

Source: Wu 2012b, 94, 95; cf. Table 8.1

2013; Wu 2012a). In 2012, the basic design of the social assessment system was finalized. It now included more than 15,000 respondents (see Tables 8.1 and 8.3), amongst them a substantial number of migrant workers and rural residents, profit-making enterprises (1000), non-profit enterprises, business associations, and other social organizations (100). Respondents were also asked to comment on 216 "adjustment promises". There were about 9500 suggestions and criticisms, 1000 fewer than the year before. Half came from ordinary citizens. Businesses took second place, with 569 concrete comments (2012 niandu 2013).

The 2014 survey involved 6000 ordinary citizens (people with *hukou* and migrant workers), and more than 2000 representatives from companies, amongst them representatives of 100 private enterprises and business organizations. The government departments also invited special "people's supervisors" (*gongshu minping* 公述民评) to participate as expert evaluators of government organizations (Hangzhou Huanbaoju 2012). The management of one of the most influential green NGOs in Hangzhou, "Green Zhejiang" (*Lüse Zhejiang*绿色浙江), was invited to work in such a capacity by the EPBs of both Zhejiang Province and Hangzhou (I-1003). All in all, 12,377 survey questionnaires were issued, of which 99.56% – with 5999 concrete suggestions – were handed in (2014 niandu 2015).

Respondents to the social survey were generally satisfied with public services, as well as with the main government projects for improving their livelihoods and enhancing public sector performance. If the three response options "satisfied", "generally satisfied", and "basically satisfied" are combined, the approval rating in 2014 came to 95.5%. Each category received about one third of the responses (2014 niandu 2015).

Whereas sustainability issues or indicators as defined here do not figure as an independent category in the social assessment, they have been increasingly highlighted in analyses of the reviews based on the responses. Respondents have focused on the need to monitor and improve air quality, as well as to monitor emissions from polluting enterprises and to track energy consumption in big energy-consuming

Table 8.4 Major environmental concerns of respondents in 2014 survey

Concern	Share of env. Concerns (%)
Management of air pollution	42.25
Water treatment, supplies, and management	14.93
Treatment of garbage	5.92
Management of "greening" (i.e. green urban areas/ spaces)	3,10
Management of scenic areas	2.54
Others	31.27

Source: 2014 niandu 2015

enterprises (I1016; Hangzhou Kaopingban 2013; Wu 2012a; Zhuyao Chengxiao n.d).

About 3% of the criticisms or proposals made in 2012 focused on environmental issues (Hangzhou Kaopingban 2013). In 2014, the figure was 6.8%. In the 2014 survey, however, the number suddenly dropped to 412 from 1034 in 2013; and environmental suggestions dropped from fourth place in 2013 to sixth in 2014, amongst the 15 issues mentioned as the most critical (2014 niandu 2015). While it is impossible to know why, there are many indications that sustainability-related issues remain important after all. Amongst the 20 most important issues to be addressed as a consequence of the 2012 social survey, the quality of water for household consumption was no. 3, the need for an air quality measurement system based on PPM 2.5 was no. 8, river pollution was no. 10, and exhaust fumes from restaurants, especially at night, was no. 11 (2012 niandu 2013). In the 2014 report, environmental measures featured as the first category of measures that had to be dealt with to address issues brought up in the 2013 survey (2014 niandu 2015).

In 2014, finally, environmental issues ranked no. 2 amongst the 18 issues of greatest concern to survey respondents. First place went to inconveniences in connection with public transport, road transport, car parking, etc. Other environmental concerns – relating to garbage sorting and food safety, for instance – also made it on to the list (2014 niandu 2015). The environmental concerns listed in Table 8.4 were the most important.

Furthermore, at the end of 2014, *Kaopingban* asked 2000 people's representatives to rank the most important issues amongst the 183 identified as serious challenges in the annual citizens' survey. Amongst the 20 top challenges, air pollution from vehicles and construction sites ranked no. 12, pollution of rivers no. 14, garbage sorting no. 16, and environmental protection in one of the city districts no. 19 (2014 niandu, 2015). The top 20 on the list all relate to the daily life concerns of citizens in Hangzhou. It was thus confirmed that the citizens of Hangzhou regard sustainability-related issues as important daily life concerns, although not the very most important.

Already in 2003, the analysis of the citizens' survey had identified "seven urgent difficulties" (*po qi nan* 破七难) affecting citizens in their daily life. They were deficient public hygiene, limited access to health services, limited access to schooling, limited access to housing, insufficient employment for destitute families, inadequate pedestrian passageways and parking facilities, and difficulty in managing affairs with the public sector (Wu 2014). The city government tasked the responsible agencies with turning these problem areas into core intervention zones for which they would be held directly accountable (I-1016). Starting around 2007, moreover, food safety and environmental issues came to feature more prominently in the responses, and they were subsequently included as additional "serious difficulties" (I-1016). Surprisingly, the 2014 survey report showed that the leaders who responded did not concern themselves with environmental issues (0%). They were more concerned with good government, management of party and state cadres, and the party's campaign against corruption (2014 niandu, 2015). These examples clearly show why public accountability reviews are important for making local leaders toe the green party line.

In connection with this, the 2014 edition of the social assessment report also focused on concerns raised by respondents about the need to improve democratic consultation. In the environmental field, the Hangzhou city government has faced some opposition to garbage incineration (cf. Burell and Almén, Chap. 10; Wang et al., Chap. 11), to restrictions on private cars, and to curtailment of polluting vehicles. According to the report, there is a general understanding of the need for such policies, but ordinary citizens are concerned that the public is not heard and that the policies pursued are not sufficiently evidence-based. It therefore suggested more public consultation and greater government transparency (2014 niandu 2015).

Finally, in the instruction to the municipal departments about the 2015 survey, the relevant departments were made responsible for following up on 15 areas of special concern that had been specifically raised in surveys over the previous years. Environmental protection was mentioned as the first such area (Yi yange 2015). The main environmental issues specified were air pollution, garbage sorting, disposal of waste water, environmental management of urban spaces, strict law enforcement in relation to management of pets, and the need to investigate violations of environmental laws and regulations in urban communities (2015 Nian Shi 2015). Interestingly, the government seems even more concerned about such issues than the respondents (I-1019).

The system for performance reviews is increasingly used to enhance public accountability. City departments are encouraged to publicize their procedures, and they must make public promises – often under instruction from the *Kaopingban* – about how they will address specific concerns and suggestions raised in the social assessments. These promises are written into the responsibility contracts of the departments, which are increasingly made available to the public in order to enhance public transparency (Hangzhou Shengtai 2015; I-1016; Hangzhou Kaopingban 2013; Wu 2012a).

A concrete example of how the pressure for public accountability can change the behaviour of government departments came up in my fieldwork. During a visit to Green Zhejiang in October 2014, a delegation from the Hangzhou city government, including the EPB and the Office of the City Government, came to visit, and I was invited to sit in at the meeting. The main point under discussion was how a social organization like Green Zhejiang could support the municipal government in its efforts to handle thorny environmental issues. The representatives of the city government pointed out that it is generally unable to expand its staff, so departments like the EPB would have to collaborate with social organizations and urban communities to address new policy issues. One of the main issues discussed related to garbage handling, and it became clear that the local government was under pressure due to some mass incidents, in which citizens had protested against plans to build garbage incineration plants in Hangzhou (see Burell and Almén, Chap. 10; Wang et al., Chap. 11). The municipal representatives emphasized that democratic consultation would be essential to solve such issues; otherwise they would not be able to proceed with plans which all of the parties at the meeting agreed would be beneficial for sustainable development in Hangzhou (I-1017).

A division chief from the DRC noted that public accountability and social participation represent a new and democratic dimension of the interaction between the city government and the public (I-1007), while the head of the Xiacheng EPB emphasized that the public is increasingly using the traditional instrument of complaints (*xinfang* 信访) to focus attention on environmental issues. His estimate was that his bureau receives more than 100 complaints annually. The EPB is required to respond quickly, and response times are important for its performance review. The department devotes considerable time and resources to meeting this requirement while also trying to avoid complaints in the first place by initiating procedures to prevent incidents from happening (I-1018). This tallies with the provisions of the EPB's performance contract discussed above.

To further incentivize government departments through public accountability, the departments and subgovernments of the city government are ranked annually on the basis of their performance record (Table 8.5).

Table 8.5 shows that the ranking of the three departments studied here has changed considerably over time. They have all been at the lower end of the ranking, as well as in the top third. The rankings do not only reflect their standing on sustainability-related criteria, but it is evident from the above analysis that sustainability indicators have become increasingly important. When reflecting on the comparatively low ranking of the municipal EPB in 2013, the head of the Xiacheng EPB – who was newly appointed – immediately noted that issues of air and water pollution were important in the assessment of his department (I-1018). The head of the climate change division in the DRC noted that staff members had actually had their bonus cut when they received a low ranking, due to the low scores on their KPIs (I-1015).

While the figures in Table 8.5 are the result of complex assessment procedures with quite narrow differences in terms of points gained, the ranking is important. The performance reviews stimulate competition amongst the departments to perform better (I-1018; I-1019), because their reputation is at stake (I-1018). The results also affect the bonus granted to the employees of a given department (I-1016). Furthermore, even if the scores are not that different, local leaders will be concerned about the ranking of their unit for personal reasons (I-1016), because the results of

Table 8.5 Ranking of sampled departments

Department	2003*	2010** (rank no./points/ category)	2012** (rank no./points/ category)	2013** (rank no./points/ category)	2014 ** (rank no./points/ category)
Development and Reform Commission	Not included	27/93.088/advanced	24/94.026/advanced	51/89.884/compliant	18/91.478/advanced
Economy and Information Commission	14/77.965	25/93.251/advanced	30/93.609/compliant	49/89.992/compliant	20/91.276/advanced
Environmental Protection Bureau	61/72.165	34/92.842/compliant	42/93.304/compliant	69/88.159/compliant	36/90.699/compliant

Sources:
*According to satisfaction only
**Result of comprehensive performance review 2003: Wu 2012b. 453–454
2010: 3 年度市直单位综合考评结果(综合考评单位)http://kpb.hz.gov.cn/showpage.aspx?nid=4688&id=217, accessed 20.9.2015
2011: 4 年度市直单位综合考评结果(综合考评单位). http://kpb.hz.gov.cn/showpage.aspx?nid=7332&id=232, accessed 20.9.2015
2012: 5 年度市直单位综合考评结果(综合考评单位). http://kpb.hz.gov.cn/showpage.aspx?nid=9064&id=217, accessed 20.9.2015
2013: 6 年度市直单位综合考评结果(参评单位). http://kpb.hz.gov.cn/showpage.aspx?nid=10059&id=217, accessed 79.2015
2014: 8 年度市直单位综合考评结果(参评单位) http://kpb.hz.gov.cn/showpage.aspx?id=205, accessed 18.9.2015

the review and the position of their department in the ranking may eventually affect their prospects for promotion (I-1019).

8.6 Narrowing the Implementation Gap?

This case study has documented how Hangzhou's party-state leadership has gradually come to use performance reviews as an instrument for incentivizing local leaders and departments to act on the city's green agenda. The system shows a breadth and depth in handling sustainability challenges at the city level that has not been documented before. The organizational setup surrounding the performance review and results management system demonstrates that the local party-state leadership is intent on making the system as unbiased and reliable as possible. This happens, primarily, through comprehensive provisions for popular participation. The scope and relative weight of the annual social assessments that form part of the performance reviews are significant. Up to this point, no other Chinese city seems to have been able to do anything similar (I-1019).

Does this system for performance review and management contribute to better green governance and better environmental management, then? Clearly, the system has increasingly incorporated and upgraded sustainability indicators, and it is evident from the analysis that these indicators have become important at both policy and operational level. The provisions for enabling the public to monitor and influence public agencies – and to hold them accountable – are also quite innovative in the overall Chinese context.

I would therefore argue that the performance review system – with its combination of performance review and results management on the one hand and with an increasing focus on transparency in governance and public accountability on the other – could well contribute to the promotion of green governance and sustainable development and to mitigating the so-called implementation gap. But the field has only just been opened. I have not looked at causal relationships, e.g. what is the actual impact of the interventions examined here on the city's prospects for sustainable development? Meeting the targets is plainly not easy, and much critical data is still kept secret (I-1020). It is also still unclear to what extent the performance reviews affects the careers of individual leaders. Furthermore, the nature of the policy agenda presupposes collaboration and coordination across departments, in order to mitigate tendencies towards fragmentation. While this is being addressed through the establishment of coordinating leading groups (Delman 2016), the performance reviews appear to accentuate competition; they may therefore lead to greater fragmentation along vertical lines. Finally, public accountability does not come without opposition: resistance from some departments and local governments has been publicly acknowledged (2014 niandu 2015), and a couple of interviewees indicated that differences of opinion could be found about the usefulness of such a complex system.

It is clear from the Hangzhou experience that there has been a good measure of policy learning through policy mobility from Western public governance theory. At the same time, there has not been a transfer of any particular package, but rather a transformation of *new public management* experiences and instruments in conjunction with their redevelopment in the local context (cf. Peck and Theodore 2010). The local authorities have taken particular account of the need to address the fragmentation of the party-state system to make implementation of green policies more efficient (cf. Delman 2016). The party-state's tradition for taking control of its own agents and partners through top-down command-and-control instruments (cf. Guan and Delman 2017) is also reflected in the approach to performance reviews and results management. Conversely, the significance of using proto-democratic mechanisms to enhance transparency in government and public accountability of the local party-state system cannot be neglected.

At this stage, we can ascertain that Hangzhou's leadership is experimenting with a new approach to improve urban governance, part of this being urban green governance. We do not know what this will do to the party-state system in the long run, and it would not be unlikely that the process could be derailed. Certainly, many lower-level party-state leaders in Hangzhou do not appear to see sustainability issues as particularly important at this stage. Given the nature of the party-state system, some leaders might also be suspected of worrying about the impact on the party-state system of too much public engagement and accountability. For the time being, however, those favouring the system do seem to have full support and attention, at both local and national level. Therefore, continued strong support from the city leadership and from citizens and non-state stakeholders will be critical for the system to be retained and refined. In this context, the ability of the *Kaopingban* to follow up on the "public promises" of the individual departments in the city's party-state system is also important for the legitimacy of the approach.

Acknowledgements This study has been undertaken with financial support from the Sino-Danish Centre for Education and Research in Beijing. I am grateful to Prof. Yu Jianxing, Zhejiang University, for his continuous support for the project and to Guan Ting, Zhang Liyan, and Stine Olsen for research assistance. Thanks are also due to Anna Lisa Ahlers and Richard Hardiman for commenting on an earlier version of the manuscript.

Literature

2012 niandu. (2013). 2012年度杭州市市直单位综合考评社会评价意见报告 [2012 report on the social assessment and comprehensive performance reviews within units directly under Hangzhou City], (17 April). http://www.hzkpb.gov.cn/showpage.aspx?nid=9028&id=194. Accessed 13 May 2015.

2014 niandu. (2015) 2014年度杭州市市直单位综合考评社会评价意见报告 [2014 report on the social assessment and comprehensive performance reviews within units directly under Hangzhou City] (24 April), http://kpb.hz.gov.cn/showpage.aspx?nid=13519&id=194. Accessed 17 Sept 2015.

2015 nian Shi. (2015). 2015年市考评办跟踪督办社会评价意见分类分组 [The City Performance Review Office on tracking and supervising categorizations and group establishment in relation to opinions from the social assessment].

2015 niandu (DRC). (2015). 2015年度杭州市市直单位绩效考核目标. [2015 results indicators for performance assessments for units directly under Hangzhou City]. http://220.191.210.153:8023/kpb/Com_allIndic2012.do?id=24267&year=2015&pUnitName=%25E5%25 B8%2582%25E5%258F%2591%25E6%2594%25B9%25E5%25A7%2594&isJxkh=1&pmgfinalgoal.fgagoaltype=0. Accessed 20 Sept 2015

Almén, O. (2016). Local participatory innovations and experts as political entrepreneurs: The case of China's democracy consultants. *Democratization, 23*(3), 478–497.

Baeumler, A., Ijjasz-Vasquez, E., & Mehndiratta, S. (Eds.). (2012). *Sustainable low-carbon city development in China*. Washington, DC: World Bank.

Burns, J. P (8). (2014). Looking West. The impact of Western ideas on public sector reform in China. In: Brødsgaard, K. E. (Ed) Globalization and public sector reform in China. Routledge

Burns, J. P., & Zhou, Z. R. (2010). Performance management in the government of the People's Republic of China: Accountability and control in the implementation of public policy. *OECD Journal on Budgeting, 10*(2), 1–28.

Dai, G. Q., & Teng, W. L. (2008). Performance Evaluation at Local Government Department Level in China – A case of People's Municipal Government of Hangzhou (PMGH). (unpublished paper)

Delman, J. (2014). Climate change politics and Hangzhou's green city making. In E. Björner & P. O. Berg (Eds.), *Branding Chinese mega-cities: Policies, practices, and positioning* (pp. 249–246). Cheltenham: Edgar Elgar Publishers.

Delman, J. (2016). Urban climate change politics in China: Fragmented authoritarianism and governance innovations in Hangzhou. In K. E. Brødsgaard (Ed.), *Chinese politics as fragmented authoritarianism: Earthquakes, energy and environment* (pp. 156–180). London/New York: Routledge.

Economy, E. C. (2004). *The river runs black: The environmental challenge to China's future*. Ithaca: Cornell University Press.

Energy Conservation Law. (2007, October 28). http://www.npc.gov.cn/englishnpc/Law/2009-02/20/content_1471608.htm. Accessed 10 Feb 2017.

Gao, J. (2009). Governing by goals and numbers: A case study in the use of performance measurement to build state capacity in China. *Public Administration and Development, 29*, 21–31.

Guan, T., & Delman, J. (2017). Energy policy design and China's local climate governance: energy efficiency and renewable energy policies in Hangzhou. *Journal of Chinese Governance, 1*, 1–23. https://doi.org/10.1080/23812346.2017.1284430.

Hangzhou CC Plan. (2014). 杭州市应对气候变化规划 (2013–2020年) [Hangzhou guidance plan for addressing climate change]. 杭州市发展和改革委员会办公室, 5 September.

Hangzhou Eco. (2011). 杭州市生态文明建设规划 (2010–2020). [Hangzhou City guidance plan for eco-civilization construction (2010–2020)]. 杭州杭州市人民政府.

Hangzhou Government (2005) 杭州市十一五节能专项规划 [11th 5-year guidance plan for special energy savings projects in Hangzhou]

Hangzhou Government. (2010). 杭州市十二五节能专项规划[12th 5-year guidance plan for special energy savings projects in Hangzhou].

Hangzhou Huanbaoju. (2010). 杭州环保局 2012年"公述公评"工作方案 [Hangzhou Environmental Protection Bureau work programme for special people's supervisors of government officials in 2012]. "公述民评"专栏 , 9 November 2012, http://www.hzepb.gov.cn/hbzx/ztzl/gsmp/gzjh/201209/20120911_17085.html. Accessed 5 Nov 2013.

Hangzhou Kaopingban. (2013). 2012年杭州市市值单位综合考评社会评价意见报告 [Report on the comprehensive assessment of opinions from society on units directly under Hangzhou City]. 杭州考评网, http://kpb.hangzhou.gov.cn/showpage.aspx?nid=9028&id=194. Accessed 16 Oct 2013.

Hangzhou Shengtai. (2015). 关于印发《杭州市2015年度生态文明建设目标责任书》的通知 [On the publication of "Hangzhou's contracts regarding target responsibilities in relation to ecological civilization for 2015] 杭生态办〔2015〕33号.

Hangzhou Shi. (2011a). 杭州市人民政府办公厅关于印发杭州市节能减排专项整治行动工作方案的通知 [Office of Hangzhou City Government on the publication of the notice on the scheme for actions to regulate the work relating to special energy savings and emission reduction projects in Hangzhou in 2015] 杭政办发电,〔2011〕34 号
Hangzhou Shi. (2011b). 杭州市"十二五"低碳城市发展规划 [Hangzhou's 12th five-year guidance plan for development as a low-carbon city], http://dtfz.ccchina.gov.cn/Detail.aspx?newsId=45742&TId=171%22%20title=%22%E6%9D%AD%E5%B7%9E%E5%B8%82%E2%80%9C%E5%8D%81%E4%BA%8C%E4%BA%94%E2%80%9D%E4%BD%8E%E7%A2%B3%E5%9F%8E%E5%B8%82%E5%8F%91%E5%B1%95%E8%A7%84%E5%88%92. Accessed 15 Aug 2016.
Hangzhou Shi Jixiao Guanli. (2015). 杭州市绩效管理条例 [Regulations for performance management in Hangzhou]. http://www.hzpolice.gov.cn/GOVHtml/201511/16/577e5378-6ac6-4851-b24b-c58a7306af79.html. Accessed 17 Feb 2016.
Hood, C. (1995). The "New Public Management" in the 1980s: Variations on a Theme. *Accounting, Organizations and Society, 20*(2/3), 93–109.
Huanbaoju Kaohe. (2014). 环保局考核内容及指标 [Contents and indicators for performance review of Environmental Protection Bureau]. (Mimeo provided by Head of *Kaopingban*, October 2015).
Jabareen, Y. (2008). A new conceptual framework for sustainable development. *Environ Dev Sustain, 10*, 79–192.
Jiang, H. Q. (蒋惠琴), Zhang, L. L. (张丽丽) (2012).低碳城市综合评价指标体系研究 [Research into the indicators for comprehensive review of low-carbon cities]. 经营与管理 11:92–94.
Kostka, G. (2014). Barriers to the Implementation of Environmental Policies at the Local Level in China. *World Bank Policy Research Working Paper*, WPS7016 (August).
Liang Bumen. (2011). 两部门要求开展节能减排财政政策综合示范工作 [Two organizations demand that comprehensive demonstration work for financial policies for energy saving and reduction of emissions is initiated]. The Central People's Government of the People's Republic of China (28.6.). http://www.gov.cn/gzdt/2011-06/28/content_1895327.htm. Accessed 9 Mar 2013.
Liu, Y. (2015). 杭州将编制 84 项 "十三五"规划 (2016–2020) [Hangzhou will elaborate 84 13th five-year guidance plans]. 360doc 个人图书馆 (7 April), http://www.360doc.com/content/15/0704/10/8507568_482527021.shtml. Accessed 20 Sept 2015.
Mai, Q., & Francesch-Huidobro, M. (2015). *Climate change governance in Chinese cities*. London/New York: Routledge.
Paul, S. (1992). Accountability in public services: Exit, voice and control. *World Development, 20*(7), 1047–1060.
Peck, J., & Theodore, N. (2010). Mobilizing policy: Models, methods, and mutations. *Geoforum, 41*, 169–174.
Ran, R. (2013). Perverse incentive structure and policy implementation gap in china's local environmental politics. *Journal of Environmental Policy & Planning, 15*(1), 17–39.
Shi Jingxinwei. (2015).市经信委 2015年度杭州市市直单位绩效考核目标 [Performance targets for 2015 for units directly under Hangzhou City, the Economy and Information Commission]. http://220.191.210.153:8023/kpb/Com_allIndic2012.do?id=2575645&year=2015&pUnitName=%25E5% 25B8%2582%25E7%25BB%258F%25E4%25BF%25A1%25E5%25A7%2594&isJxkh=1&pmgfinalgoal.fgagoaltype=0 and http://220.191.210.153:8023/kpb/Com_jinduJxkh2012.do?id=47221698. Accessed 21 Sept 2015.
Streb, C. K. (2010). Exploratory case study. In A. Mills, J. G. Durepos, & E. Wiebe (Eds.), *Encyclopedia of case study research* (pp. 372–374). Thousand Oaks: SAGE Publications.
Van Rooij, B. (2006). Implementation of Chinese environmental law: Regular enforcement and political campaigns. *Development and Change, 37*(1), 57–74.
Wang, A. (2013). The search for sustainable legitimacy: Environmental law and bureaucracy in China. *Harvard Environmental Law Review, 37*, 367–440.

Wang, G. P. (王国平) (2010).杭州如何建设低碳城市 [How should Hangzhou establish a low-carbon city]. 城市建设 (5:1):7–11
Wu, B. (伍彬) (2012a). 公民参与和重述"我们"的价值观 [Participation of the ordinary people and the remoulding of the values of "We"]. Zhonggong Hangzhou Shiwei Jiguan Kanwu "Hangzhou Women", 11 Nov., http://kpb.hangzhou.gov.cn/showpage.aspx?id=264&nid=7313. Accessed 16 Oct 2013 (Originally published in: 行政管理改革 2010/12)
Wu, B. (伍彬) (Ed.). (2012a). 综合考评与绩效管理 / 杭州的实路和探索 [Comprehensive performance review and management/the practice and explorations in Hangzhou]. 人民出版社, 北京.
Wu, B. (伍彬) (2012b).中国地方政府绩效管理中的民意价值和治理创新以杭州市位列 [The popular will and governance innovation in performance management at local government level in China with Hangzhou as a case]. Lecture at Vienna University, 11 June, http://kpb.hz.gov.cn/showpage.aspx?tid=259&nid=12717. Accessed 3 July 2014.
Wu, B. (伍彬), Chen, G. Q. (陈国权) (Eds.). (2014). 创新性政府 - 杭州的探索与实践 [Innovative governments – Exploration and practice in Hangzhou]. 浙江大学出版社,杭州.
Xie, G. D., et al. (2012). China ecological footprint report. In *Consumption, production and sustainable development*. World Wildlife Foundation: Beijing etc.
Yi Yange. (2015). 以严和实的要求加强跟踪督办推进整改落实.市考评办召开跟踪督办意见整改对接会 [Strengthen the tracking of supervision and monitoring with serious and practical requirements to push the implementation of reforms. The Performance Review Office of the City conducts liaison meeting to get opinions on reforms]. 考评版, 3 July, http://kpb.hz.gov.cn/showpage.aspx?id=211&nid=13739. Accessed 18 Sept 2015.
Yin, R. K. (2009). *Case study research. Design and methods*. London: Sage.
Yin, B. Q. (殷宝庆), Yan, Q. (颜青) (2011).低碳经济发展与杭州发展方式的转变[Low-carbon economic development and the transformation of Hangzhou's development model]. 湖北函授大学报. *Journal of HUBEI Correspondence University*. 24/6: 61–62.
Zhonggong. (2009). 中共杭州市委杭州市人民政府关于建设低碳城市的决定 [Decision by CPC Committee of Hangzhou and the People's Government of Hangzhou on the establishment a low-carbon city]. 市委[2009]3. 杭州,中共杭州市委办公厅 31 December
Zhonggong. (2013). 中共杭州市委杭州市人民政府 - 关于优化综合考评强化绩效管理的意见.市委 (6) 15号 [CPC Committee of Hangzhou and the People's Government of Hangzhou – opinions on how to improve comprehensive performance reviews and strengthen performance management] (mimeo provided by Head of *Kaopingban*, I-2016).
Zhou, N., He, G., & Williams, C. (2012). China's development of low-carbon eco-cities and associated indicator systems. Ernest Orlando Lawrence Berkeley National Laboratory (July) identifier: LBNL-5873E.
Zhuyao chengxiao (n.d.). 主要成效 [Main results]. 杭州考评网, http://kpb.hz.gov.cn/showpage.aspx?id=224. Accessed 26 Oct 2013.

Interviews

I-1003 (15.10.2013). Director of Secretariat of Green Zhejiang.
I-1007 (18.10.2013). Head of Division, Deputy Head of Division, DRC (Climate Change).
I-1015 (22.10.2014). Head of Division, Deputy Head of Division, DRC (Climate Change).
I-1016 (21.10.2014). Head of Department, *Kaopingban*.
I-1017 (2014.10.23) Meeting with Hangzhou government representatives from EPB and Office of Government at the premises of Green Zhejiang.
I-1018 (22.10.2014). Head of Xiacheng EPB.
I-1019 (11.11.2015). Head of Department and Senior Researcher, *Kaopingban*.
I-1020 (13.11.2015) Head of Division, Deputy Head of Division, DRC (Climate Change).

Part IV
Society Knocking on the Door

Chapter 9
Digital Media, Cycles of Contention, and Urban Governance in China: Anti-PX Protests as an Example of the Sustainability of Environmental Activism

Jun Liu

Abstract Since the emergence of the environmental crisis in China in the early 1990s, environmental activism has emerged as a key type of digital activism as well as a major challenge to urban governance. While scholars have studied issues arising from digital activism in general and environmental activism in particular, most have failed to scrutinize possible interconnections among different instances of digitally mediated political contention. To advance such an understanding, this chapter employs the concept of "cycles of contention" – a concept helpful for investigating recurrent mechanism of protest in contemporary society. The study takes seven anti-petrochemical (anti-PX) environmental protests in China between 2007 and 2014 as cases. During the time, 54 in-depth interviews were conducted. The study finds that coverage in the traditional media, on the one hand, served to legitimize and to modularize these anti-PX protests, thereby facilitating the adoption of digital media as part of the repertoire of contention and helping to make political contention sustainable over the long run. The use of digital media, on the other hand, enabled protestors to diffuse contention widely and quickly and to learn from past experiences. The chapter concludes that, as a new challenge to urban governance, digitally mediated environmental activism is shaped by the specific communication ecology in China.

Keywords Digital media · Political activism · Cycles of contention · Repertoire of contention · Environmental activism · Urban governance · Anti-PX protests · China

---

This chapter is a revised version of "Digital Media, Cycle of Contention, and Sustainability of Environmental Activism – The case of anti-PX protests in China", which is published in *Mass Communication and Society* [19(5), 604–625]. An earlier version was presented at the conference "Governance for Urban Sustainability in China: Challenges and Practices", University of Copenhagen, 31 October–1 November 2015.

J. Liu (✉)
Department of Media, Cognition and Communication & Centre for Communication and Computing, University of Copenhagen, Copenhagen, Denmark
e-mail: liujun@hum.ku.dk

J. Delman et al. (eds.), *Greening China's Urban Governance*, ARI - Springer Asia Series 7, https://doi.org/10.1007/978-981-13-0740-9_9

At the end of March 2014, a message went viral both through the Internet and through mobile phone networks in China, calling for an environmental protest against a paraxylene ("PX" for short) project in Maoming City, in southeast China's Guangdong Province.[1] Following the diffusion of this message, thousands of residents took to the streets, holding signs protesting the proposed project. On-the-spot images and videos of protests proliferated on the Internet and on mobile social networks, as did expressions of concern about the implications of the project in terms of the environment and public health. Widespread news coverage ensued in both national and international media (e.g. BBC 2014).

The environmental protest in Maoming was more than just one of the thousands of political contentions in contemporary China that have caught public attention and manifested the increasing adoption of information and communication technologies (ICTs) as a means of protest. It was also the latest in a series of protests against PX projects in China between 2007 and 2014, following actions in Xiamen (Xie and Zhao 2007), Kunming (Chang 2013), and Shifang (FlorCruz 2012). In 2012, for example, a similar protest took place against a PX project in Ningbo, in southeast China's Zhejiang province (Liu and Yan 2012). In that case too, the wide and rapid diffusion of a message[2] via the Internet and mobile phone networks led to an environmental protest (which in this case lasted for 4 days and involved more than 5000 participants). Besides calling on recipients to spread the news further, the message explicitly recalled the earlier protests in Xiamen and Dalian. In short, earlier anti-PX protests have spawned subsequent ones, underpinning *continuity* in the urban movement against PX projects in China.

How to explain these protests – especially how to explain *the continuity* among these digitally mediated anti-PX protests? What are the factors, furthermore, that account for the continual adoption of digital media for environmental activism here? A number of thorough studies have documented the increasingly prominent role of digital media in political activism (e.g. Anduiza et al. 2012; Castells 2012; Earl and Kimport 2011; van de Donk et al. 2004; for a review, see Garrett 2006). However, most of this scholarship tends to focus either on the analysis of discrete contentious events (e.g. McCaughey and Ayers 2013; van de Donk et al. 2004) or on the comparative analysis of similar events (e.g. Bennett and Segerberg 2012; Howard and Hussain 2013). Such approaches fail to analyse possible interconnections among different instances of (digitally mediated) political activism (i.e. the anti-PX protests) or to consider their possible influence on urban governance. To advance such an understanding, I employ the concept of "cycles of contention"

[1] Paraxylene is a petrochemical feedstock used in plastics, polyester, and other synthetic manufacturing. A key part of the message read: "…Pollution: Xiamen residents do not want it! Kunming residents do not want it! Shifang residents do not want it!…We Maoming residents do not want it either!"

[2] The message read: "A PX project that has been forbidden in Xiamen and discarded in Dalian will be moved to Zhenhai, Ningbo, which is only 20 km away from the old city of Ningbo and 10 km away from the new city…PX is paraxylene, which easily leads to cancer and pathological changes of the reproductive system; please forward this message if you live in Ningbo!"

(Tarrow 1993a, 1993b) in this study – a concept well-suited for exploring *the recurrent mechanism* of environmental activism in contemporary urban society. I take the *recurrent* anti-PX protests in six Chinese cities between 2007 and 2014 as my case, in order to examine the factors contributing to the *emergence* and *sustainability* of *assimilative* digitally mediated environmental activism. I hope thereby to shed light on the enduring mechanisms of environmental activism in contemporary China and on the impact thereof on urban governance in that country.

I begin with a critical review of current studies of ICTs and political activism in China. Second, I develop the concept of *cycles of contention* as a theoretical framework for analysing the factors that shape the mechanisms of recurrence and sustain digitally mediated political contention. Third, I set out my selection of cases, my collection of data, and my methods of analysis. Fourth, I examine the mechanism that produces the recurrence of certain strategies of digitally mediated environmental activism in urban China. I then conclude with some thoughts on the mechanism that underpins the dynamics of digitally mediated environmental activism and sustains it beyond a single contentious event.

9.1 Digital Media and Political Activism in China: A Review

The impact of digital media on political activism has emerged as a prominent and enduring theme in studies of ICTs in China (e.g. Esarey and Xiao 2011; Yang 2009; Yang and Calhoun 2008; for critical reviews, see Herold and de Seta 2015, 75–76; Qiu and Bu 2013, 126). Nevertheless, research on this questions remains dominated by case studies of *discrete* – and in some cases *isolated* – contentious events, the emergence of which often reflects unique internal dynamics (e.g. Hassid 2012; Wu 2012; Yang and Calhoun 2008). Such an approach fails to explore possible *interconnections* among different cases of digitally mediated political activism or to recognize their *long-term* effects. In this sense, as Huang and Sun suggest, "…activism in China fails to meet the definition of a 'social movement' because it is usually *localized* and falls short of *sustained contention*" (2014, 87, emphasis added).

However, existing scholarship does point out that earlier political struggles affect, facilitate, or even legitimize subsequent contentious movements. This enables some such movements to gain a long-term influence over society. For instance, several studies on digitally mediated environmental activism indicate that earlier protests not only serve as the blueprint for later political movements; they also lend "legitimacy" to later struggles (Lee and Ho 2014, 34–35; Hu 2012, 104; Zeng 2015). Yang reminds readers of the "gradual revolution" which is taking place in China under the influence of online political activism. Already, he writes, "these popular political forms have…left their imprints on Chinese politics. The changes may be slow and piecemeal…but they add up and their cumulative effects are profound" (2011, 1044–1045). In this sense, failing to recognize the impact of early riser activism on subsequent movements, or the *interconnections* among movements and activism, leaves significant lacunae in our understanding of the sustained influence of political activism on Chinese society.

Students of social movements have long argued that contentious events are by no means discrete or mutually independent phenomena (McAdam 1995, 218, also see Minkoff 1997; Traugott 1995a). We must go beyond a single contentious event as our unit of analysis and instead canvass the *interconnections* among different political activism. The concept of "cycles of contention" can be useful here.

9.2 Cycles of Contention: Beyond the Single Contentious Event

Sidney Tarrow coined the term "cycles of contention" (1993a, b, 2010) to refer to recurrent protests and the long-term changes which they generate in collective action and the larger society (see also McAdam et al. 2001, 65–68; Traugott 1995b). A "cycle of contention" is:

> …a phase of heightened conflict across the social system, with rapid diffusion of collective action from more mobilized to less mobilized sectors, a rapid pace of innovation in the forms of contention employed, the creation of new or transformed collective action frames, a combination of organized and unorganized participation, and sequences of intensified information flow and interaction between challengers and authorities. (Tarrow 2011, 199)

Moreover, "cycles of contention" acts as a key term to explicate the translation of forms of collective action during *moments of contention* – or, as Zolberg (1972) describes them, "moments of madness" – into *the historical development* of means and methods for making claims and engaging in struggles in a society. These means and methods form the "repertoire of contention" – "…a limited set of routines that are *learned*, *shared*, and *acted out* through a relatively deliberate process of choice" (Tilly 1995, 42, emphasis added). The repertoire of contention is an array of methods for contentious claim-making and performance – methods which are always rooted in *prior* societal experience. People learn from the history and experience of contention, often finding "inspiration elsewhere in the ideas and tactics espoused and practiced by other activists" (McAdam and Rucht 1993, 58). The consequent borrowing or imitation leads to cycles wherein certain types of protest recur in a society.

A key contribution made by cycles of contention to what becomes the eventual repertoire of contention is the *modularization* of emerging forms of struggles, which facilitates the diffusion of these forms and ultimately leads to their translation into permanent changes in the repertoire of contention. The outcome of recurrent struggles influence the shape of subsequent ones by providing an array of possible tactics and, more importantly, "modular repertoires" (Tarrow 1993a; Tilly 1986, 6). These modular repertoires manifest themselves in forms of collective action used in a number of different conflicts by a range of social actors and against a variety of opponents (Tarrow 1993b, 77). In short, cycles of contention encompass models of collective action that enable a single instance of political activism to have an impact that transcends itself, nourishing collective challenges by subsequent movements.

One key element facilitating cycles of contention is the diffusion of contention (Tarrow 1993b, 82). Some studies portray the distribution of information about contentious events by organizations operating within established networks of communication as a precondition for diffusion (e.g. McAdam 1995, 232; Minkoff 1997). Others note the significance of the mass media for disseminating news about protests (e.g. Norris 2002, 208).

Nevertheless, current studies are largely based on protests that take place in Western democracies, and they fail to consider the distinctive shape taken by the diffusion of contention in highly controlled societies like that of China. The Chinese government is working hard to tighten state control over both traditional and digital media, and information on protests is severely suppressed (e.g. King et al. 2013). Meanwhile social organizations, such as environmental NGOs, have been missing from the process of political activism, in large part due to "...daunting political pressure, prevalent surveillance and [the] hidden constraints they endure on a daily basis" (Wu and Wen 2015, 113). How, then, can further cycles of contention develop in a highly controlled and politically sensitive context, where protest information is highly restricted and social organizations are under close government control? This context makes the Chinese case uniquely relevant for studying the *recurrent dynamics* of environmental activism mentioned at the beginning of this chapter. The overarching research question is accordingly: *What specific factors and processes account for the recurrent dynamics of environmental activism in China?*

More specifically, I address the following questions in this study: How do people get information about protests – including information about environmental activism (i.e. anti-PX protests) of the past and present? What are the factors contributing to people's perception of digital media as part of the repertoire of contention? What are the processes – e.g. diffusion of knowledge about protests – by which earlier movements facilitate follow-up activism, generate cycles of contention, and further shape urban governance in China?

9.3 Methods

In this study, I examine seven anti-PX protests in six cities between 2007 and 2014. Starting with the anti-PX protest in Xiamen in 2007, PX has become a common target for public protest and environmental activism in China. Since then, anti-PX protests have broken out in several cities, including Chengdu in 2008, Dalian in 2011, Ningbo in 2012, Kunming and Chengdu in 2013, and Maoming in 2014. In short, anti-PX protest has grown into one of the most prominent and *recurrent* forms of environmental activism in contemporary China.

The anti-PX protests have seen the appropriation of digital media as an indispensable means of environmental activism. As early as in 2007, during the anti-PX protest in Xiamen, people relied largely on mobile communication to organize their protests (Liu 2013). In subsequent actions, they have continued to exploit various digital media, including Internet forums, QQ, *Weibo*, text messaging, and WeChat.

They have employed such methods to share information, to plan protest actions, and to facilitate collective mobilization generally (e.g. Goldkorn 2008; Sevastopulo 2014). The anti-PX protests thus allow us to observe the integration of digital media into environmental activism and to assess their possible contribution to its sustainability.

I used snowball sampling to gather information on these cases, as well as in-depth interviews with protest participants. My purpose was to explore both the diffusion of protests over time and the perception of digital media as part of the repertoire of contention. More specifically, in order to access the hidden populations (Salganik and Heckathorn 2004) of protestors in these cases, I identified three to five protest participants in six cities and then asked them to specify three to five persons they had met during protests or from whom they had received protest information or to whom they had delivered such information. The purpose of this process was to reveal as many protest participants as possible. I then repeated the procedure with an eye to identifying and recruiting the remaining participants, until the process came to a halt of its own accord. This network-based sampling procedure guaranteed the privacy of interviewees and ensured that the profile of their protest participation would be kept low. Of the initial 85 participants contacted for the study, 54 agreed to be interviewed.[3] Fifty had participated in protests physically, while four had done so "virtually": i.e. they had not taken part locally but instead helped to distribute protest information via online forums and social networking sites (e.g. Facebook and *Weibo*).

I conducted semi-structured in-depth interviews, following the direction and general framework of the interview[4] while probing further and eliciting more detailed responses when clarification was necessary. Questions in the interview guide thus focused not only on the use of digital media in protests but also on how interviewees *perceived* digital media as a means of struggle, on what kind of experience they had had with such media that could be useful for carrying out protests, on where they had gained this experience, and on how such experience was *justified* and *validated*.

The interviews typically lasted 1.5 h. Of the 54 interviews, I conducted 48 face to face, having travelled to the cities where the interviewees resided. I carried out the remaining six via Gmail, due to a concern about government surveillance of social media and e-mail communications. I conducted all of the interviews in person

---

[3] The 54 interviewees had an average age of 35.7 years. Eighty-three percent (45 out of 54) had a bachelor's degree or higher. All were Internet users who were familiar with digital media in various forms. Professions among the interviewees encompassed a considerable range: journalist, editor, graduate student, high school student, lawyer, sales representative, consultant, university lecturer, taxi driver, IT professional, mobile phone salesperson, barber, and owner of a small clothing store.

[4] The interview framework includes (1) basic demographic information, (2) the availability of different types of media during activisms, (3) the information exchanged via digital media in protests, (4) the interpretation of both communication practices and information via digital media (e.g. how interviewees perceive the mobilizing message on digital devices, whether they follow it or not and why), and (5) how interviewees deal with messages (e.g. do they disseminate the message or respond to it, to whom do they forward the message, via which channels, and why).

and in Chinese, after which I transcribed them and then translated them into English. I only documented interviews where the interviewee was comfortable and anonymity could be guaranteed.

After collecting the data, I used thematic analysis and cross-case synthesis (Yin 2009, 18) to identify the factors contributing to the diffusion and recurrence of environmental protests.

9.4 Findings

In the next section, I present findings from the interviews about the diffusion of contention and the perception of digital media as a means of contention. As the results demonstrate, diffusion takes place through social- and group-based applications of digital media and subsequently through other digital media platforms. After that, traditional media such as newspapers and TV start to publish news about protests. Both traditional and digital media contribute to the diffusion of information about anti-PX protests (digital media especially), but they do so in different ways.

9.4.1 Diffusion of Contention via Digital Media

When both social organizations and the mass media are subject to highly repressive control, digital media act as a crucial channel through which people can convey and receive information about (1) politically sensitive issues (e.g. PX in this study), (2) collective mobilizations, and (3) collective actions of the past (which enables people to learn from earlier experiences).

According to the persons interviewed, social- and group-based applications of digital media, such as QQ, WeChat, and mobile phones, served as the primary source of relevant but politically sensitive information (in this case PX-related information). Fifty-one of 54 interviewees (94%) said they received information about the PX project from their social networks via mobile call, text messaging, QQ, or WeChat. The rest got their information via face-to-face interaction, as well as through their digitally mediated social networks. For instance, several participants in the Xiamen case recalled that, around 3 months before the protest, they started receiving "a warning text message regarding the PX project"[5] from their friends, colleagues, and relatives. (This was the very first time they had heard the word "PX".) The text message spread widely within a very short time. Against this backdrop, the phrase "Did you get the SMS?" reportedly became the opening remark when Xiamen residents met each other during this 3-month period (Zhu 2007). People got their information on the matter through digitally mediated social networks, and a wide-ranging discussion about the project began.

[5] For detailed information on the text message, see Xie and Zhao 2007.

If Xiamen residents largely relied on text messaging to receive and spread PX information, later protestors adopted various social network-based digital media for the same purpose. In Kunming, Chengdu, and Maoming, residents mostly used QQ and WeChat to disseminate PX information, to inform their social networks, and to initiate discussions – even when the traditional media did not cover the issue at all. In this sense, people no longer had to depend on the traditional media as a key information source.

Second, in addition to diffusing PX-related information, people spread mobilizing messages for a "stroll (*sanbu*)", a euphemism used by Chinese people for street protests. They used social- and group-based applications of digital media to do this. In the case of Xiamen, the mobilizing message had been disseminated via mobile communication. Similarly, as we saw at the beginning of this chapter, the messages calling for protests in Maoming and Ningbo proliferated via digital media, which turned into a key part of the mechanism of mobilization.

More specifically, the easy-to-use quality of digital media allowed people to invite members of their social networks to collective actions without much effort. A 23-year-old graduate involved in the Kunming mobilization said that she had forwarded the message via WeChat to her relatives and her entire school class (a total of 75 people) by "…just twiddling my thumb". In Xiamen, the anti-PX movement reached its zenith as "millions of Xiamen residents frenziedly forward[ed] the same [mobilizing] text message" (Lan and Zhang 2007), urging each other to join a street protest opposing the government's PX project. Interviewees who took part in other protests confirmed that they too had forwarded such mobilizing messages to social groups with which they were affiliated. They did so by "copying", "pasting", and "sending" the message or simply by pressing the "forward" or "retweet" button. To summarize, easy-to-use digital devices with low fees for the conveyance of information became the key facilitator of protest mobilization – be it through text messaging, QQ, or WeChat.

Third, after receiving the message from their digitally mediated social networks, people moved on to search engines, microblogging sites, and online forums to look for more information about PX. These digital platforms became major channels through which they could acquire more detailed information about the issues they were concerned with (i.e. PX) and more importantly through which they could retrieve information about previous protests on such issues (enabling them to learn from earlier events). Interviewees reported finding a lot of information about anti-PX protests in other cities as soon as they searched with keywords like "PX" via digital media. They were then able to read about these past experiences and to learn from them. A 33-year-old from Ningbo, owner of a small clothing store, described the situation thus:

> …You can find web pages, online forums, and tweets, including photos and videos of the anti-PX protest in Xiamen, the very first one in 2007. This information expands our awareness about the PX project. Most importantly, it shows us how we can adapt in order to stop the project.

While the protest in Xiamen remained "the most renowned anti-PX protest", according to a 27-year-old mobile phone saleswoman from Chengdu, news about later protests in Kunming, Dalian, and so on could be easily retrieved (despite censorship) from web pages, microblogging sites, blogs, and online forums. She explained that:

> …after checking the internet, we saw several anti-PX protests around the nation. If the project is not toxic or problematic [as the government claimed], why would people in other cities oppose it? Why do we need a project that people elsewhere discarded? If they [people in other cities] succeeded in forcing the government to give up the project by carrying out protests, we should do the same!

Although such reasoning did not convert into offline protest action *directly*, it did encourage people to pass information on within their social networks via digital media. In this way, information about other anti-PX protests proliferated within digitally mediated social networks, enabling many people to learn from experiences elsewhere.

To sum up, social- and group-based digital media, web pages, online forums, and microblogs became a crucial channel for the diffusion of protest information. This included both mobilizing messages and information about past protests. Furthermore, social- and group-based applications of digital media enabled people to convey information of an urgent kind – information demanding immediate attention and action – extremely quickly and widely. Meanwhile web pages, online forums, and microblogs provided abundant information, especially about previous protests. Such information was critical in shaping people's perceptions of the PX project and in laying the foundations for further contention.

9.4.2 Coverage and Diffusion of Contention via Traditional Media

People also read about the PX project, the ensuing debate, and anti-PX protests in the traditional media, such as newspapers and TV. In general, such organs tried to persuade the public to accept the PX project, by citing professional opinion or by claiming that ill-intentioned protesters had falsified the data or exaggerated the toxicity of PX in order to mislead the public. But campaigns of this kind failed to change attitudes towards the PX project. Instead they increased general knowledge about the protests, highlighted successful examples that could be learned from and followed, and encouraged the adoption of digital media for protest purposes.

To begin with, coverage in the traditional media acted as a *catalyst* igniting widespread contention against the project, as people perceived the coverage as meaning both that censorship on the topic was being removed and that the danger of the PX project was being admitted by the authorities. In practice, PX-related news did not appear in the traditional media until the information had already proliferated within social- and group-based digital media. For instance, local mass media in Xiamen only started covering the PX project after controversy about it had spread via mobile phones for about 3 months. Meanwhile, the authorities worked hard from the beginning to censor sensitive words like "PX" and "petrochemical", shutting down online forums that featured arguments against the project and blocking access to overseas coverage of the issue. Nevertheless, faced with the spread of PX-related news via digital media, as well as coverage by international news organs, the traditional media in China also started covering the debate. This significant change was perceived as meaning that censorship of the debate over PX had ceased. The interviews showed that people believed that, following the open coverage of the issue by the traditional media, they were allowed to argue against the PX project *publicly*. A 38-year-old IT professional said that "…even the government had to admit the damaging nature of the PX project in the [mass] media! Why can't we oppose it?"

In addition, coverage of the issue by the traditional media encouraged people to learn from and to imitate earlier collective actions against the project – particularly the successful "stroll" in Xiamen. According to the persons interviewed, anti-PX protests were covered in the *People's Daily* (the official newspaper of the Communist Party of China), in *China Newsweek* (which is published by China News Service, the second-largest state news agency), and in national and regional newspapers such as *Southern Metropolis Daily*. People perceived this coverage as implying that "… the government had accepted 'the stroll,' or street protest, as a *legit* way of opposing the PX project" (in the words of a 36-year-old taxi driver from Chengdu, emphasis added). The protest in Xiamen became not just a benchmark for anti-PX protests generally but also *a successful* and *politically accepted* example for people to learn from and to follow in later cases.

Finally, and most importantly, coverage by the traditional media played a key role in encouraging the adoption of digital media for protests. The way the traditional media covered anti-PX protests underlined the importance of digital media for enabling people to learn from past examples and helping them initiate and organize protests. Several interviewees, for instance, mentioned a report by *China Newsweek* entitled "The Power of Mobile Messaging" (Xie and Zhao 2007). This report detailed how Xiamen residents used text messaging to organize their anti-PX protest. People interpreted this report as meaning that the authorities had given *tacit permission* for the use of mobile phones for *organizing protests*. A 36-year-old accountant from Chengdu recalled that:

> …even the authorities acknowledged *the power of mobile phones* for successful protest. We can just *copy* the "stroll" in Xiamen by using our mobile phones to distribute protest information and to organize similar protests against the PX project. This successful example [of using digital media such as mobile phones] shows *a way to oppose the project which is recognized by the government* [emphasis added].

Similarly, in later coverage of the anti-PX protests, the traditional media pointed out that various digital media were being used as a means of protest. Such coverage became a major driving force inspiring and encouraging the "replication" of successful past examples involving the use of digital media for protest purposes.

To sum up, the traditional media played a key role in spreading knowledge about past protests. This served to facilitate digitally mediated protests against the PX project at later stages as well. While *none* of these reports said that protests against the PX project were legal, coverage by the traditional media was taken to mean that censorship of opposition to PX had ceased – that the government had given a go-ahead signal for protest activity. Similarly, the way these reports covered the protests played a crucial role in shaping attitudes and actions on the issue. People took this coverage as an encouragement to follow "the successful model" of the past, i.e. to adopt digital media for purposes of political contention.

9.5 Discussion

In a series of recurring protests against PX projects between 2007 and 2014, both digital and traditional media in China facilitated the diffusion of contention and contributed to the mechanism of recurrence. They did so, however, in two distinct ways. In this section, I explore the different contributions of digital and traditional media in terms of the diffusion, legitimation, and modularization of contention. I then discuss digital media as a repertoire of contention and the anti-PX protests as an example of the sustainability of environmental activism in contemporary China.

9.5.1 Digital Media, Diffusion of Contention, and Mass Self-Mobilization

Due to its fear of the contagion effect of protests, the Chinese government works hard to detect and eradicate information on collective action as it emerges. Against this backdrop, digital media serve as a key resource against censorship and for the diffusion of contention. More specifically, the unique advantages of digital media – availability, accessibility, and affordability – facilitate the proliferation of contention information outside the traditional media, generating a process of what I call "mass self-mobilization".

Low-end ICTs have become a simple yet central necessity for everyday life. Digital media have thus been adopted as a quotidian tool for political activism. In 2014, the population of mobile phone users in China surpassed 1.2 billion, or 90.8% of the country's total population (CNNIC 2014). WeChat, the mobile messaging app, had over 500 million active users each month. There were more than 290 million *Weibo* users in 2014, accounting for 45.9% of the total number (632 million) of Internet users in China (CNNIC 2014). Due to the authoritarian conditions in the country, resources and institutional support for political contention are lacking, prompting people to rely largely on digital media – *a mundane communication tool* – when undertaking collective action (e.g. Yang 2009). Similarly, as the cases examined here show, the Chinese public resorts to digital media as its most readily available resource for distributing information and organizing protests.

Low cost and ease of use contribute to the accessibility and affordability of digital media. In particular, social- and group-based applications of digital media permit the instantaneous dissemination of protest information, thereby greatly facilitating collective mobilization. These advantages make digital media a key resource both for informing and for uniting individuals, including those with little technological know-how. Digital media allow people to spread politically sensitive messages, to initiate collective actions, and to resist the authorities with a simple method – e.g. twiddling their thumbs on a mobile phone. They make it possible to release contentious information at the flick of a button, be it through mobile phones or personal computers. This enables contentious efforts to transcend geographical boundaries, facilitating their large-scale impact. If a protest gets international news coverage, for example, the government will be unable to cover up either the contested issue (i.e. the PX project) or the politically sensitive development (i.e. the protest).

Furthermore, digital media convey calls to act on the basis of social networks; in other words, they incorporate social relationships as the agent of mobilization. The latter, in turn, becomes the driving force in both information dissemination and collective mobilization. As has been observed elsewhere (e.g. Bennett and Segerberg 2012; Liu 2015), digital technology activates social networks for political contention. In short, digital media lower the threshold for initiating, promoting, supporting, and engaging in political action.

These advantages contribute to a process of "mass self-mobilization" – a term I have adapted from Castells' phrase "mass self-communication" (2007, 248). This process involves *quasi*-mass communication on the basis of horizontal networks *outside* the traditional mass media and *without* an organizational foundation in the usual sense. In the case of the anti-PX protests, digital media first enabled people to distribute protest information and mobilizing messages *without* an organizational basis. Then, in the next stage, they facilitated collective action and mobilization. "Self-mobilization" thus involves one further step than does "self-communication", since it refers to the process by which individuals, largely through their digital media, *mobilize* others within their social networks. This process occurs when large numbers of people circulate mobilizing messages spontaneously within their social networks. Thus, digital media drive *mass mobilization* through interpersonal interactive practices even in the absence of social organizations, transforming a means of interpersonal interaction into a mass-mobilizing tool (see also Liu 2015). This point is relevant for understanding digitally mediated processes of collective action in the absence of social organizations, including social movement organizations, under authoritarian regimes specifically.

9.5.2 Traditional Media and the Legitimation and Modularization of Contention

In a repressive political environment such as that in China, political contention normally entails great risk and provokes harsh repression by the state (e.g. O'Brien 2009). Against this backdrop, the mere diffusion of contention – the simple spread of knowledge about a protest – is not enough to inspire political activism. Instead, people need to exploit openings in the official ideology. For example, through "… the innovative use of laws, policies, and other officially promoted values" (O'Brien

1996, 32), they can *legitimate* their resistance and protest (see also O'Brien and Li 2006). In the case of the anti-PX protests, digital media facilitated the diffusion and mobilization of contention, but it was the traditional official media that *legitimized* and *modularized* the contention, i.e. allowed certain types of protest repertoire to be transferred or to be employed across different contexts. The result was to encourage an assimilation of past protest and to promote a challenge to prevailing patterns of urban governance in the long run.

Coverage in the traditional media served to increase the influence of the anti-PX protests, by legitimizing them and thus encouraging their widespread duplication. The government censored PX-related information to begin with but then removed the censorship and allowed the official media to cover the issue, including the protests. To the average person, coverage of the anti-PX protests in the traditional media implied that controversy concerning the PX project was no longer *politically taboo*. Instead the authorities had recognized its existence and allowed people to discuss the issue openly. More importantly, the government had accepted protests as *a legitimate response* to the PX project. In other words, the anti-PX protests had been recognized as "a form of officially legitimate public action" (Thireau and Linshan 2003, 87). This "schemata of interpretations" (Zhao 2010, 42) among the public expanded *political opportunities for contention* and encouraged others elsewhere to adopt the same type of collective action against PX projects.

It is also necessary to address the anti-PX protests as a specific form of activism in contemporary China. Such protests – and environmental activism generally – are tolerated by the authorities because they largely concern environmental issues and do not challenge the authoritarian system (see also Zhao 2010, 45). This fundamental fact serves to lower the political risks involved – both in connection with such protests themselves and in connection with coverage thereof in the traditional media.

In addition, the traditional media highlighted the use of digital media in protests. This accounted for the modularity of anti-PX activism and encouraged imitation by others at a later stage. Many of the persons interviewed, for instance, recalled clearly that *China Newsweek* had dedicated its cover story to the use of text messaging in the anti-PX protest in Xiamen. Reports described in detail how people used their mobile phones to distribute "millions of messages" for the mobilization of protest (Xie and Zhao 2007; Zhu 2007). Similarly, later reports called attention to how people employed various digital media, such as QQ, online forums, and WeChat, to organize protests successfully (The Center for Public Opinion Monitor 2014; Zeng 2015, 112–113). The adoption of digital media was portrayed in traditional news organs as a unique and successful aspect of these protests. As news about this form of digitally mediated collective action spread, it generated a demonstration effect. According to the persons interviewed, the anti-PX protests became a typical and, more importantly, successful model for digital activism (or digitally mediated public engagement). This model has been followed in many struggles against the authorities, who have been forced in some cases to change their decisions. In this way, the coverage by the traditional media *modularized* the digitally mediated anti-PX protests as *a successful example* of protest organization and as *an established pattern* of political engagement in urban areas. The modularity of the anti-PX protests resulted in the establishment of patterns of digitally mediated contentious behaviour that could "...be learned, adapted, routinized, and diffused from one group, one locale, or one moment to another" (Traugott 1995a, 7). The tactics, strategies, and organizational means employed in earlier anti-PX protests proved useful in later struggles. Advantages included the mundaneness of the means of protest (i.e. digital media), the rapidity of transmission, and the low costs of mobilization that such methods made possible. Thus, coverage by the traditional media allowed people to adopt modes of contention quite easily – to learn from past experiences and to duplicate them – and furthered the popular perception of *digital media as a repertoire of contention*, especially in connection with environmental activism.

9.5.3 Digital Media as a Repertoire of Contention and the Sustainability of Anti-PX Protests

Repertoires of contention involve "...not only what people *do* when they are engaged in conflict with others but *what they know how to do* and what others expect them to do" (Tarrow 2011, 39, emphasis in original). Such repertoires spring from experience with varied means of contention in the past. For instance, a key contribution by traditional media to the recurrent mechanism of anti-PX protests was to establish *the perception of digital media as a repertoire of contention*. More specifically, the emphasis on digital media, be it text messaging or WeChat, in reports in the traditional media gave people the impression that anti-PX protests and the use of digital media were bound up together. As the interviews made clear, coverage of protests by the traditional media established a framework in which digital media – as a flexible and well-understood means for making claims and opposing government policy – became a key component of these repertoires of contention. Coverage of anti-PX protests in the traditional media supplied the "guideline" for people to duplicate previous protests, by using digital media to engage members of their social network and to spread political contention. To the public, "the power of mobile messaging" is a slogan to encourage individuals to employ their digital media (e.g. mobile phones) for protests. In this way, the traditional media played a key role in establishing digital media as a repertoire of contention and in setting the recurrent mechanism of digital activism into motion.

9.6 Conclusion

Most current studies on political activism focus on a singular contentious event. In this study, by contrast, I underline the interconnections among different instances of contention. Utilizing the concept of "cycles of contention", I investigate the recurrent mechanism of anti-PX protests in cities in contemporary China. I argue that digital media play a key role in informing people about contentious questions. More specifically, the advantages of digital media, such as ease of use, immediate transmission, and the use of embedded social networks, generate a process of *mass self-mobilization*. In this process, people are able to disseminate protest information quickly and widely and to engage rapidly in collective action. Past experiences of collective action also become accessible to the public, who can then – censorship notwithstanding – learn from and duplicate them.

Meanwhile, and more importantly, it is the traditional media that legitimize contention, thereby encouraging the emergence of repeated patterns of collective action. Indeed, coverage of contention by the official media not only legitimizes contention; it also modularizes it, further encouraging the assimilation of protests. The traditional media play a crucial role in shaping perceptions of digital media as a repertoire of contention, which in turn increases the sustainability of digitally mediated political contention. Following Lund's discussion in this volume, I show in this study that environmental activism and urban governance form a complex, interrelated system. In this sense, the intertwining of digital and traditional media generates the recurrent dynamics of contention considered here, thereby sustaining the long-term influence of digitally mediated environmental activism in China.

Acknowledgements The author deeply appreciates comments from Klaus Bruhn Jensen, Jørgen Delman, Mattias Burell, Oscar Almén, and Ran Wei. This study has been funded by grants from the Carlsberg Foundation (CF14–0385), S. C. Van Fonden (reference numbers 1267, 1503), and Zhejiang Provincial Philosophy & Social Science Research Base (15JDCB05YB).

References

Anduiza, E., Jensen, M. J., & Jorba, L. (2012). *Digital media and political engagement worldwide*. New York: Cambridge University Press.

BBC. (2014, April 2). *China Maoming environmental protest violence condemned*. Available via BBC. http://www.bbc.com/news/world-asia-china-26849814. Accessed 22 Sep 2016.

Bennett, W. L., & Segerberg, A. (2012). The logic of connective action. *Information, Communication & Society, 15*(5), 739–768.

Castells, M. (2007). Communication, power and counter-power in the network society. *International Journal of Communication, 1*, 238–266.

Castells, M. (2012). *Networks of outrage and hope*. Cambridge: Polity.

Chang, M. (2013, May 17). *Thousands protest Kunming PX plan*. Available via Global Times. http://www.globaltimes.cn/content/782252.shtml. Accessed 22 Sep 2016

CNNIC. (2014, July). *The 34th statistics report on internet development in China (July 2014)*. Available via CNNIC. http://www1.cnnic.cn/IDR/ReportDownloads/201411/P020141102574314897888.pdf. Accessed 22 Sep 2016

Earl, J., & Kimport, K. (2011). *Digitally enabled social change*. Cambridge: The MIT Press.

Esarey, A., & Xiao, Q. (2011). Digital communication and political change in China. *International Journal of Communication, 5*, 298–319.

FlorCruz, J. A. (2012, July 6). *People power a sign of times in China's internet age*. Available via CNN. http://edition.cnn.com/2012/07/06/world/asia/china-shifang-protest-florcruz/. Accessed 22 Sep 2016.
Garrett, R. K. (2006). Protest in an information society. *Information, Communication and Society, 9*(2), 202–224.
Goldkorn, J. (2008, May 6). *NIMBY protest hits Chengdu*. Available via DANWEI. http://www.danwei.org/environmental_problems/nimby_protest_hits_chengdu.php. Accessed 22 Sep 2016.
Hassid, J. (2012). Safety valve or pressure cooker. *Journal of Communication, 62*(2), 212–230.
Herold, D. K., & de Seta, G. (2015). Through the looking glass. *The Information Society, 31*(1), 68–82.
Howard, P. N., & Hussain, M. M. (2013). *Democracy's fourth wave?* New York: Oxford University Press.
Hu, Y. (2012). The internet and social mobilization in China. In L. FLF, L. Leung, J. L. Qiu, & C. DSC (Eds.), *Frontiers in new media research* (pp. 93–114). London: Routledge.
Huang, R., & Sun, X. (2014). Weibo network, information diffusion and implications for collective action in China. *Information, Communication & Society, 17*(1), 86–104.
King, G., Pan, J., & Roberts, M. E. (2013). How censorship in China allows government criticism but silences collective expression. *American Political Science Review, 107*(2), 1–18.
Lan, Y., & Zhang, Y. (2007, May 29) *Millions of Xiamen residents spread crazily the same SMS against high-pollution project*. Available via Southern Metropolis Daily. http://news.hsw.cn/2007-05/29/content_6311487.htm. Accessed 22 Sep 2016.
Lee, K., & Ho, M. (2014). The Maoming anti-PX protest of 2014. *China Perspectives, 3*, 33–39.
Liu, J. (2013). Mobile communication, popular protests and citizenship in China. *Modern Asian Studies, 47*(3), 995–1018.
Liu, J. (2015). Communicating beyond information. *Television & New Media, 16*(6), 503–520.
Liu, D., & Yan, S. (2012, October 29). *Ningbo backs down from PX project*. Available via Global Times. http://www.globaltimes.cn/content/740943.shtml. Accessed 22 Sep 2016.
McAdam, D. (1995). "Initiator" and "Spin-off" movements. In M. Traugott (Ed.), *Repertoires and cycles of collective action* (pp. 217–239). Durham: Duke University Press.
McAdam, D., & Rucht, D. (1993). The cross-national diffusion of movement ideas. *The Annals of the American Academy of Political and Social Science, 528*(1), 56–74.
McAdam, D., Tarrow, S. G., & Tilly, C. (2001). *Dynamics of contention*. Cambridge: Cambridge University Press.
McCaughey, M., & Ayers, M. D. (2013). *Cyberactivism*. London: Routledge.
Minkoff, D. C. (1997). The sequencing of social movements. *American Sociological Review, 62*(5), 779–799.
Norris, P. (2002). *Democratic phoenix*. Cambridge: Cambridge University Press.
O'Brien, K. J. (1996). Rightful resistance. *World Politics, 49*(1), 31–55.
O'Brien, K. J. (2009). *Popular protest in China*. Cambridge: Harvard University Press.
O'Brien, K. J., & Li, L. (2006). *Rightful resistance in rural China*. New York: Cambridge University Press.
Qiu, J. L., & Bu, W. (2013). China ICT studies. *China Review, 13*(2), 123–152.
Salganik, M. J., & Heckathorn, D. D. (2004). Sampling and estimation in hidden populations using Respondent-Driven Sampling. *Sociological Methodology, 34*(1), 193–239.
Sevastopulo, D. (2014, April 1). China PX plant protests spread to provincial capital Guangzhou. *Financial Time*. Available via http://www.ft.com/intl/cms/s/0/cf6594fa-b98d-11e3-957a-00144feabdc0.html - axzz3ZeGWgZ3k. Accessed 22 Sept 2016.
Tarrow, S. G. (1993a). Cycles of collective action. *Social Science History, 17*(2), 281–307.
Tarrow, S. G. (1993b). Modular collective action and the rise of the social movement. *Politics & Society, 21*(1), 69–90.
Tarrow, S. G. (2010). Dynamics of diffusion. In R. K. Givan, K. M. Roberts, & S. A. Soule (Eds.), *The diffusion of social movements* (pp. 204–219). Cambridge: Cambridge University Press.

Tarrow, S. G. (2011). *Power in movement* (3rd ed.). New York: Cambridge University Press.
The Center for Public Opinion Monitor. (2014). To guide the public opinion toward PX project in the Maoming protest. *Public Opinion on Politics and Law, 114*(14), 3–8.
Thireau, I., & Linshan, H. (2003). The moral universe of aggrieved Chinese workers. *The China Journal, 50*, 83–103.
Tilly, C. (1986). *The contentious French*. Cambridge: Harvard University Press.
Tilly, C. (1995). *Popular contention in Great Britain* (pp. 1758–1834). Cambridge: Harvard University Press.
Traugott, M. (1995a). Recurrent patterns of collective action. In M. Traugott (Ed.), *Repertoires and cycles of collective action* (pp. 1–14). Durham: Duke University Press.
Traugott, M. (1995b). *Repertoires and cycles of collective action*. Durham: Duke University Press.
van de Donk, W., Loader, B. D., Nixon, P. G., & Rucht, D. (Eds.). (2004). *Cyberprotest*. London: Routledge.
Wu, C. (2012). Micro-blog and the speech act of China's middle class. *Javnost-the Public, 19*(2), 43–62.
Wu, F., & Wen, B. (2015). Nongovernmental organizations and environmental protests. In P. G. Harris & G. Lang (Eds.), *Routledge handbook of environment and society in Asia* (pp. 105–119). London: Routledge.
Xie, L., & Zhao, L. (2007). The power of mobile messaging. *China Newsweek, 326*(20), 16–17.
Yang, G. (2009). *The power of the Internet in China*. New York: Columbia University Press.
Yang, G. (2011). Technology and its contents. *The Journal of Asian Studies, 70*(4), 1043–1050.
Yang, G., & Calhoun, C. (2008). Media, power, and protest in China. *Harvard Asia Pacific Review, 9*(2), 9–13.
Yin, R. (2009). *Case study research* (4th ed.). London: Sage.
Zeng, F. (2015). 环境抗争的扩散效应 (Diffusion effects of Chinese environmental contentions). *Journal of Northwest Normal University, 52*(3), 110–115.
Zhao, D. (2010). Theorizing the role of culture in social movements. *Social Movement Studies, 9*(1), 33–50.
Zhu, H. (2007, May 30). 百亿化工项目引发剧毒传闻 厦门果断叫停应对公共危机(Xiamen calls an abrupt halt to the million-cost petrochemical project to deal with the public crisis). Southern Weekend, p. A1.
Zolberg, A. R. (1972). Moments of Madness. *Politics & Society, 2*(2), 183–207.

Chapter 10
The Role of Social Protests in Environmental Governance in Hangzhou

Mattias Burell and Oscar Almén

Abstract This chapter examines the consequences of social protests for environmental governance in China. Environmental protection is a policy area where the party state has allowed civil society to play an increasingly important role. Chinese citizens can take part in state-sanctioned forms of political participation, but one of the most important ways they can influence policy is by protesting. If we are to understand environmental governance in China, we must study the impact of such protests. Students of social movements in China have made great efforts to understand what explains protest success in that country. The authors contend that, in order to understand environmental governance, we must look at all the different outcomes of a social protest. In most cases a protest elicits at least some concessions from the authorities, even when the protesters themselves perceive it as a failure. This study examines urban environmental governance by looking at the varying outcomes of three different waves of environmental protest in Hangzhou. Protests may lead to the development of new institutional tools such as dialogue platforms, or involve external actors like courts or NGOs. Through these mechanisms, citizen protests form a new and important element in China's fragmented governance system.

Keywords Environmental protests · Social movements · Waste incineration · Environmental governance · Hangzhou · NGOs

10.1 Introduction

This book is concerned with understanding urban sustainability governance in China. Sustainability can refer to environmentally friendly practices like the reduction of pollution, as well as to citizens' quality of life and economic viability in the

M. Burell (✉)
Hankuk University of Foreign Studies, Seoul, South Korea (Republic of)

O. Almén
Uppsala University, Uppsala, Sweden

J. Delman et al. (eds.), *Greening China's Urban Governance*, ARI - Springer Asia Series 7, https://doi.org/10.1007/978-981-13-0740-9_10

community. However, sustainability can also have social and political dimensions. In that sense, it signifies a constructive relationship between civil society and local political authorities. Sustainability, in other words, can denote both environmental and political outcomes. Such environmental and political outcomes can be either sustainable or not, which in turn is often linked to prevailing patterns of local governance. Some patterns of governance involve adapting to and dealing with new challenges, but others do not. The resulting feedback loops can have an effect on environmental and political sustainability.

In the social sciences, governance theory has developed as a reaction to the limitations of state-centric approaches to explaining modern governing processes and as an attempt to overcome the dichotomy between state and society (Pierre and Peters 2000; Bellamy and Palumbo 2010). As Lund (Chap. 1) discusses in this volume, three governance paradigms prevail in Western democracies: traditional public administration, new public management, and network governance. Although some aspects of all three paradigms can be found in China too, the Chinese case does not fit easily into any specific paradigm. The most obvious difference lies in the lack of democratic mechanisms in China. Instead, Chinese governance has typically been labelled as some hybrid form of authoritarianism. Examples of such labels include the influential *fragmented authoritarianism* model first described by Lieberthal and Oksenberg (1988) and later revisited by other scholars (Mertha 2009). Similar concepts such as *consultative authoritarianism* (Harding 1987; He and Thøgersen 2010; Teets 2013) or *consultative Leninism* (Tsang 2009) emphasize the importance of including a plurality of actors and allowing greater leeway for grassroots participation. This way, civil society forms a part of the policy-making process – but within the limits of an authoritarian system. These labels distinguish the Chinese governance model from its Western counterparts, but they also involve a departure from traditional simplifications that stress top-down authoritarian rule. In this chapter, we examine the Chinese governance model in environmental protection from a bottom-up perspective. Specifically, we study how grassroots social protests become a part of local environmental governance. We argue that the recurrent interplay between social protests and government responses contributes to new governance patterns which may be more sustainable – both environmentally and politically – over the long run.

The dominance of the Communist Party and the state sector in China has traditionally left limited space for society to be involved in governance. However, relationships in this area are changing in China too, notwithstanding the pronounced state-centrism in that country. These changes are perhaps more obvious in environmental governance than in other policy sectors. After all, the state has recognized the importance of environmental non-governmental organizations (ENGOs), and it has officially encouraged citizens to play a greater role in this area. Public participation in environmental protection is a recurrent slogan, and the number of ENGOs is increasing. According to a review in 2013 by the China Development Brief, ENGOs were the largest group among Chinese NGOs at the time (Shie and Brown-Inz 2013,14). Official calls for citizen participation in environmental governance are issued repeatedly, and local authorities are seeking new methods to involve local citizens in these issues

(Kostka and Mol 2013). In recent years the political space for NGOs has decreased, and ENGOs have also been affected by this. However, many grassroot ENGOs still find ways to influence policy-making at the local level (Dai and Spires 2017).

Citizens can take part in state-sanctioned forms of society-based environmental governance. However, another important way they can influence policy is by protesting. Social protests like collective petitions and demonstrations are common and have proven fairly effective for expressing grievances and getting them addressed. According to the common expectation, authoritarian regimes respond to protests with repression (Koopmans 2007, 39). Yet Chinese social protests often result in government concessions and policy adjustments (Cai 2010; Tong and Lei 2010). If we are to understand environmental governance in China, then, we must study the role and impact of environmentally related protests. Students of social movements in China have made great efforts to understand what explains the varying levels of protest success (Cai 2010). We contend that, in order to understand environmental governance, we must look at *all* the different consequences of a social protest. We cannot restrict ourselves to examining the obviously successful cases. One could certainly argue that, when citizens protest and the government makes some concessions, this entails a shift in governance or at least an adjustment in policy. However, even when social protests are met with resistance from political authorities, the simple fact that there is a tense, recurrent, and significant interaction between social actors and the government may contribute directly or indirectly to changes in the pattern of governance. Previous research has shown, for example, that protests in China sometimes result in improved policy implementation (O'Brien and Li 2005) and that disorganized popular contention can prompt the state to create case-by-case mechanisms for the resolution of conflict (Chen and Kang 2016). Even social protests which look like failures from a citizen perspective may have a positive long-term effect on local governance in the form of policy learning, initiatives for increased consultation with the public, greater government anticipation of negative social reactions to certain policies, etc. We must accordingly dispense with the idea that only successful social protests – for instance, in the area of environmental protection – have an impact on local governance.

In this study, we examine sustainable local governance by looking at the varying outcomes of three different waves of environmental protest in Hangzhou. We define sustainable governance as *the coordinating mechanisms and processes used by state and societal actors to steer environmental development*. We ask the question: how do social protests affect the sustainability of local environmental governance?

The study is based on 21 in-depth interviews with local officials, NGO leaders, activists, and scholars conducted between 2009 and 2015 but mainly in 2013–2014 (see list of interviews). In addition, we have studied documents and media reports in relation to the cases to triangulate this data with the information given by informants. Case 1 was first introduced to us in an interview with the Hangzhou EPB, while cases 2 and 3 were selected based on media reports. We conducted the interviews personally in Chinese, sometimes with a Chinese research assistant and sometimes alone.

10.2 The Role of Social Protest in Chinese Governance

Protest performs an important function in the Chinese political system. It puts pressure on local authorities from below to do their job. The political careers of local leaders are managed through the so-called cadre responsibility system, which evaluates the performance of all party cadres in China in relation to policy targets such as economic growth, family planning, social stability, and the reduction of environmental pollution (Whiting 2001; Edin 2003; Heberer and Trappel 2013; Gao 2015). A violent protest in any given policy area means that the responsible cadres will not be eligible for promotion. Thus, there are strong incentives for them to ensure that protests do not get out of hand. Hence, social protests have direct consequences for the careers of local leaders, and they can lead indirectly to sanctions against local cadres. In the Chinese context, we contend the concept of social accountability (Peruzzotti and Smulowitz 2006) can help us understand the feedback mechanism between social protests and the monitoring and sanctioning of local officials.[1] While vertical and horizontal accountability mechanisms, such as elections and legislative oversight, are weak in China, informal channels such as protest and pressure from traditional and social media can play an important role. This is because they can plug into the machinery whereby cadres are evaluated, rewarded, and sanctioned by higher levels of the party state. Of course – and this is important – such access to the machinery in question only works indirectly, tenuously, and with the approval and oversight of higher political authorities.

In fact, it can be argued that the central government uses social protest as a way of monitoring the behaviour of local officials to ensure that they actually do their job. Citizen grievances can be a sign of bad governance, which in turn threatens the legitimacy of the political system. While the central authorities do not actively encourage protests, neither do they suppress them. As much as possible, the party state tries to uncover the root causes of the protest and to fix the problem, in order to secure its own legitimacy. In other words, protests perform an important signalling function in the Chinese governance system. They can ring the alarm that something is seriously wrong and that public policies or their specific implementation may need revision and adjustment. Alternatively, they may signal that something has gone badly wrong in the local administration and that this needs to be examined further. In other words, while social protests are almost always bad news for local authorities, they may actually entail some governance benefits for the central government. Environmental protests are also more likely to be treated leniently than protests on other issues, due to the straightforwardness of the moral issues involved. The environment is something around which most Chinese can unite – in contrast to other policy areas, where the moral rights and wrongs are hard to distinguish. Most

[1] In another paper (Almén and Burrell forthcoming), we analyse various social protests in Hangzhou from a social accountability perspective.

political leaders and state officials would hesitate to defend a polluting factory or to ignore the obvious worries and anger of victims of environmental disasters.

Earlier research on social protest in China has shown that local authorities tend to meet citizen protest and resistance with some concessions, in order to avoid social unrest and prevent intervention from higher-level governments (Cai 2008a; Chen 2009). At other times, local leaders may consider the economic or political costs of yielding to citizen demands (*concession costs*) to be too high and decide to suppress the movement in question (Cai 2008b, 2010). Observers agree, however, that irrespective of whether repression or concession is chosen, contentious politics in China consist of an interplay between citizen actions and responses by local authorities (Diamant et al. 2005; Lee and Zhang 2013; Deng and O'Brien 2014). In other words, the conflict usually takes the form of reciprocal and strategic actions and counteractions by citizen groups on the one hand and local authorities on the other. Such conflicts also often involve other actors, such as lawyers, the mass media, and various NGOs (Fu and Cullen 2008; Ru and Ortolano 2009; Wu and Chan 2012; Wu 2013).

Several aspects of environmental protests in China distinguish them from many other protests. Tong and Lei (2013, 56) note that, in contrast to the more common protests related to labour conflicts or land rights, antipollution protests mainly involve the urban middle class. Improving the environment and fighting pollution are also in line with official policy at both central and provincial levels, in contrast to protests that directly challenge government policy and are more politically sensitive, such as demanding religious freedom or the right to form independent trade unions. In O'Brien and Li's terms (2006), environmental protesters can successfully frame their actions as a form of "rightful resistance", simply by citing central government regulations against polluting factories or the lax implementation of environmental standards by local authorities. The central government has thus acknowledged, accepted, and encouraged the participation of civil society in fighting environmental pollution (Mertha 2008; Johnsson 2010). Usually, however, the demands of environmental protesters conflict with other government concerns; and there is a sharp tension between environmental issues and the need for economic growth and employment creation. For the sake of social stability and economic progress, economic issues are often given higher priority. Obviously, this dilemma is part of the incentive structure of local government officials, since their careers are affected by whether or not economic growth targets are achieved. Recently, environmental protection targets have gained in importance, but as Ran (2013) notes, this does not always translate into better implementation of environmental policy.

This study is concerned with understanding how protests affect sustainable governance. All social protests with an environmental focus – as soon as they are nontrivial in size and organization, vocal in their demands, or successful in getting media attention or in an alliance with like-minded NGOs or other sympathetic actors – will gain enough strength to pose a serious challenge to local authorities. As such, the actions and claims of the protesting groups will pose a dilemma for local governments and will surely have an impact on local environmental governance. The *inputs* and *impacts* flowing from the actions of the protesting groups will force local authorities to respond in one way or another. More significantly, recurrent protests will have a long-term impact on the governance process. Certainly, such social protests are unwelcome from the local government's point of view; nevertheless, they occur and must be dealt with. Hence, we argue repeated cycles of social protest and government response will produce an altered style of environmental governance. Such changes are hard to predict, but they may include a more anticipatory style of policy-making and implementation, a heavier use of opinion polls and of consultation of citizen groups, and redefined roles for NGOs, the media, the judicial system, etc. We further argue that, while their exact outcomes vary, the social protests seen in these case studies contribute directly and indirectly to changes in the system of environmental governance – and this is true whether a specific protest is successful or not.

10.3 Environmental Politics in Hangzhou

If environmental governance in Hangzhou is to be understood, a brief introduction of the particular circumstances of the city is needed. Hangzhou currently profiles itself as a "green city", and gives official priority to protection of the environment. It is a wealthy region, and Hangzhou citizens are generally satisfied with the performance of the municipal government. In national surveys comparing the major municipalities of China, Hangzhou often ranks among the top three cities in terms of citizen satisfaction.[2] However, it does not stand out as vibrant with regard to the media situation or civil society in general. Local officials sometimes contrast Hangzhou with Guangzhou, the capital of Guangdong province. Guangdong province and Guangzhou city are known for their fairly independent media, active NGO movement, and assertive legislature. This has been labelled the Guangdong phenomenon (*Guangdong xianxiang*广东 现象). One official at Zhejiang Provincial People's Congress claimed that, in contrast to the open conflicts in Guangdong, politics in Hangzhou are based more on consensus and cooperation.[3] Hangzhou's leadership has been at the forefront in developing a set of local governance reforms. One Chinese scholar (Lin 2012) has described the Hangzhou model of governance as "composite democracy" (*fuhe minzhu*复合 民主). The name comes from the special type of performance evaluation that the municipal party committee developed

[2] There are many different surveys on citizen satisfaction in China. One example is the Chinese Institute for Economic Experiments (中国经济实验研究院) that recently reported that Hangzhou residents are the second most satisfied in China: http://news.sohu.com/20160703/n457484600.shtml. Accessed 29/8 2016.

[3] Interview (HZ-020) with provincial people's congress official, April 2009

in the 1990s. Since the late 1990s, Hangzhou's government has developed methods for evaluating the performance of local public agencies. The so-called comprehensive evaluation (*zonghe kaoping*综合 考评) combines internal (bureaucratic) evaluation with external citizen satisfaction ratings to produce an annual appraisal of more than 100 municipal government and party departments. Many other cities in China have been inspired by Hangzhou and have set up similar systems. In recent years, the system has spread to include district governments in Hangzhou as well. The result of the evaluation is made public at the end of the year. Each department receives a score between 1 and 100. Half the score is based on a satisfaction survey of 15,000 participants from different groups in society, including 9000 randomly selected citizens. The score affects the annual bonus of the department staff and leaders. More importantly, the result is directly connected to the cadre responsibility system. A department head that fails to receive the highest grade in the evaluation will not be eligible for promotion.[4] The impact of Hangzhou's evaluation system should not be underestimated. A social protest which involves many citizens and gains attention in the media can negatively affect the score of a public agency. For example, some departments that often experience conflicts with citizens, such as the City Administration Bureau (*chengguan*城管局), consistently end up at the bottom of the list.[5]

Let us look at how the comprehensive evaluation system works in practice. How did the protests in question affect the grades given to the responsible government department? It is difficult to evaluate the extent to which individual incidents affect the annual evaluation of different government departments. Only once in the history of the evaluation office has a government department received the grade labelled "not up to standard" (*wei daibiao danwei*未代表单位). This was in December 2008, when a serious accident occurred in the construction of the new subway system in Hangzhou. The accident was reported as major news in the media, and since the year-end evaluation was going on at the same time, the accident seems to have had a direct and tangible effect on the evaluation of the responsible government department, i.e. the department of safety supervision.[6] In the three cases we have studied, however, the effect of protests and incidents – if there is one – is less obvious. We have looked into the figures for the Hangzhou Environmental Protection Bureau (EPB), which is the key agency tasked with environmental protection in the city. The main incident that was reported in the media was the violent protest in Yuhang district in 2014 (case-study 3). Table 10.1 below shows that the evaluation results for 2014 were unusually good for the Hangzhou EPB – far better than the

[4] Interviews (HZ-021) with the head of the Hangzhou city evaluation office, June 2009 and November 2013

[5] Hangzhou's performance evaluation system is described in detail in Almén 2017. See also Delman Chap. 8 in this volume.

[6] Interview (HZ-021) with the head of the Hangzhou city evaluation office, November 2013

Table 10.1 Evaluation results for the Hangzhou Environmental Protection Bureau 2008–2016

Year	Ranking (total number of depts.)	Points (out of 100)
2008	60 (72)	89.6605
2009	48 (74)	91.1967
2010	34 (73)	92.8420
2011	61 (69)	91.8365
2012	42 (71)	93.3040
2013	69 (71)	88.1590
2014	36 (66)	90.6990
2015	37(65)	89.3330
2016	17 (63)	91.8890

Comments: The Hangzhou EPB had really low rankings in 2008, 2011, and 2013, but better rankings were recorded in the other years; thus there is no clear pattern across time in the ratings of the city EPB

Sources: Authors' compilation based on the Office of Hangzhou Comprehensive Performance Measurement Committee (杭州市综合考评委员会办公室). The reports are available online at: http://kpb.hz.gov.cn/showpage.aspx?id=181

year before, when the EPB ended up almost at the bottom of the list compared with other agencies.

While the annual evaluation does not indicate that protests had a direct impact on the score and rating of the EPB, we cannot rule out the possibility that responsible people in the EPB and other departments were held accountable for the communication failures that led to the Yuhang protests. An official in the performance-measurement office argued that the mere presence of the rating system and the threat of receiving low grades put a lot of pressure on officials to avoid mistakes.[7] As Delman (Chap. 8) shows in this volume, there are clear rules in the performance evaluation system which link social protests that attract unwelcome media attention to negative sanctions for the responsible department. We do not know, however, which department was ultimately deemed responsible for the protests in Yuhang in 2014.

In the next couple of sections, we turn our attention to the three cases of social protest. In our analysis, we focus on the role of social protests in environmental governance in Hangzhou. We discuss the input factors involved in each protest, such as the methods and channels used by citizen groups to express their demands, the role of other actors such as environmental NGOs, and the government's responses to these challenges. Finally, we discuss the policy changes and governance outcomes of these events.

[7] Interview (HZ-021) with the head of the Hangzhou city evaluation office, June 2009.

10.4 Case 1: Homeowners Protest Against Industrial Pollution in Gongshu District (拱墅区): Conflict-Solving Through a Residents-Factory Dialogue Platform

10.4.1 The Gongshu City District

Gongshu, located in the northern part of Hangzhou municipality, is a central city district. The Hangzhou municipal government is located there. In 2013 it had 560,000 residents, a sizeable part of Hangzhou's total population (8.9 million). Over the last 20 years, this district has gone through big demographic and industrial changes – with peripheral farming activities and suburban industries gradually becoming fully urbanized and included within the expanding residential and commercial development of the city centre. These processes of industrialization and urban sprawl, together with the shifting priorities of city planning, constitute a background to the citizen protests which erupted in Gongshu district in 2008 and 2009 and continued until 2011 (Interview HZ-006, HZ-012). The protests had their root in environmental concerns and a rising dissatisfaction expressed by local homeowners' associations.

In several interviews, the dilemma in question was described as a conflict between "old industries" and a "new residential area" (*laochang xinqu*老厂新区). In other words, it has been seen as a problem of developing new housing projects in a formerly suburban and industrial zone (Interview HZ-006, HZ-012). Old factories in the area were operating according to environmental standards that were out of date or inadequately enforced, or which simply did not meet the expectations of modern urban citizens. Described in this way, the problem can also be seen as a failure of city planning. After all, the residential developments were approved and completed before the old polluting industries had been relocated to other areas.

10.4.2 The Evolution of Events in Gongshu

Protests by local residents against noise and air pollution in Gongshu district started in late 2008. By that time many people had already moved into their newly built apartments, the complex was quickly filling up with residents, and the homeowners' association (HOA) had been formed.[8]

The HOA and the local community branch office (*yuanjiaqiao shequ*缘佳桥社区) took the initiative to bring these issues to general attention and to organize the

[8] The area consisted mainly of commodity housing (*shangpin zhuzhai*商品住房) of an upscale and expensive kind. Thus, the people moving in, among them well-educated university staff, were fairly well-off. They saw their housing as an investment offering better living standards. There were also many young families with pregnant mothers and small children, and they were keen to ensure a healthy living environment (Interview HZ-012, HZ-013, HZ-014).

residents' complaints.[9] The complaints at the beginning were disorganized and individualized, including calls to the mayor's office and the 12345 hotline. In 2009, however, they became more organized, focused, and intense.

There were a number of complaints and petitions on this issue in 2009, as well as some media attention. The residents wrote a letter to the Zhejiang provincial governor; they called in complaints to 12345 and 96666 hotlines; they formed QQ groups to discuss the issue on the Internet; and they brought banners to protest in front of the district office of environmental protection. Staff at the district EPB clearly stated that they were worried about falling satisfaction ratings and a rising number of complaints (Interview HZ-006, HZ-012). This confirms Delman's finding that the speed at which government departments handle complaints and respond to satisfaction ratings is central to how their performance is assessed. The protests also attracted the interest of several local media outlets in Hangzhou and even caught the attention of a China Central TV talk show (Interview HZ-006, HZ-012). According to central and local regulations, protests reported in the media should lead to sanctions for the responsible government cadres.[10] The pressure thus increased on the district government, especially the district and city-level environmental protection bureaus (EPBs). Thus, in 2009 it was locally suggested (but approved at higher levels) that the district EPB set up a "residents-factory dialogue platform" (RFD) to handle complaints, manage Q&A sessions, offer clarifications, negotiate solutions, and so on (Interview HZ-012, HZ-013). It seems that an overall motivation was to bring the conflicting parties to a common table and somehow, in a more organized manner, to reduce or defuse the tensions of the situation.

The key participants in the RFD were people from the homeowners' association, from the community branch office, from the local EPB, and from the main factories. However, due to high levels of distrust surrounding the situation in 2009, there were many angry bystanders watching over these meetings. Moreover, the first meetings between residents and factory representatives were so heavily influenced by discontent and mistrust that they often ended in shouting matches and mutual accusations. For example, members of the homeowners' association would not listen to the arguments offered by representatives of the EPB, since they suspected them of being on the same side as the factories. Therefore, in the beginning stages, the dialogue platform took on the function of a "punching bag", an outlet for the expression of strong feelings. In sum, it was a kind of safety valve for letting off the steam of citizen discontent (Interviews HZ-012, HZ-013, HZ-014).

Gradually, over the course of 2009, the RFD's work procedures grew more orderly, and the atmosphere calmed down a bit despite continued feelings of mistrust. There were repeated talks about noise levels and hours of factory operation

[9] Commodity housing residents have formed homeowners' associations to protect their legal and economic interests vis-à-vis real estate management or the local community. Their significance as a platform for residents' grievances in urban China has been previously documented by Tomba (2004, 2005) and Cai (2005).

[10] This is regulated both in the central cadre responsibility system and in the local performance assessment system.

(night and day shifts), environmental standards for air quality, and the need to instal additional air filters. The RFD's discussions also focused on the need to relocate factories away from Gongshu district. On this key issue, there was a great deal of disagreement between the local residents and the factories (Interviews HZ-012, HZ-013).[11] Finally, by 2011–2012, most of the polluting factories had moved away from Gongshu; they are now located in Jiande county outside Hangzhou. It should be noted that the Hangzhou city authorities had already intended to relocate these industries, and the main question was whether to do it sooner or later (Interview HZ-012, HZ-015).[12] The relocation programme proceeded according to the original schedule and did not speed up markedly because of citizen complaints. We can still say, however, that the residents' request for recognition in their complaints was met through the "residents-factory dialogue platform".

10.4.3 Protest Input and Sustainable Governance Outcomes in Gongshu

This protest featured a well-organized residents' group set up by a homeowners' association and a neighbourhood committee. Its members used social media to organize, and they managed to put pressure on the district EPB by protesting and by giving it low satisfaction ratings. This seems to have been a crucial trigger for the EPB's initiative to form the residents-factory dialogue platform, with an eye to reducing the level of conflict. In terms of sustainable governance, this case has several implications. First, the net result of the conflict was that the factory finally had to change location. To what extent the protests prompted this decision is unclear, but they are likely to have been at least one important factor. Whether or not the relocation represented a move towards sustainable development is debatable, but at least the environment for Gongshu residents improved. In terms of governance, one clear result of the protest was the creation of the dialogue platform – a governance innovation that helped to improve communication between citizens, local authorities, and businesses. Such dialogue platforms fit well into the models of

---

[11] Many old factories in this area had been established 20 years earlier. They produced steel, chemicals, pharmaceuticals, processed paper, and rubber tyres. Thus, even if they had not broken any environmental standards in a strict sense, they were bound to produce a lot of noise and an unpleasant air quality in the community (Interview HZ-012, HZ-015).

[12] According to the city planning department, the relocation of these polluting factories to nearby rural counties was already decided as part of a 5-year schedule. The relocation, however, was a slow and time-consuming process. The totality of industries in Gongshu district employed maybe 30,000–40,000 workers, and it was difficult to transfer their work and housing to other locations. In addition, there was the issue of determining the levels of compensation (for land, etc.) for the factories that had to move away. On top of that, the factories had to ensure that the proper infrastructure (water, power) was in place in their new location of Jiande. Otherwise, they would be unable to start production and would lose money, as confirmed by representatives of these companies (Interview HZ-015).

governance – fragmented authoritarianism and consultative authoritarianism – mentioned earlier in this chapter.

The dialogue platforms are typical conflict management mechanisms, and their aim is to strengthen government legitimacy. They serve to reduce social tensions by including citizen participation in the governance process. Such mechanisms have become more and more common for handling social protests in China (Chen and Kang 2016).

10.5 Case 2: Garbage Incineration Plant in Binjiang District (滨江区): Residents Engage a Beijing-Based NGO to Sue the Hangzhou Municipal EPB in Court

10.5.1 The Binjiang City District

Binjiang is a city district of Hangzhou municipality. It is located across the Qiantang River from the Hangzhou city centre, at a distance of some 20 km. Similar to Pudong in Shanghai, the Binjiang district is a rapidly developing suburban area on land that, until 10–15 years ago, consisted mainly of farmland or undeveloped land. The district has a population of 351,000 (2017) people. Binjiang residents living near the southern banks of the Qiantang River have been concerned about the environmental effects of a refuse incineration plant in their district. Many residents contacted the local EPB and complained about foul smells coming from the plant. They also had concerns about the detection of a large number of cancer cases in the nearby area (Interview HZ-001, HZ-002).

The construction of the garbage plant started in 2002; test processing and incineration began in 2002; and normal dump station operations started in 2004.[13] Thus, the factory has been in Binjiang district for over 10 years; however, according to residents, the negative effects on health and the environment have only become evident in recent years. The plant is located close to a mountain in the vicinity of Shan'Er village, and since 2004 it has been surrounded by more than ten residential communities (Interviews HZ-001, HZ-002). The size of the local population is considerable. In Shan'Er village alone, there are more than 3000 inhabitants, and in the area around the waste station, there are over 50 communities with companies and schools. There are 17 day-care centres for pre-school children, 14 elementary schools, and 7 high schools with over 80,000 pupils. There are also more than

[13] We do not have any precise information about this factory (*Binjiang luneng laji fenshao fadianzhan*), or about the exact relationship between it and the city government of Hangzhou. However, our impression is that this is a state-owned enterprise. We also know that the original decision to place it in Binjiang district was a point of controversy between the Hangzhou city government and the Binjiang district government. It is worth noting that Hangzhou has several other incineration plants for handling large volumes of municipal garbage and that the siting of their operations is always a sensitive and contentious issue (Interview HZ-011).

50,000 college students in the area. In the Shan'Er village sub-district closest to the waste station, comprising 720 residents, more than 30 cancer cases have been recorded, and an estimated 10 new cases are detected each year. According to investigative journalists and other observers who have visited the area, this higher incidence of cancer could be connected with the fact that local villagers grow vegetables on plots in the area (Interviews HZ-004, HZ-009; Can 2014).

Increasingly fearful and discontent with the garbage incineration plant, villagers in Binjiang submitted suggestions for inspection and improvement to the local government. But they did not receive any answers. Local residents also demanded that the plant be moved and that the responsible authorities stop the plans for a second development stage of the existing plant. They then filed a petition with a formal request for disclosure of data regarding the level of toxic particles in the groundwater, the nearby river, and the surrounding land (Interviews HZ-004, HZ-005). Furthermore, in February 2013, a group of eight municipal people's deputies from Binjiang, led by the local entrepreneur Zhang Fangxiang, submitted a suggestion (*jianyi* 建议) to the Hangzhou municipal people's congress for "accelerating the relocation of the Hangzhou Lüneng factory". Like the petitioners, the people's deputies argued mainly from a health perspective, but they also pointed out that moving the factory would be beneficial for social stability and for the industrial development in Binjiang.[14]

In connection with these actions, the villagers organized and mobilized themselves, held meetings and discussions about the issue, and created an online social forum for exchanging views and ideas. In October 2013, this online forum used a QQ chat room comprising 1500 members, the aim of which was to find ways to solve the problem with the garbage incineration plant. In addition, activists contacted local environmental NGOs in Hangzhou, seeking their support and asking for advice on how to articulate their demands vis-à-vis the local government. One of these NGOs was instrumental in helping Binjiang residents to get in touch with environmental activists and NGOs from other provinces (e.g. Guangdong and Beijing) with experience on these issues (Interview HZ-009). Local activists in Binjiang tried several times to get attention from journalists and the mass media in order to gain sympathy for their cause. In 2013, however, local media grew increasingly reluctant to cover this story; it was seen as too sensitive. Apparently, the Hangzhou propaganda department had instructed local media outlets to keep a low profile in reporting this case (Interviews HZ-003, HZ-009).

Meanwhile, the attempts of Binjiang activists to organize both online and in real life were placed under surveillance by the local security police, and both journalists and NGOs were warned not to get involved in the case (Interview HZ-009). The intimidation – as well as other aspects of the case – has been described in a book by one of the key activists with whom we were in contact (Can 2014). Nevertheless, the

[14] A suggestion carries less weight than a motion (*yian* 议案), but it still requires swift action from the responsible department. The suggestion can be read on the Hangzhou People's Congress webpage. (http://60.191.40.91/rd_yajy_cx.html), which states that the Binjiang district government had been designated as the responsible government department for handling the suggestion.

Binjiang activists managed to contact a Beijing-based NGO called *Nature University*, which helped them submit a request for disclosure of factory operations and data on the emission of toxic particles.[15] In May 2013, *Nature University* filed such a request with the Hangzhou EPB, but it was refused with the argument that such information contained sensitive technical details and business secrets. Next, *Nature University* engaged a Hangzhou-based lawyer, Mr. He Rongming, with experience of similar cases (Interviews HZ-004; HZ-005). The aim was to sue the Hangzhou EPB for not disclosing data in line with paragraphs 22 and 23 of the National Regulations on Information Disclosure. In March 2014, the first court hearings on the case took place in a Hangzhou district court. The event was surrounded by extensive security, and public access was restricted. Journalists were also still being asked to keep a low profile in reporting on the sensitive case (Interview HZ-018). On June 15, the Hangzhou Intermediate People's Court dismissed the appeal from Nature University for information disclosure and upheld the original verdict.[16] In legal terms, the protesters did not succeed in their demands. On the other hand, the decision to expand factory operations in a second stage was postponed.

The disputes surrounding the Binjiang garbage plant are complex and politically sensitive. The stakes were high, and local residents and the Hangzhou authorities came into sharp conflict. The latter have clearly been very reluctant to enter into any dialogue with citizen groups protesting against the plant operations. Their refusal to communicate, to explain, or to disclose information – even when it became an open court case – is a sign that local authorities view any concessions to citizen demands as costly or unacceptable. This impression is accentuated by the tactics of police harassment and "demobilization" of activists, journalists, and NGOs involved in the case. There are signs, however, that the EPB has tried to probe citizen claims indirectly – via talks with local environmental NGOs occupying a midway position in the conflict (Interview HZ-009, HZ-019). Protesters could not force the EPB to disclose information about emissions. However, we should note that plans for stage two of the Binjiang plant – i.e. an expansion of the facilities for garbage incineration – have been postponed. In this sense, local residents have scored some success. Since the dispute was handled as a court case in Hangzhou, it can be seen as a partial success and a recognition of the right of citizens to challenge the position of the authorities.

[15] One problem is that the waste products have not been properly sorted, which results in more toxics during incineration. In addition, while the recommended incineration temperature is 900 degrees Celsius, it seems the Binjiang plant only operates at 700 degrees Celsius (and burning waste at a lower temperature result in higher dioxin levels). The decision to burn at lower temperatures is probably based on economic calculations. However, in March 2013, the Hangzhou EPB claimed that the toxic substances emitted by the Binjiang garbage plant were still far below hazardous levels (Interview HZ-003, HZ-004, HZ-009).

[16] Cai Xin, June 17, 2014, NGO诉杭州环保局信息不公开再败 [NGO that sued Hangzhou EPB for not disclosing information lost again] accessed at http://china.caixin.com/2014-06-17/100691456.html

10.5.2 *Protest Input and Sustainable Governance Outcomes in Binjiang*

In contrast to their counterparts in case 1, the protesters in Binjiang were met with silence and hostility from the local government. The residents had used all the channels available to them, such as collective petitions and demonstrations, but the response of the authorities consisted of threats, police surveillance, and severe harassment. It was only when higher-level actors became involved that the protesters managed to put some real pressure on the local government. By engaging the Beijing-based NGO and a local lawyer, they were able to use the court system and to sue the Hangzhou EPB. At this stage, the national media were already writing about the case. As for governance outcomes, stage two in the building process for the incineration plant was stalled, because the judicial system was reviewing the case. This example shows how, in China's fragmented authoritarian system, national non-state actors – and to some degree local NGOs – can enter the game of environmental governance and change the policy process.

10.6 Case 3: Residents Protests Against Plans for a Waste Incineration Plant in Yuhang (余杭区): Violent Protest and Media Attention Force the Government to Postpone Its Plans Temporarily

Yuhang is the second largest suburban district of Hangzhou. With a population of over one million, it extends to the northeast of the city centre. Zhongtai township, which was the focus of social protests in this case, is located 50 km to the west of Hangzhou.[17] The conflict started with an announcement in February 2014 by the district government that construction would begin on a new waste incineration plant in Jiufeng village, 10 km from Zhongtai township on highway G-56, east of Hangzhou. Citizens fearing negative environmental consequences were frustrated by the lack of information from the authorities. As a result, violent social protests took place in May 2014, attracting national and international media attention.[18] The protests involved several thousand people, and some of the clashes with the police were so harsh that about 40 injuries (on both sides) resulted.

[17] The management of domestic waste is a huge problem in China. In 2010–2015, about 300 waste incinerators were planned to be built to meet the national target of 35% of household waste to be incinerated. (See Shanghaiist.com, 11 May 2014: "Hangzhou halts waste plant construction following bloody protests". Accessed at http://shanghaiist.com/2014/05/11/hangzhu-halts-waste-plant-construction-following-bloody-protests.php.)

[18] See BBC online: "China waste plant protest in Hangzhou 'injures dozens'." Accessed on the internet at http://www.bbc.com/news/world-asia-china-27360779.

10.6.1 The Evolution of Events in Yuhang

In February 2014, following the local government's announcement of construction plans for the new waste incinerator plant, local citizens were apprehensive. They feared the effects on their health of such a plant and suspected it would worsen air pollution in the area. They were also upset about the lack of any government consultation with local interests. In a meeting with the local government, some residents submitted a collective petition signed by 20,000 people, with the complaint that due legal procedures had not been respected in this case. No public hearing had been held, and no proper environmental impact assessment (EIA) had been made, although both are required before projects of this kind are started. Residents also questioned the location of the plant, since it was so close to residential areas. In that respect, this was a typical "Not in my backyard" (NIMBY) kind of protest.

The lack of information and of any response from the local government led to heightened frustration, and soon local citizens were taking to the streets in large numbers. The authorities were forced to close off a major highway for 90 minutes when demonstrators occupied it, and protesters damaged cars and set fire to police vehicles.[19]

Apparently, demonstrations against the construction project by residents and activists went on for 2 weeks at the end of April and beginning of May 2014. The overt expression of public discontent may have been triggered by an official news article in the *Hangzhou Daily* on Friday 9 May. This article defended the construction of the incineration plant, saying that the "technology used was safe and up to standard".[20] As a result the demonstrations escalated, becoming full-blown riots on Saturday 10 May. Several thousand protesters on the streets confronted several hundred policemen, and the situation became chaotic and violent. In the end, 29 policemen and 10 citizens were severely injured. In the process, over 30 vehicles were overturned, 2 police cars were set on fire, and 4 police cars were smashed up.[21]

Initially, local police officials were hesitant to intervene in the conflict. Eventually, the Hangzhou city SWAT team was sent in to stop the violent demonstrations. As part of their strategy for handling the conflict, the local authorities backed off from their initial policy decision. They said they would postpone the construction of the plant and conduct discussions with local citizens. On Saturday 10 May 2014, immediately after the large-scale demonstrations, the Yuhang authorities issued a statement calling for more consultation with the public. On its website, the local government announced that all work on the incinerator plant had stopped and that

---

[19] Xu Wen, "Waste incinerator generates hot question", 13 May 2014. Accessed at http://www.hicenter.cn/columns_detail.asp?id=436232

[20] Reuters online, 10 May 2010 (via Voice of America): "Protesters in East China clash with police". Accessed at http://www.voanews.com/content/protesters-in-east-china-clash-with-police-over-waste-incinerator-plan-/1911828.html

[21] Reuters online, 11 May 2014, "China says no waste plant without support after protests" Accessed at
http://www.reuters.com/article/2014/05/11/us-china-protests-idUSBREA4904320140511

the public would now be invited to participate in the eventual decision. It stated that "We will invite the local people to participate [...] and seek everyone's opinions, guarantee people's right to know and right to participate". It also expressed the hope that people would end their protests, that public order would be restored, and that social stability would be maintained. Finally, it declared that it recognized the protests were a reflection of people's concern.[22]

Despite all this, plans to build the waste incineration plant in Yuhang have remained the same.[23] The general policy of Hangzhou municipality is quite firm in this regard: the growing mountains of domestic waste must be handled in one way or another. Meanwhile, dialogue platforms with the public were initiated in the summer of 2014, and the plan is to continue to use them to enhance public consultation. However, according to a scholar who interviewed local residents, people were distrustful of the local government, and few showed any interest in participating in dialogue platforms (Interview HZ-016). Local and international environmental NGOs were also involved in the resistance. According to one local ENGO, the government has had serious and ongoing problems communicating with citizens. This ENGO is now being approached by the government to take part in citizen dialogues (Interview HZ-017). Another NGO representative noted that all observers of the Yuhang protests, including Hangzhou's political authorities, were surprised by the quick rise and organizational force of the social protests in 2014. This may have increased their willingness to accommodate citizen demands (Interview HZ-018).

Waste incineration is an important aspect of waste management in Hangzhou, and it is not likely that protests will change that policy priority. A complete abandonment of the project would be very costly for the local government. On the other hand, the intensity, organizational capacity, and surprisingly quick build-up of the protests, and the far-reaching attention from the national media they received, meant that potentially high disruption costs were associated with the project (Interview HZ-018). It was imperative that the conflict be handled in an effective manner. The local government used force to stop the violence and arrested several protesters, including people involved with a local environmental NGO. At the same time, it was quick to announce that the project would be temporarily halted and that the dialogue with citizens would be improved. One reason the conflict became so violent was that the local government was so incommunicative. It did not follow its own guidelines for involving citizens in the process. The seriousness of the protests forced it to respond quickly. Then, once the national media got involved, the pressure rose dramatically on local officials to address the protest properly.

[22] Reuters online, 11 May 2014, "China says no waste plant without support after protests". See footnote 20.

[23] According to *The Telegraph* (11 May 2014), Chinese cities produce 160 million tons of waste per year, and the country is planning 300 waste incinerators within the coming 3 years to handle the problem. Accessed at http://www.telegraph.co.uk/news/worldnews/asia/china/10822917/Bloodshed-as-Chinese-environmental-protest-sparks-riot.html

10.6.2 Protest Input and Sustainable Governance Outcomes in Yuhang

The numbers of people involved, the degree of violence, and the media attention attracted had the effect of putting heavy pressure on the local government to respond to the protesters' demands. Clearly, large-scale social discontent forced the local government to change its position. It had to postpone the project and to establish dialogue platforms to involve citizens in the process. The government also invited local NGOs to serve as go-betweens, in the hope of improving communication with citizens. As for the effects on governance, public hearings were held and a more thorough environmental impact assessment (EIA) was carried out. However, the project to build the incineration plant continued, and the protesters failed in their fundamental demands. Yet, in spite of this apparent failure, the protests clearly added a new dynamic to the governance process. Among other things, they forced the authorities to pay close attention to EIA procedures. They also reinforced and legitimized the role of environmental NGOs as pivotal actors in the governance process.

10.7 Conclusions

The three cases of protest that we have described represent different examples of how social contention and protests can influence the patterns of local environmental governance. In Gongshu it resulted in a governance innovation, with negotiations between the homeowners' association, the neighbourhood committee, and the polluting factories, under the monitoring eyes of the local EPB. The Binjiang case involved external NGO actors and lawyers. In the end, appeals were made to the judicial system, a court case was opened, and the expansion of the incineration plant was stalled. In Yuhang, on the other hand, the judicial system was never involved. Instead, violent street protests forced the authorities to make concessions, to consult with residents (with the mediating help of NGOs), and to follow existing legal procedures (like proper EIAs). In all three cases, the local government had to make big or small adjustments. It had to change its policy substantially or at least to postpone the implementation of its envisioned project.

Through these mechanisms, social protest can be an important society-driven input into the cadre responsibility system, which is a key feature of local governance in China. We have no clear information about whether local officials actually faced sanctions in these cases. However, according to existing regulations on cadre management, as well as previous research (Cai 2015), local officials are often punished when social protests turn violent and attract unwelcome media attention – especially when the events catch the attention of provincial or central authorities.

Social protests constitute an important element in the process of environmental governance in China today. They have an impact on the pattern of local governance, regardless of whether or not they produce the final results sought by protesters. They tend to have a constraining effect on local authorities. At times they succeed in changing government policy, as in the Gongshu case. At other times, they delay policy implementation, which leads to new windows of opportunity when political or economic circumstances change. Protesting is usually a method of last resort for Chinese citizens, and it entails a fair amount of risk for participants. But it also signals that the issue is of great importance. The three cases in Hangzhou vary in how far the protesters achieved their demands. In each case, however, the protestors forced the city government to respond, to make concessions, and to address the contentious issue in some manner. Therefore, all three protests had an undeniable impact on the policy process and should therefore be seen as a constitutive element of the city's environmental governance. The result is a fragmented governance structure, which can be sustainable as long as conflict management mechanisms are developed to handle the tensions between social protesters and local political authorities.

References

Almén, O. (2017). Participatory innovations under authoritarianism: Accountability and responsiveness in Hangzhou's social assessment of government performance. Journal of Contemporary China, online version, https://doi.org/10.1080/10670564.2018.1389003

Almén, O., & Burell, M. (forthcoming). Social accountability as social movement outcome: Protests in a Chinese city. *Social Movement Studies*.

Bellamy, R., & Palumbo, A. (Eds.). (2010). *From Government to Governance*. Burlington: Ashgate.

Cai, Y. (2005). China's moderate middle class: The case of homeowners' resistance. *Asian Survey, 45*(5), 777–799.

Cai, Y. (2008a). Social conflicts and modes of action in China. *The China Journal, 59*(January), 89–109.

Cai, Y. (2008b). Local governments and the suppression of popular resistance in China. *The China Quarterly, 193*(March), 24–42.

Cai, Y. (2010). *Collective resistance in China: Why popular protests succeed or fail*. Stanford: Stanford University Press.

Cai, Y. (2015). *State and agents in China: Disciplining government officials*. Stanford: Stanford University Press.

Can, N. (2014). 美丽山水是忧伤 (Sad Beautiful Landscape), Book manuscript, on file with the authors.

Chen, X. (2009). The power of 'Troublemaking': Protest tactics and their efficacy in China. *Comparative Politics, 41*(4), 451–471.

Chen, F,. & Kang, Y. (2016). Disorganized popular contention and local institutional building in China: A case study in Guangdong, Journal of Contemporary China (online publication).

Dai, J., & Spires, A. (2017). Advocacy in an Authoritarian State: How Grassroots Environmental NGOs Influence Local Governments in China. The China Journal 79. Online version.

Deng, Y., & O'Brien, K. (2014). Societies of senior citizens and popular protest in rural Zhejiang. *The China Journal, 71*(January), 172–188.

Diamant, N., Lubman, S., & O'Brien, K. (Eds.). (2005). *Engaging the Law in China: State, society, and possibilities for justice*. Palo Alto: Stanford University Press.

Edin, M. (2003). State capacity and local agent control in China: CCP cadre management from a township perspective. *The China Quarterly, 173*, 35–52.

Fu, H., & Cullen, R. (2008). Weiquan (Rights Protection) lawyering in an authoritarian state: Building a culture of public-interest lawyering. *The China Journal, 59*(January), 111–127.
Gao, J (2015) Pernicious manipulation of performance measures in China's cadre evaluation system. The China Quarterly, Available on CJO 2015 doi:https://doi.org/10.1017/S0305741015000806
Harding H (1987) China's Second Revolution: Reform after Mao. Brookings, Washington.
He, B, Thøgersen S (2010) Giving the People a Voice? Experiments with Consultative Authoritarian Institutions in China, Journal of Contemporary China 19 (66), pp. 693–717.
Heberer, T., & Trappel, R. (2013). Evaluation processes, local Cadres' behaviour and local development processes. *Journal of Contemporary China, 22*(84), 1048–1066.
Johnson T (2010) Environmentalism and NIMBYism in China: Promoting a rules-based approach to public participation, Environmental Politics, 19(3), 430–448.
Koopmans, R. (2007). Protest in time and space: The evolution of wages of contention. In D. Snow, S. Soule, & H. Kriesi (Eds.), *The Blackwell companion to social movements* (pp. 19–46). Oxford: Blackwell Publishing.
Kostka, G., & Mol, A. (2013). Implementation and participation in China's local environmental politics: Challenges and innovations. *Journal of Environmental Policy and Planning, 15*(1), 3–16.
Lee, C. K. & Zhang, Y. (2013). The power of instability: Unraveling the microfoundations of bargained authoritarianism in China. *American Journal of Sociology 118*(6), 1475–1508.
Lieberthal, K. & Oksenberg, M. (1988). *Policy Making in China*. Princeton: Princeton University Press.
Lin, S. (2012). *Fuhe minzhu:renmin minzhu zujin minsheng jianshe de hangzhou shijian [Composite democracy:the practice of promoting people's livelihood by people's democracy in Hangzhou]*. Beijing: Central Compilation and Translation Press.
Mertha, A. (2008). *China's Water Warriors*. Ithaca: Cornell University Press.
Mertha, A. (2009). "Fragmented authoritarianism 2.0": Political pluralization in the Chinese policy process. *The China Quarterly, 200*, 995–1012.
O'Brien, K., & Li, L. (2005). Popular contention and its impact in rural China. *Comparative Political Studies, 38*, 235–259.
O'Brien, K., & Li, L. (2006). *Rightful resistance in Rural China*. Cambridge: Cambridge University Press.
Peruzzotti, E., & Smulowitz, C. (Eds.). (2006). *Enforcing the Rule of Law: Social accountability in the new Latin American democracies*. Pittsburg: University of Pittsburg Press.
Pierre, J., & Peters, G. (2000). *Governance, politics and the state*. Basingstoke: Macmillan Press.
Ran, R. (2013). Perverse incentive structure and policy implementation gap in China's local environmental politics. *Journal of Environmental Policy and Planning, 15*(1), 17–39.
Ru, J., & Ortolano, L. (2009). Development of citizen-organized environmental NGOs in China. *Voluntas, 20*(2), 141–168.
Shie, S., & Brown-Inz, A. (2013). A special report mapping China's public interest NGOs. In: Chinese NGO directory: A Civil Society in the Making, China Development Brief Publication. http://chinadevelopmentbrief.cn/wp-content/uploads/2014/10/Special-Report-Mapping-NGOs.pdf
Teets, J. (2013). Let many civil societies bloom: The rise of consultative authoritarianism in China. *The China Quarterly, 213*(March), 19–38.
Tomba, L. (2004). Creating an Urban Middle Class: Social Engineering in Beijing. *The China Journal, 51*(January), 1–26.
Tomba, L. (2005). Residential space and collective interest formation in Beijing's housing disputes. *The China Quarterly, 184*(December), 934–951.
Tong, Y., & Lei, S. (2010). Large-scale mass incidents and government responses in China. *International Journal of China Studies, 1*(2), 487–508.
Tong, Y., & Lei, S. (2013). War of position and microblogging in China. *Journal of Contemporary China, 22*(80), 292–311.

Tsang, S. (2009). Consultative leninism: China's new political framework. *Journal of Contemporary China, 18*(62), 865–880.
Whiting, S. (2001). *Power and wealth in rural China: The political economy of institutional change*. New York: Cambridge University Press.
Wu, F. (2013). Environmental activism in provincial China. *Journal of Environmental Policy and Planning, 15*(1), 89–108.
Wu, F., & Chan, K. (2012). graduated control and beyond: The evolving government-NGO relations. *China Perspectives, 3*, 9–17.

Field Interviews

HZ-001, Interview in Hangzhou (June 2013) with environmental NGO*
HZ-002, Interview in Hangzhou (June 2013) with environmental NGO**
HZ-003, Interview in Hangzhou (June 2013) with newspaper journalist**
HZ-004, Telephone interview (Oct 2013) with Lawyer He Rongming*
HZ-005, Interview in Beijing (Oct 2013) with environmental NGO*
HZ-006, Interview in Hangzhou (Nov 2013) with Gongshu district EPB*
HZ-007, Interview in Hangzhou (Nov 2013) with Hangzhou city evaluation office*
HZ-008, Interview in Hangzhou (Nov 2013) with environmental NGO**
HZ-009, Interview in Hangzhou (Nov 2013) with environmental NGO**
HZ-010, Interview in Hangzhou (Nov 2013) with social science researcher**
HZ-011, Interview in Sweden (Dec 2013) with Hangzhou-based environmental NGO**
HZ-012, Interview in Hangzhou (March 2014) with Hangzhou and Gongshu EPB officials**
HZ-013, Interview in Hangzhou (March 2014) with Gongshu, Dabeiqiao neighbourhood committee**
HZ-014, Interview in Hangzhou (March 2014) with Gongshu, Dabeiqiao homeowner association**
HZ-015, Interview in Hangzhou (March 2014) with Gongshu, Dabeiqiao Miles Industrial Corp.**
HZ-016, Interview in Hangzhou (Nov 2014) with university academic at ZU*
HZ-017, Interview in Hangzhou (Nov 2014) with environmental NGO*
HZ-018, Interview in Beijing (Nov 2014) with environmental NGO**
HZ-019, Interview in Hangzhou (March 2015) with environmental NGO**
HZ-020, Interview in Hangzhou (April 2009) with a Zhejiang provincial people's congress official*
HZ-021, Interview in Hangzhou (June 2009; Nov 2013) with the head of Hangzhou city evaluation office*

Chapter 11
Green Justice Approach to the Environmental Governance Dilemma: A Case Study of Jiufeng Environmental Energy Project in Yuhang District, Hangzhou

Jiangli Wang, Youxing Lang, Feifan Huang, and Biao Wei

Abstract This chapter reviews a public protest against the construction of a waste incineration power plant during April and May 2014 in Yuhang, Hangzhou. Based on the theory of green justice, the article analyses the core reasons of the protest and discusses the absence of justice between the residents and the local government from the perspectives of procedural justice and spatial justice with a special focus on the NIMBY (not-in-my-backyard) phenomenon. It suggests that the significant cause of such conflicts lies in the dominance of spatial justice of a private nature combined with the closed governmental decision-making process, which lacks procedural justice. To build a waste incineration power plant, it is argued, is not only a precise measure of environmental governance but also meant to protect the quality of civic lives. This is a dominant requirement of citizens during urbanization. Therefore, all the parties, i.e. the government, residents, and interest groups, should adhere to green justice as a holistic approach to deal with public bads like municipal waste, in this case based on common and long-term interests, and subsequently share the environmental public goods equally.

Keywords Green justice · Spatial justice · Procedural justice · NIMBY · Hangzhou · Jiufeng Protest

J. Wang (✉) · Y. Lang · F. Huang
Department of Political Science, School of Public Affairs at Zhejiang University, Hangzhou, People's Republic of China
e-mail: wangjiangli@zju.edu.cn

B. Wei
Certified Planner of Hangzhou City Planning and Design Academy, Hangzhou, People's Republic of China

J. Delman et al. (eds.), *Greening China's Urban Governance*, ARI - Springer Asia Series 7, https://doi.org/10.1007/978-981-13-0740-9_11

11.1 Introduction

In the process of contemporary urban development and urbanization, people around the world often find themselves in a dilemma, in which they usually cannot strike a balance between development and environmental protection and between public demand and the interests of individuals or certain interest groups. China's urbanization has witnessed this dilemma too. Chinese citizens with environmental awareness and citizen consciousness now devote increasingly more attention to the need for public environmental facilities and services, while many others still do not concern themselves about it. At any rate, the construction of such facilities easily brings up the NIMBY (not-in-my-backyard) phenomenon which frequently leads to protests and bottom-up obstruction. This kind of public concern and the protests put stress on local governments, thus challenging the process of urbanization in China.

When faced with NIMBY-related protests, local governments in China mostly choose to discontinue the construction and then proceed with the construction process later, if at all possible, after calming down the public to maintain their authority. The most common method to cope with the protestors is to provide compensation for their potential loss. Meanwhile, the media tend to blame the participants of NIMBY movements for their selfishness, because they do not put the public and national interests first.

In academia, research on the NIMBY phenomenon and the social protests mainly focus on the underlying reasons of NIMBY movements and the measures to solve the problem. Previous analysis has emphasized the local government and the residents as the main actors, but seldom talks about justice. Most research thus fail to realize the justice issue in the dilemma of local environmental governance as it materializes in the conflicts between the public interest and the private benefits.

In this chapter, we will examine the case of such a NIMBY incidence, the Jiufeng Environmental Energy Project of Yuhang in Hangzhou (JEEP; earlier called the Jiufeng Waste Incineration Power Plant Project). We aim to elaborate and discuss the topic of justice behind this type of environmental governance dilemma. The specific case provides us with an example with which to analyse conflicts of environmental justice. The JEEP was built in response to increasing demands for waste-handling arising from urbanization. The local government regarded the project as its legally imposed responsibility (Ma 2014). Hangzhou citizens were supposed to enjoy the JEEP's benefits as a public good, while nearby residents would suffer from its negative externalities. Naturally, such a situation could easily give rise to a sense of injustice among the people living around the plant. As a matter of fact, the public protest against the JEEP turned into both a movement for protecting the interests of local residents and a social action in pursuit of social justice. The local residents believed that the government's decision-making lacked transparency and that it violated the right of citizens to information about the project. The government's decision on the project was also thought to lack the necessary legal foundation. Furthermore, the local government controlled all information during the process of planning and construction. The ensuing protest led to a series of negative consequences. Local residents lost trust in the government; the protest eventually turned violent; and the JEEP was caught in an impasse. The Jiufeng Protest was not extraordinary. Many similar protests have taken place elsewhere, in response to similar environmental projects. In recent years, public protests over environmental issues in urban areas have been increasing at an average annual rate of 29% (Wu 2013).

In general, environmental projects face tough opposition as soon as they are proposed (Yu and Zhang 2014). The ensuing protests often lead to a change in project location, as happened with similar cases in Beijing and Guangzhou (Wong 2016); or they experience a delay in construction, so that the necessary facilities remain lacking; or they may face to more serious conflict, in the midst of which construction must proceed under governmental protection. The contradiction between the need to build environmental facilities and the emergence of social protests in response has made urban environmental governance in China increasingly difficult.

To address these questions, we take green justice as our theoretical starting point. We examine some of the key issues from the standpoint of procedural justice and spatial justice, respectively. We consider the absence of justice in these matters critically, especially in connection with cases that embody the NIMBY phenomenon. The most significant cause of such conflicts, we suggest, lies in spatial justice of a private nature on the one hand and in the government's closed approach to decision-making, which fails to provide procedural justice, on the other hand.

In the first part of the chapter, we briefly discuss green justice focusing on its components, i.e. spatial justice and procedural justice, to establish our analytical framework. We point out that, by embracing holism, the agencies responsible for urban development planning and environmental governance could facilitate the elaboration of detailed development and construction plans that consider the long-term and common benefits of achieving harmonious development between human beings and nature, between the individual and society, and between each person and himself/herself. In the second section, we describe the Jiufeng Protest in brief. In the next two sections, we examine the actions of two critical actors—the public and the government—according to the spatial justice and the procedural justice approach, respectively. In the final section, we conclude that the key to resolve the NIMBY dilemma in the process of urbanization, *in casu* building a waste incineration power plant, is to strike a balance between public and private interests as well as overall interests and partial ones.

11.2 Green Justice: Analytical Framework

It is well acknowledged by governments and academia that the core of the new type of urbanization is the urbanization of people. However, what does the 'urbanization of people' mean specifically? What does the concept refer to under Chinese conditions? There is no common or mainstream view on this question as yet. We think that, given the political-structural framework of China, the new round of urbanization is aimed at realization of fairness and justice, compared to the past few rounds of urbanization in China since 1949 which focused more on urban growth at any cost. Nevertheless, not everyone is clearly aware of this. Therefore, to summarize our position, urbanization is not only a significant project of industrialization and modernization but also a tangible issue that relates to fairness and justice.

Whether the need we face is to solve or avoid environmental protests, or to ascertain the characteristics and requirements of current environmental governance, we must adopt an integrated perspective on the rights and interests of all species on earth. In this section, we analyse the Jiufeng Protest from the standpoint of green justice. This is a new kind of justice that is increasingly being pursued by human beings. It has emerged on the basis that the human race needs to seek for a new kind of global ecological civilization that embodies 'self-conscious' justice in a global ecological security community within which human beings are dependent on a strong consciousness of the rights and interests of all species. It goes beyond the limits imposed by patterns of life in previous periods. It stresses sustainable development and takes intergenerational equity into account (Yu and Wang 2009, 75).

Green justice does not originate from a certain civilization or a specific religion. It does not express the global expansion of any theory or -ism. It should be seen, rather, as representing the accumulation of human wisdom over the course of history and as an innovation which this wisdom has made possible. With his principle of justice, J. Rawls sought to regulate the distribution of certain social goods, among them rights, freedom, power, income, wealth, dignity, and recognition (Rawls 1999, 54). With which social good, then, is green justice concerned? Fundamentally it is concerned with ecological security. The arrangements required to ensure such security involve two major policies: on the one hand, the provision of public environmental goods like beneficial ecological conditions and the sustainable use of environmental resources and, on the other, a just distribution of the social benefits and burdens arising from these ecological conditions, including rights and freedoms, duties and responsibilities, interests and actions, and dignity and sharing. Green justice is based on a critique of the unilateral pursuit of material growth that is so characteristic of industrial civilization and the serious ecological crises resulting therefrom. The emergence of the contemporary ecological crisis and the ensuing need for global ecological security guarantees have raised considerations of justice to the level of multiple dimensions (see Fig. 11.1).

In Fig. 11.1, we show these dimensions in a conceptual framework for explaining the nature of green justice. Green justice reflects the dualism of human beings as ecological and social beings. It combines environmental justice, which focuses on Dimension III, with ecological justice, which pays more attention to Dimension I. It starts with a low-key (*didiaode* 低调的) consciousness of the rights and interests of all species (as opposed to anthropocentrism). It understands human beings as forming a part of nature and as being on a par with the other species with whom we share this earth (see Dimension I). In vertical terms, it stresses the need not just for intragenerational justice but for intergenerational justice as well. It guarantees the sur-

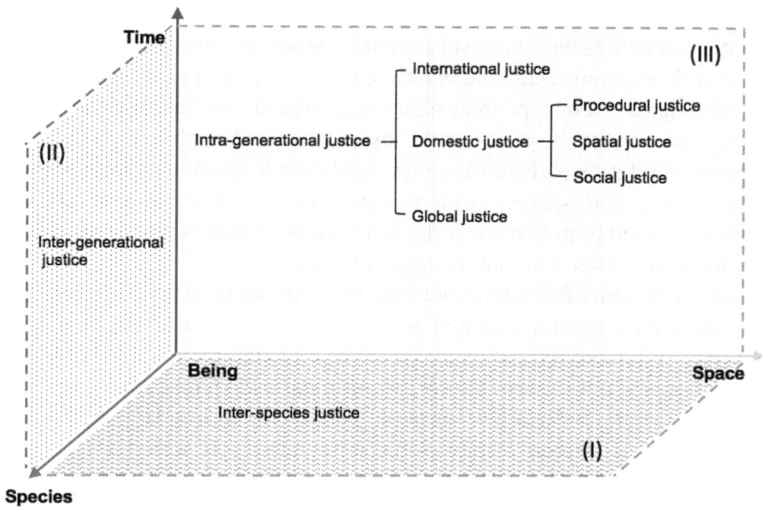

Fig. 11.1 Dimensions of green justice (Yu and Wang 2009 (modified))

vival rights and ecological interests of other species and of future generations of human beings (Dimension II), and it seeks to maintain the harmony and balance of the ecosphere on earth. In horizontal terms, it transcends the regional or national environmental justice of 'others' (i.e. 'no rubbish or toxic-material dumping in my backyard'), and it applies to the international or global level as well—the environmental justice of 'all of us' (i.e. 'no rubbish or toxic-material dumping in anyone's backyard') (Dimension III). Moreover, green justice involves more than just historical understanding; it also entails an awareness of the global community across regions and nations. As a category with abundant connotations, it integrates previous notions of environmental justice and ecological justice and transcends them as well.

Green justice relates to and incorporates the concepts of 'environmental' justice and 'ecological justice', in different yet essential ways.[1] In his book *Defining Environmental Justice*, David Schlosberg (2007) discusses environmental and ecological justice. The focus of environmental justice, he suggests, is on relations

[1] Environmental justice was first discussed in the works of some US scholars and activists (e.g. Wenz 1988; Hofrichter 1993; Bryant 1995), who focused on issues of environmental rights among different ethnic groups in different regions within the USA. Environmental justice in this sense concerns the extension of ethnic and regional equality, in terms of both content and range (Hayden 2005: 133–134; Wenz 1988; Dobson 1999; Bullard 1999; Sandler and Pezzullo 2007; Zeng 2007). In our view, the concept of environmental justice is based on the attributes of human beings as social beings, and it is focused on the realization of social justice. The concept of ecological justice is often equated with that of environmental justice; however, ecological justice is based on the biological attributes of human beings. It therefore puts more emphasis on the preservation of natural habitats, the rights of non-human species, and the relationship of other species to our own (Baxter 2004; Wulfhorst and Haugestad 2004; Pepper 2005; Li 2005, 2007; Huang 2006; Li 2008; Li 2002).

among members of our own species, whereas that of ecological justice is on relations between human beings and the rest of nature. However, similar conceptions can be applied to both environmental and ecological issues. Green justice, for its part, is a more comprehensive concept than either 'ecological' or 'environmental' justice, because it has a much broader scope: more than just a philosophical notion, it encompasses regulation and coordination in relation to the three general dimensions of species, time, and space. From the perspective of ecological existence, it emphasizes the need both to respect the full variety of beings and to restore balance and coordination between human beings and nature.

It also pays considerable attention to the many multi-level issues of social justice posed by the environmental challenges we face. Green justice thus has a broader meaning than either ecological justice or environmental justice. It is both a survival philosophy and a set of criteria for socioecological and spatial ethics and values. It calls for a brand-new civilization—an ecological civilization—beyond the industrial one that now exists. Green justice is the cornerstone on which a new civilized, ecological society can be built. Ecological civilization is not just a Chinese development concept; it is also a historical stage which all countries and all human beings are entering, due to globalization.

The advancement of globalization and the deterioration of the environment have influenced people's awareness of the planet on which they are living. To defuse the ecological crisis and enact effective governance, all human beings—old and young, male and female—will have to readjust their relationship with the environment (Vogt 1981, 289). People need to restore a harmonious relationship between economic development and the natural world. If injustice on the above-mentioned three dimensions is to be avoided, the global societies of the future will have to make green justice a priority (Jamieson 2003, 426–237). It is important to maintain the holistic approach of green justice and to show foresight in governance; otherwise it will not be possible to solve the environmental problems that arise in the course of urbanization. Green justice is also a basic value in relation to local issues: it can guide people to 'think globally and act locally', enabling them to contribute to effective environmental governance and to a more science-based process of urbanization.

With green justice as our theoretical foundation, we discuss and analyse the Jiufeng Protest. We do so in terms of two of the three aspects of domestic justice contained within the concept of green justice (see Fig. 11.1): that is, spatial justice and procedural justice. The Jiufeng Protest reflects the process of environmental governance that is emerging in the course of Chinese urbanization.

Spatial justice functions on various spatial levels: planet, continent, country, city, etc. Residents of these different spatial communities may enjoy a fair or an unfair distribution of environmental resources, in the form of either public goods or public bads. Spatial justice also implies that poor areas should be compensated by rich areas. In the case of China, spatial justice requires that rural areas be compensated by urban areas. The development of the urban economy and environment often comes at the cost of deteriorating living conditions in the countryside. The government must therefore devise efficient mechanisms to balance environmental benefits between urban and rural areas during the process of urbanization. The NIMBY phenomenon is a very general problem in spatial justice all over the world. It is a form of opposition to external influences, and it can materialize into protest activity by residents when they fear that damage may be inflicted on their health, their property, or their living environment by the construction of public facilities like public toilets, power lines, garbage collection systems or transfer stations, etc. Residents often express their unwillingness to accept the establishment of such facilities close to their homes (Hall 2009, 430; He 2013, 51).

Procedural justice requires that a coherently internally integrated legal system ensures a just distribution of environmental resources (Wang 2012). Spatial justice and procedural justice are both essential components of green justice, and we use them both to ground our analysis theoretically when analysing the Jiufeng Protest. We argue that both are helpful for understanding the motives for this protest by local residents, as well as for explaining how and why the local government lost the trust of the local people in this case.

11.3 The Jiufeng Protest: Conflict Process

Zhongtai Sub-district, located in the western suburbs of Hangzhou, is 25 kilometres away from the city centre. It borders Lin'an and Fuyang to the southwest and Xianlin Sub-district and Yuhang Sub-district to the northeast. Zhongtai has an area of 71.05 km^2 and a population of 25,700, of which the agricultural population constitutes 77% (cf. Introduction to this volume). It has a beautiful green environment and a forest coverage rate of 68%. Epithets for the area include 'the home of bitter bamboo' and 'the home of the Chinese bamboo flute'. Zhongtai's economy is dominated by agriculture, tourism, and the tea and flute industries.

With its huge population of nearly 9 million, Hangzhou lacks the capacity to dispose of its waste properly. Tianziling Refuse Landfill, on which the city relies, has been overloaded for years. The government is thus confronted with an acute waste treatment problem, and the urgent need to build new waste treatment facilities is generally recognized.

The JEEP has therefore been considered one of Hangzhou's most significant livelihood projects. The bidding announcement for the proposed waste incineration power plant was released by Hangzhou Solid Waste Disposal Company on 5 June 2012. Before long, local residents started opposing the project. On 27 December 2013, the Hangzhou Communist Party Committee approved the JEEP proposal. According to the plan, the plant would incinerate 3200 tons of rubbish per day during the first phase of the project and 5600 tons per day during the second phase. In the *Revised Draft of Professional Environmental Sanitation Planning of Hangzhou (2008–2020)* (the Bureau of Urban Planning of Hangzhou 2014), the whole city was divided into five disposal areas, each with a plant able to incinerate 3000 tons of rubbish per day. The public was kept informed of the project from then on, and local residents began to oppose it strongly, due to the serious harm they believed it would pose to their health and their living environment.

Over the course of the publicity period, citizens from western Hangzhou and Zhongtai Sub-district made their resistance to the project clear. They made calls, sent letters of protest, talked to government officials, and left messages on government websites. In April, residents organized specific groups on QQ, a popular social media platform, and the number of members in the groups quickly increased. Opponents of the project also wrote petitions, collected signatures, handed out leaflets, and carried out other offline campaigns, as well as mobilizing mass gatherings via online social networks.

On 22 April 2014, the Housing and Construction Department of Zhejiang Province announced that the construction of the JEEP would proceed. Locals were infuriated. Two days later, at the Bureau of City Planning, a small hearing was held at which citizens' representatives presented a petition against the project signed by over 20,000 citizens. They also submitted a request, signed by 52 leading protestors, for notarization of the *Revised Plan*. The Bureau of City Planning issued a written response, but the public did not regard it as plausible. People furthermore felt that subsequent opportunities to express their opinions were blocked, so they resorted to other approaches. Residents of the area adjoining the construction spot were not the only ones opposing the project either: people from other sub-districts and from the downtown area joined the opposition as well.

Then, on 26 and 27 April, opponents of the project from Zhongtai and western Hangzhou protested at shopping centres, at the Alibaba Group, and at Xixi National Wetland Park. Some of the protest leaders were detained by local police, and the marchers were asked to cancel the demonstration. The attitude of the government and the intervention by the police resulted in increasing public anger, so residents and their supporters planned an even larger protest. People's concerns had switched from one type of framing ('the waste-incineration power plant may cause health risks') to a new one ('the government has ignored basic human rights'). On 7 May, a message that the JEEP was to be launched spread secretly among the protestors, who got even more agitated. From that evening on, an angry crowd of over 10,000 people gathered at Jiufeng to stop the measuring work, and more protestors arrived during the following days and helped to block the way to Jiufeng. On 8 May, the government arranged a meeting among experts, and the media introduced the project's security assessment to the public, promising that there would be no dioxin emissions and that the waste gas would be processed under standards stricter than those applied by the EU (Wei 2014).

On 9 May 2014, the government of Yuhang District issued an announcement clearly stating that the work at Jiufeng would be stopped immediately and that it would not be started up again unless legal procedures and a public consensus allowed it (the Government of Yuhang District 2014). But this announcement failed to calm the public, who felt that their concerns about health, environmental quality, and housing prices had not been sufficiently addressed. The protest then continued for several days. On 10 May, residents in Zhongtai mistakenly assumed that some high-level government leaders were going to visit them, and they flocked around the toll gate of the highway. Public security forces were called out to disperse the crowd, and a fierce clash broke out at nightfall. Several vehicles were damaged and 39 people were injured, including some policemen. On the afternoon of 11 May, the city government held a news briefing to report on the conflict. Deputy Mayor Xu Liyi said that people's right to know would be protected and that the JEEP would reflect common interests. The most advanced equipment would be used in the project. The government of Yuhang District promised again that the project would not be started until the legal process had run its course and public expectations had been fulfilled (the Government of Yuhang District 2014). Earlier on that same day, moreover, a public notice had been issued by the People's Court, the People's Procuratorate, the Public Security Bureau, and the Judicial Bureau of Yuhang District. It urged protestors who had blocked traffic, attacked people, spread rumours, or damaged public or private property to turn themselves in. According to reports from the city police, 53 persons were detained and 11 surrendered themselves to the police. From 12 May on, a number of teams formed by public officials went to every village in Zhongtai, in an attempt to stabilize the situation and persuade local residents. Furthermore, a news programme on CCTV (*News 1 + 1*) reported and commented on the incident on 12 May, which helped to calm down the public.

On 11 September 2014, the government of Zhejiang Province released further information about the JEEP. The printed materials for the project were discussed, and detailed information was posted online for public inspection. The new proposal was to build a waste incineration power plant with a capacity of 3000 tons per day, consisting of four incinerators with a daily capacity of 750 tons each (Wei 2014). After a further round of siting analysis and environmental assessment, construction of the JEEP started on 14 April 2015. Only government officials, construction workers, and the police force were there. No opponents were present.

11.4 Spatial Justice and NIMBY Movements

The Jiufeng Protest in Yuhang is a typical NIMBY case, in which considerations of spatial justice are central. David Harvey tries to construct a theory of social justice for regulating questions of spatial or territorial allocation. He raises two questions: What are we distributing? And among whom or what are we distributing it (Harvey 2009, 99–100)? His concern is to examine the justifications that people provide when they make claims on the product of the society in which they live and work (Harvey 2009, 100). These are also the core problems connected with the NIMBY phenomenon. NIMBY explains the structural imbalance of interests that obtains in the distribution of social resources: that is, those who enjoy the benefits are not the same as those who pay the costs. Citizens accordingly find the distribution of benefits and burdens to be unfair and unequal. As an aspect of spatial justice, NIMBY reflects a conflict between individual or small-group interests on the one hand and public or collective interests on the other.

Citizens were no doubt fully aware of the urgent need for new waste disposal facilities. The question, however, was not whether to build such facilities, but instead where to locate them. The disagreement in the Jiufeng case arose despite the consensus obtained about the urgent need for the project. Effectively, there was a difference within the public, rather than a difference between the government and the public. For the government, it was an issue of location: that is, whether to build the facility at location A or at location B. But for the public it was a NIMBY phenomenon: that is, the object was to avoid having garbage dumped in my backyard when it could be dumped in other places. Effectively, this was a case, then, of environmental justice with narrow self-interest at its core. The purpose in such a case is simply to protect the relative environmental safety of certain groups based on the narrow interests of the location or region in question.

Some scholars have analysed the Jiufeng Protest in terms of certain presumed general patterns of environmental protest and come up with what appears to be a series of predetermined conclusions: the protest starts—the government represses—the protest escalates—the government compromises—the protest ends (To End NIMBY Movements 2013). Yet projects are still being initiated and plants are still being built, which is obviously not the ideal response to a NIMBY-type social movement. Some projects are carried out at the site initially chosen, but in a revised form after negotiation and compromise; some are moved elsewhere. But such siting and resiting are not done purely in accordance with considerations of spatial justice, for an underlying logic of power is still present. The more that people protest in a specific area, for example, the less is the likelihood that the facility will be built there. Even in the West, landfills are often placed in relatively weaker communities; and hazardous NIMBY facilities are often placed in areas with less political or economic strength. It is a common misconception, when public facilities for dealing with public bads are being built, that the government considers the interests of most people and assumes that the minority will compromise.

In China, public facilities like the JEEP are usually built in remote areas with few inhabitants. Remote places are always more suitable for the placement of hazardous facilities in terms of technology and risk assessment, as well as in a moral sense. But why are remote and sparsely populated places the best choice? Is it the case that remote areas with few inhabitants—in the extreme case with just a single resident—ought to bear the risks or costs associated with such facilities? In point of fact, we see a pattern here whereby disadvantaged social groups are oppressed in the name of the 'public good'.

However, Jiufeng Village is neither remote nor lightly populated (Yu 1 July 2014). There are four villages very close to the planned incineration plant: viz. Nanfeng Village, Nanhu Village, Zhongqiao Village, and Baiyun Village. They have a total population of about 10,000. Zhongqiao Village is only two kilometres from the site. The area, then, is not as sparsely populated as the media have reported. The centre of Zhongtai Sub-district, which is even more densely populated, is five kilometres from the site. There are many residential buildings in the area, and news of the planned construction of the incineration plant influenced the purchasing behaviour of buyers, causing losses to real estate development companies operating in the area (Yu 1 July 2014).

As to why Jiufeng Village was chosen as the site for the incineration plant, an official from the Bureau of City Planning explained that this was the choice that had to be made and that residents should take the overall situation into account. In the context of relations between officials and the public in China, then, taking the overall situation into account means conceding that the interests of the minority must be sacrificed in order to satisfy major collective interests. Jiufeng Village in Hangzhou's Zhongtai Sub-district was a relatively suitable place, and there were no other choices available. After the Jiufeng Protest broke out, Xu Liyi, Hangzhou's Deputy Mayor, openly told the public—meaning especially the residents around Jiufeng—that: 'We cannot get the garbage out of Hangzhou to dispose of it. To deal with this problem, we need a joint effort from the government and society' (Wei 13 September 2014).

Apparently, the sense of injustice felt by the residents around Jiufeng Village cannot be assuaged by a purely propagandistic appeal to consider the public need for waste disposal and the common interests and obligations entailed in urban environmental governance; nor can such an approach be conducive to the effective realization of spatial justice. Instead it will likely give rise to another round of environmental protests following the temporary suppression by the government. From the government's point of view, then, appropriate measures must be taken to compensate residents from around the location of the environmental facilities. Only in this way will it be possible to avoid the NIMBY dilemma and to fulfil the requirements of spatial justice in environmental governance. Procedural justice also calls for such an approach, as we shall discuss later.

NIMBY movements in China reflect an enhanced public awareness of questions of environmental protection and environmental rights. They also register the rise of a civil society in the country. For a long time, hardly anyone in Chinese society dared express dissatisfaction publicly, due to direct or indirect political pressure, even when people thought their interests had been violated (Chen and Bao 2013). But, as Feng (2007) points out, the people of China have dispensed to some extent with the traditional concept of hierarchy (see Introduction to this volume) and begun to exercise the modern rights of citizenship. On the other hand, we must stress that participants in Chinese NIMBY movements are not citizens (using the term 'citizen' here not in a legal but rather a political sense). Instead they are individuals or a loose group of people. Some scholars have described the NIMBY phenomenon as a kind of self-interested ideological or political tendency (Hunter and Leyden 1995). In essence and in most cases, however, the Chinese NIMBY movement is not a movement of civil society, but instead a display of private-property interests. It is not based on the idea of green justice and environmental protection.

As befits persons organized in defence of private property-based interests, NIMBY protestors depict people as individuals focused on their own interests and emotions. The concept of the so-called minority is basically a synonym for this. A citizen is of course a natural person. In this case, however, he/she is a natural person who is not involved in public affairs and who only takes part in activities together with others for the purpose of securing group or private benefits. He/she is not a 'public' individual—only a 'private' one. Persons engaged in defending their private interests do not engage in public discourse on public political issues; nor do they put forward reasons for their position that can be shared by all (Tan 2007).

This discussion of the role of private interests in Chinese NIMBY movements is not a moral criticism; even in Western democratic societies with a representative system, 'citizens' are influenced by interests rooted in 'private property'. What the Jiufeng Protest has shown is the particularistic value of individual freedom. Only public opinions were involved, not the general will. The general will focuses solely on public interests, whereas public opinions are based on private interests. Public opinions reflect the aggregation of individual wills; the general will, by contrast, is what remains after different individual wills have offset each other (Roussau 1985). There are only 'public opinions' in private society; there is no 'general will'. NIMBY movements do not seek environmental protection, or ecological justice, or the welfare of society as a whole. They may display flashes of the general will, but such flashes are not the general will itself.

Social development, urbanization, and environmental degradation have brought the common interests of human beings to the fore. People seek more and more to achieve a harmonious relationship with nature. It is clear from the Jiufeng Protest, however—and indeed from the pattern of environmental governance seen in the whole world—that human beings have not embraced a 'common life' built on a 'general will'. It is true that the ecological crisis and the deterioration of the environment are leading to a wider awareness that human beings are a community of destiny—and all the more so in densely populated cities like Hangzhou. But while people live a common life, they want to protect their own interests and to meet their own needs, rather than securing the common good. The common life is not based on public goals or public goods or the public interest. It comprises, rather, a collection of individual wills and desires (J. J. Feng 2011).

It cannot be denied that, notwithstanding a gradual increase in people's civic and environmental awareness, public consciousness and common spirit are still weak. Private interests are central in NIMBY movements: their supporters seek a reduction of garbage 'in my back yard only'. Due to this lack of consideration for other people's interests and demands, NIMBY movements can hardly be regarded as champions of environmental justice. Justice for the sake of self-interest is actually pseudo-justice, and pursuing it compounds the injustice that already exists. Realizing the goal of green justice—in which no one's backyard will be a dump for garbage and toxic substances—may be difficult, but there is no other way. People must embrace holistic attitudes; they must come to see human beings as a community of destiny.

Spatial justice has still not been realized in China. A major reason is that within the framework of the current political system, both government officials and the public have put collective interests above all and refuse other demands on the excuse of common interests. In the case of Jiufeng, the protests launched by the residents who would live in the pollution radius are not based on justice. Their actions were motivated by two factors: to get compensation or to force the government to change the construction spot. They did not care about the rights of the residents that might be affected due to a possible change of location. The main trouble in this case lies in China's current reality that the private rights are always buried and forgotten due to the emphasis on common interests or collective benefits. In fact, the selection of sites for 'NIMBY facilities' is still dominated by the logic of the strongest interests. Even in Western countries, waste disposal facilities are usually located in the relatively weak communities, and the final site of the unpopular NIMBY facilities is always in areas that are undeveloped or weak in politics (Lang and Xue 2015). During the construction of most NIMBY facilities in China, the government often considers the interests of the majority and takes the sacrifice of the minority as granted. Chinese mainstream thinking is to always treasure collective benefits while being indifferent to the opinions of the minority or individuals (Cui 2011). There are many successful cases of NIMBY movements in urban areas, but the resistance by rural victims that suffer from environmental pollution mostly seems powerless. The urban-rural difference in treatment of NIMBY movements leads to the tendency that waste disposal projects are moving from the city to the country or the suburb (Cui 2011). An official of the Bureau of City Planning of Hangzhou informed us during an interview that it was a choice that had to be made and that residents there should consider the overall situation:

> There is only one waste incineration plant in the west of Hangzhou at present, and this old facility is overloaded due to the increasing population in western Hangzhou. Since the capacity of the waste incineration plant is way less than the waste output of the western area, and since there is no room for extension, we are badly in need of a new waste incineration plant. Jiufeng Village of Zhongtai Sub-district is a relatively suitable place. Other than that, we have no choice (Yu 2014).

11.5 Procedural Justice and Governmental Decision-Making

Procedural justice has not yet been clearly defined in the literature. Generally speaking, it refers to fairness in the processes employed for the resolution of disputes and the allocation of resources. According to Rawls (1999), procedural justice has three forms: pure procedural justice, perfect procedural justice, and imperfect procedural justice. The main requirement of justice, as he saw it, is that society be structured in such a way that basic rights and obligations are properly allocated. In order to achieve this, the social system can be designed in accordance with the concept of pure procedural justice, thereby ensuring that the outcomes produced within it are always just. Rawls defines pure procedural justice as follows:

> There is no independent criterion for the right result: instead there is a correct or fair procedure such that the outcome is likewise correct or fair, whatever it is, provided that the procedure has been properly followed. (Rawls 1999, 73)

Obviously, Rawls's analysis of justice cannot be applied to legal processes only. It must cover all aspects of politics and the economic structure of society as well.

Questions about the legitimacy of procedures or about the procedures themselves are profound and of wide concern to the public. Such questions also reflect the core content of procedural justice and of green justice as well. That is to say, when people address activities or decisions that have a positive or negative impact on some people but not on others, they cannot concern themselves solely with the legitimacy of the results. They must also consider the legitimacy and rationality of the process that led to those results.

When studying environmental protests in China, such as the Jiufeng Protest examined here, we can acknowledge the positive aspects of the movements in question from the perspective of the public, despite the deficiencies noted above regarding the lack of proper spatial justice animating such movements, due to their pursuit of 'private' interests. For one thing, NIMBY movements show that Chinese citizens have strengthened their demands for environmental policies, while their consciousness of the need for environmental protection has grown ever stronger. For another, such movements reflect an enhanced awareness on the part of the Chinese public of both environmental imperatives and the need to protect rights. People start learning how to mobilize environmental movements in order to protect and pursue their own environmental rights. These phenomena testify to considerable improvement in Chinese civic society in regard to social participation and the protection of rights. As civic consciousness awakens, social activities are spreading.

As for the government, its main problem lies in the general absence of transparency in its decision-making process, which leads to a lack of procedural justice. The government still takes a 'top-down' administrative approach for the most part, rather than a public participatory approach engaging with multiple agents or stakeholders. Although Chinese governments from the central level down to the local have already adopted the latter approach to governance in part, the traditional top-down management pattern still dominates in practice (c.f. Delman, Chap. 8). Improving environmental governance also requires a long-term process. In the case of the Jiufeng Protest, the Hangzhou government's determination to build a new waste disposal facility was a specific example of a broader motivation. It wanted to pursue environmental protection for the benefit of all citizens and to enhance the city's capacity to handle its environmental challenges. However, the decision gave rise to heavy opposition from local residents in Zhongtai Sub-district and even led to a large-scale environmental protest which had an enormous influence in China. In fact, the government's plan to build such a facility was reasonable, as was the ultimate result. But rationality and political legitimacy were absent from the process. In the absence of procedural justice—with its deliberative mechanisms, transparent communication channels, well-established supporting policies, a wide dissemination of information about environmental protection, and the like—local citizens acted to protect their environment and their private interests through violent confrontation. Under such conditions, namely, they believe their environmental rights will not otherwise be guaranteed.

Clearly, the processes of environmental governance in relation to the JEEP were marked by a lack of procedural justice. During the early stage of planning, the government and the construction company failed to submit the project to public deliberation or approval, and local residents had no access to the decision-making process. Similar deficiencies are found in other places. In contemporary China, a lack of specific procedures for public participation makes processes of this kind highly formalistic. Furthermore, legal provisions in this area fail to specify how citizens can truly participate in environmental governance, supervision, and impeachment. The Jiufeng Protest arose in response to a lack of public participation before, during, and after the decision-making process. On the other hand, the protest itself can be seen as a form of public participation.

In the terms of Sherry R. Arnstein's Step Theory (1969), we may say that the Chinese public can participate in environmental governance in three ways: through operation, through notification, and through consultation. However, an inadequate legal system, a resort to administrative operations in lieu of processes of law, and an overwhelming tendency towards governance by men and not by laws still pose a challenge to the democratization of the environmental decision-making process.

In most cases, a legitimate policy-making process requires professional arguments, consultations, and public hearings to prove that a facility will be damage-free environmentally. Spreading positive publicity beforehand leads to better results than lobbying people afterwards to persuade them that the project will be strictly managed. A transparent channel of communication between the government and the public is also necessary in order to mitigate information asymmetry. When there is positive interaction between the government and the public, a consensus about the project or a deal for compensation is more likely to be achieved. In reality, however, distrust and conflict tend to arise, due to inefficient communication and an authoritarian model of management.

In the case of Jiufeng, people developed a deep distrust of the government. According to residents of Zhongtai Sub-district whom we interviewed and surveyed, the key point was that people had no trust in the government in this case. In our survey of citizens in Nanfeng Village, Zhongqiao Village, Baiyun Village, and so on, 100% reported opposing the construction of the waste incineration power plant, and 45% reported refusing any form of compensation offered by the government or the waste disposal company, on the grounds that such compensation could not make up for the damage to the environment and to their quality of life. Moreover, in response to the following question: 'Why do you oppose the plan to build trash incinerators?', 100% responded that 'the emission of toxic materials such as dioxin will do harm to the environment and to our health'. 85.5% answered that 'the government can be no longer trusted', and 73.7% believed that 'local housing prices will be affected'.[2]

This distrust was focused on three questions: technological standards, management capabilities, and public credibility. The government promised that the technology and equipment used in the project would exceed the EU 2000 Standard. But the public remained highly suspicious, and the Jiufeng Protest badly undermined the government's credibility. According to our small survey, villagers felt dissatisfied with government at all levels. In terms of the satisfaction of our respondents,

[2] Based on an interview survey of 20 residents at Zhongtai Sub-district on 10 July 2014

provincial governments, municipal governments, district-level governments, and township governments ranked from ninth to twelfth among 13 candidates.[3]

Meanwhile, it may be worthwhile to look at attempts in other countries to cope with problems relating to procedural justice in this area. There is a close link between democracy and environmental protection. Some scholars find that democracy reduces all five types of environmental degradation (Li and Reuveny 2006). In France, large construction projects with a substantial impact on the environment must be debated in public. In the USA, a system of information disclosure on environmental questions encourages public participation. In Japan, the process by which sites are selected for the construction of waste incineration power plants helps to reduce public resistance and to maintain open channels of communication between the government and the public (Hang and Zhou 2011).

The government's closed system of decision-making is a major cause of NIMBY movements in China today. Closed decision-making is regarded as partisan: the government has the final say, and there is no effective channel for communication between the government and the public. Chinese policy-making has always followed a top-down pattern. The public has been accustomed to having no other option than to obey. In connection with the construction of public environmental facilities, decision-making of this kind fails to take proper account of the vital interests of the residents involved. It does not give adequate consideration to the environment either. Environmental public facilities like garbage disposal plants do impose some negative externalities. By their nature, the benefits and burdens arising from such facilities fall unevenly on the residents of different regions. Therefore, reasonable and appropriate compensation should be considered at the initial stage, rather than being taken as an emergency action when protests arise (or afterwards).

It is true that joint communication platforms have been set up in some cities (cf. Schlæger and Zhou, Chap. 7), but they are not always designed to serve as bottom-up information channels through which the public can be heard. The Jiufeng Protest reflected the *truth absence* when it came to public participation in decision-making and the lack of sufficient opportunities for people to present their demands. Even if public participation in relation to environmental assessments, hearings, consultations, dialogue, etc. in cases like this one is possible in some sense, it does not function properly. Certain key procedures are largely ignored, and at times the government pretends to go through the motions in order to fool the public. The Jiufeng project was an example. The project proposal was published online, and a hearing was held. But much of this only took place after public resistance. Local residents were not respected and their rights were not protected. They could only express their dissatisfaction and influence decision-making by resorting to extreme measures.

Furthermore, closed decision-making makes it hard to address people's perception of the risks involved in different projects, which is an important psychological factor contributing to the rise of NIMBY movements. Residents of areas close to hazardous facilities, like waste disposal plants, will tend to feel unsafe in that environment. People sense instinctively that garbage disposal facilities will

[3] Based on an interview survey at Zhongtai Sub-district

have a negative impact on their physical health and quality of life. And, indeed, environmental facilities of this kind do carry a certain risk. But what if the government were to inform the public in plain language about the incineration technology to be used, the methods by which the facility will be managed, the monitoring techniques to be employed, and the international practices which are followed in this area? If the government were to level with the public in this way before making decisions, the worries of the latter could be relieved, and the likelihood of the emergence of a NIMBY movement could be reduced.

Neither moralism nor idealism can help explain or address the NIMBY phenomenon in China. The realistic practical path is to understand, to compromise, to reach consensus, and to cooperate. Then, how is it possible to reach common recognition and achieve cooperation? To answer this, we need to go back to the basic approaches, that is, the transparency of consultation and public participation. Public participation is essential to effective environmental governance. However, the public's role has been limited in China. But, clearly, NIMBY movements can promote more inclusive decision-making processes and help consolidate governance reform (Johnson 2010). The government is the key to the problem, and it holds the initiative. There is a need for institutions and a legal system to enforce transparency and public participation. For example, information disclosure and transparency should be clearly expressed in the law, as is the case in many countries. There was, for example, a successful case of public participation in New York. In 1991, the Department of City Planning of New York issued a set of *criteria city facilities* and then executed a *fair share sitting process*. This process brought diverse parties into the process of deliberation to reach the most desired results.

11.6 Conclusion

The Jiufeng Protest is a typical example of the kind of public resistance that arises as urbanization proceeds within a country, and attempts are made to develop new types of local environmental governance. By developing and using our analytical framework of spatial justice and procedural justice, we have identified an absence of justice in relation to the JEEP which resulted in the residents' protest with support from other civic groups. Environmental protests and NIMBY movements reflect the unequal and unjust distribution of economic and environmental benefits and burdens among different interest groups, as well as the play between self-interest and the interest of others, between private interest and the common interest. The probability of protest is especially high in connection with the construction of environmental infrastructure like waste incineration power plants, which provide environmental protection and positive externalities for most of the public. But they also entail negative externalities for local communities. NIMBY movements are principally driven by the self-interest and private interest of members of the public. Therefore, the key to resolving the problem is to balance public and private interests—overall interests and partial ones—and to ensure that externalities are internalized. It is also important for residents to express their demands through rational methods. As in similar protests against Waste to Energy (WTE) incineration in Guangzhou, for example, urban homeowners use rational resistance to distinguish themselves from both villagers and the state. Their focus on rationality is a critique of the government's reliance on technology to resolve the social problem of waste management (Zhang 2014).

After all, building public facilities like a waste incineration power plant is intended to satisfy certain people's demands for a better life for themselves, but it will inevitably undermine the interests of other people. Therefore, all of the parties involved—the government, residents, interest groups, and others—might find it useful to adhere to green justice and holism when dealing with public bads like municipal waste. They should do this on the basis of common and long-term interests in order to share environmental public goods in a rational way.[4]

References

Arnstein, S. R. (1969). A ladder of citizen participation. *Journal of the American Institute of Planners, 35*(4), 216–224.

Baxter, B. (2004). *A theory of ecological justice*. New York: Routledge.

Bryant, B. (1995). *Environmental justice: Issues, policies, and solutions*. Washington, DC: Island Press.

Bullard, R. D. (1999). Environmental justice challenges at home and abroad. In N. Low (Ed.), *Global ethics and environment*. New York: Routledge.

Bureau of Urban Planning of Hangzhou. (2014). *The revised draft of professional environmental sanitation planning of Hangzhou (2008–2020)*. Retrieved May 20, 2014, from http://www.hzplanning.gov.cn/DesktopModules/GHJ.PlanningNotice/PublicityInfoGH.aspx?GUID=20140328143144609

Chen, Z. J., & Bao, Z. M. (2013). Zhidu bianqian, liyi fenhua yu nongmin huanjing kangzheng—yi hunansheng X shi Z diqu wei ge'an [Institutional change, interest differentiation and farmers' environmental resistance—A case in Z district, X city of Hunan Province]. *Zhongyang Minzu Daxue Xuebao (Zhexue Shehui Kexue Ban)* [Journal of Minzu University of China (Philosophy and Social Science)], (4), 50–61.

Cui, Z.(2011). Linbi Yundong Zai Zhongguo[NIMBY in China]. *Xinshiji Zhoukan(Century Weekly)*. (43), 97–101.

Dobson, A. (1999). *Justice and the environment: conceptions of environmental sustainability and theories of distributive justice*. New York: Oxford University Press.

Feng, J. J. (2011). Gonmin jiaoyu mubiao de dangdai jiangou [The contemporary construction of civic education goals]. *Jiaoyu Xuebao* [Journal of Educational Studies], 07(3), 34–40.

Feng, S. Z. (2007). Kuaisu zhuanxing shiqi de gonggong canyu yu shehui maodun [Public participation and social conflicts in the rapid transformation period]. In Zheng, H. S. (Ed.) *Zhongguo shehui fazhan yanjiu baogao* [Research reports on the development of Chinese society] (pp. 181–203). Beijing: Zhongguo Renmin Daxue Chubanshe [China Renmin University Press].

[4]As of the day that this chapter is finished, the JEEP has been completed and put into use. The local government is developing tourism and enhancing the infrastructures in nearby villages as compensation.

Government of Yuhang District. (2014). *Guanyu jiufeng huanjing nengyuan xiangmu de tonggao* [The announcement on jiufeng environmental energy project]. Retrieved May 14, 2014, from http://www.yuhang.gov.cn/xxgk/gggs/qt/201405/t20140511_931367.html
Hall, P. (2009). *Cities of tomorrow: An intellectual history of urban planning and design in the Twentieth Century* (3rd Edn., Trans. Tong Ming), Shanghai: Tongji University Press.
Hang, Z. F., & Zhou, M. L. (2011). Riben laji fenshao changsuo de quwei jueding yu zhengce boyi [Geographical determination and policy game-playing of waste incineration location in Japan]. Diyu Yanjiu Yu Kaifa [Areal Research and Development], *30*(4), 11–15.
Harvey, D. (2009). *Social justice and the city*. Athens: University of Georgia Press.
Hayden, P. (2005). *Cosmopolitan global politics* (Vol. 79). Aldershot: Ashgate.
He, M. J. (2013). *Urban planning in the spatial constitutionalism*. Nanjing: Southeast University Press.
Hofrichter, R. (Ed.). (1993). *Toxic struggles: The theory and practice of environmental justice*. Philadelphia: New Society Publishers.
Huang, M. J. (2006). Lun zuowei zhengti gongping de shengtai zhengyi [Ecological justice as overall fairness]. Dongnan Xueshu [Southeast Academic Research], (5), 131–137.
Hunter, S., & Leyden, K. M. (1995). Beyond NIMBY: Explaining opposition to hazardous waste facilities. *Policy Studies Journal, 23*(4), 601–619.
Jamieson, D. (2003). *A companion to environmental philosophy*. Malden: Blackwell Publishing Ltd..
Johnson, T. (2010). Environmentalism and NIMBYism in China: Promoting a rules-based approach to public participation. *Environmental Politics, 19*(3), 430–448.
Lang, Y. X.and Xue, X. J. (2015). 'Simin Shehui': Jieshi Zhongguoshi 'Linbi' Yundong De Xinkuangjia['Private Society': A New Framework to Explain NIMBY in China] , *Tansuo Yu Zhengming*(Exploration and Free Views)(12), 37–42.
Li, B. R. (2008, January 22). Huanjing lunli guanzhao shengtai zhengyi [Environmental ethic concerns of ecological justice]. *Zhongguo Jiaoyubao* [China Education Newspaper], p. 3.
Li, H. R. (2002) Shengtai zhengyi lun [A theory of ecological justice]. Taiyuan Shifan Xueyuan Xuebao [Journal of Taiyuan Normal University], (2), 11–13.
Li, P. C. (2005, March 15). Lun shengtai zhengyi [A theory of environmental justice]. *Guangming Daily*, p. 10.
Li, P. C. (2007). Duowei shijiao xia de shengtai zhengyi [Ecological justice in multi-dimensional perspective]. *Daode yu Wenming [Morals and Civilization], 2*, 10–14.
Li, Q., & Reuveny, R. (2006). Democracy and environmental degradation. *International Studies Quarterly, 50*(4), 935–956.
Ma, Y. (2014). Hangzhou huiying xinjian laji fenshaochang xuanzhi Yuhang Jiufeng: Biewu taxuan [Response to the siting of the waste incineration power plant in Yuhang Jiufeng: There is no other choice]. *Zhejiang Daily*, April 28, p. 2.
Pepper, D. (2005). *Eco-socialism: from deep ecology to social justice* (Trans. Liu, Y.). Jinan: Shandong University Press.
Rawls, J. (1999). A theory of justice (Rev. Edn.). Cambridge, MA: Belknap.
Roussau, J. J. (1985). *The social contract. (He, Z. W., Trans.)*. Shanghai: The Commercial Press.
Sandler, R., & Pezzullo, P. C. (2007). *Environmental justice and environmentalism: The social justice challenge to the environmental movement*. Cambridge, MA: The MIT Press.
Schlosberg, D. (2007). *Defining environmental justice: Theories, movements, and nature*. New York: Oxford University Press.
Tan, A. Q. (2007). Siren, gongmin, zhexuejia: Zhengzhi zhexue zhong de sanzhong siwei fangshi [Private person, civics and philosopher: Three modes of thinking in political philosophy]. Zhongshan Daxue Xuebao (Shehui Kexue Ban) [Journal of Sun Yat-sen University (Social Science)], *47*(5), 71–76.
Vogt, W.(1981). *Shengcun zhilu [Road to survival]*. (Zhang, Z. M, Trans.). Beijing: Shangwu Yinshuguan [The Commercial Press].

Wang, L. L. (2012). Jingji zhengyi yu huanjing zhengyi—zhuanxingqi woguo chengxiang guanxi de lunli zhiwei [Economic justice and environmental justice—The ethical dimension of the urban-rural relationship in the transition period]. *Lunlixue Yanjiu*[Studies in Ethics], (6):37–41.
Wei, H. F. (2014, September 13). Jiufeng laji fenshao fadian gongcheng guihua xuanzhi gonggao [The announcement on planning and siting of Jiufeng waste incineration power plant]. *Today Morning Express*, p. 2.
Wenz, P. S. (1988). *Environmental justice*. New York: State University of New York Press.
Wong, N. W. M. (2016). Environmental protests and NIMBY activism: Local politics and waste management in Beijing and Guangzhou. *China Information, 30*(2), 143–164.
Wu, C. H. (2013). Huanjing quntixing shijian de guonei yanjiu zongshu. [A review of studies on environmental public protests in china]. *Qingchun Suiyue*[Youth], (6), 447–449.
Wulfhorst, J. D., & Haugestad, A. K. (Eds.). (2004). *Future as fairness: Ecological justice and global citizenship* (p. 10). Amsterdam/New York: Rodopi.
Yu, F. (2014, July 1). Hangzhou Yuhang laji fenshao kunju [The predicament of waste incineration in Yuhang, Hangzhou]. Beijixing Dianliwang [Polaris Power Web]. Retrieved April 10, 2015, from http://news.bjx.com.cn/html/20140701/523250.shtml
Yu, X. F., & Wang. J. L. (2009). Lvse zhengyi: shengtaianquan yu shengtaiwenming de jiazhi zhiwei [Green justice: ethical dimensions of ecological security and ecological civilization]. In S. Liang & L. Y (Eds.), *Quanqiu dabiange yu zhongguo duiwai dazhanlue* [Global transformation and strategies in Chinese foreign relations] (74–86). Beijing: Shijie Zhishi Chubanshe [World Affairs Press].
Yu, Y. H., & Zhang, Y. L. (2014). Woguo huanjing 'linbi yundong' kunjing neiyin yu huajie. [Internal causes and solutions to environment 'NIMBY movement' in China]. *Huangjing Baohu* [Environment Protection], (18), 49–51.
Zeng, J. P. (2007). *Huanjing zhengyi—fazhanzhong guojia huanjing lunli wenti yanjiu* [Environmental justice: a research on environmental ethics in developing countries]. Jinan: Shandong Renmin Chubanshe [People's Press of Shandong].
Zhang, A. (2014). Rational Resistance: Homeowner contention against waste incineration in Guangzhou. *China Perspectives, 2*, 45–52.
Zhongjie linbi yundong [To End NIMBY movements]. (2013). *Wanxia [Sunset Glow]*, (2), 41.

Chapter 12
Civic Engagement and Sustainable Development in Urban China: Policy Lobbying by Social Organizations

Zhang Changdong

Abstract Sustainable development requires that China's governance structure be changed: from a state-led model to one of state-society collaboration, in which government initiatives work with civic engagements. Over the last decade, which saw rapid economic growth together with rising concerns about unsustainability, the Chinese government started to address the problem of sustainability. At the same time, many non-governmental organizations (NGOs) are trying – regardless of their varying degrees of dependence on the state – to promote sustainable development, among other things by lobbying the authoritarian government to bring a variety of stakeholders into the policy-making process. Using comparative case studies and a quantitative analysis of policy lobbying by NGOs, this study finds that many NGOs are able – regardless of their varying degrees of autonomy from the state – to influence government policy-making through various channels. At the same time, there is quantitative evidence suggesting that sustainability-related NGOs – i.e. NGOs dedicated to the defence of the environment and of vulnerable social groups – have a stronger degree of policy influence compared to other NGOs.

Keywords State-society collaboration · Non-governmental organization · Civic engagement · Environmental protection · Cities

After more than three decades of economic development and market transition, the Chinese government in the early 2000s started to emphasize sustainable development as an important challenge and to make efforts to deal with it. Certain guiding ideas, such as "harmony" and "scientific development", have been introduced as a new ideology of national development. The purpose is to deal with the negative effects of rapid economic growth, among them severe pollution, increasing instability, and rising inequality – all of which undermine the sustainability of further

Z. Changdong (✉)
School of Government, Institute of State Governance and Center for Civil Society Studies, Peking University, Beijing, People's Republic of China
e-mail: zhangchd@pku.edu.cn

J. Delman et al. (eds.), *Greening China's Urban Governance*, ARI - Springer Asia Series 7, https://doi.org/10.1007/978-981-13-0740-9_12

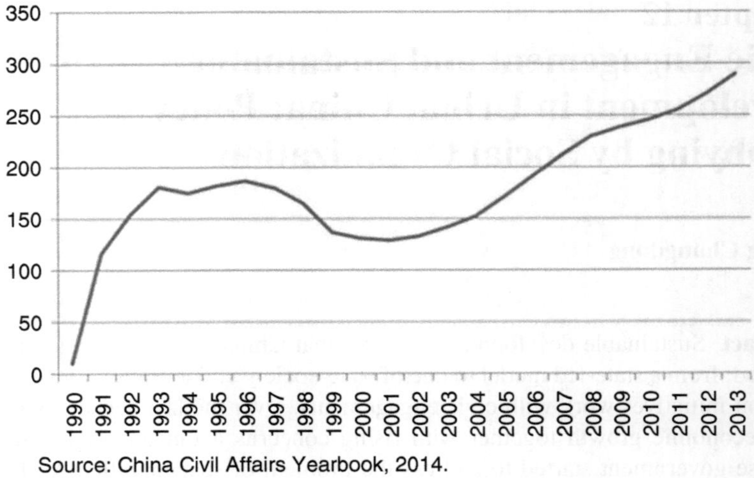

Fig. 12.1 Growth of NGOs 1990–2013. Unit: 1000. (Source: China Civil Affairs Yearbook, 2014)

development. These new ideas also entail a number of specific policies: for reducing pollution, maintaining stability, improving social welfare programmes, and the like.

At the same time, as shown in Fig. 12.1, a great many officially registered NGOs have emerged in China. Beyond these registered NGOs, moreover, there are many unregistered grassroots NGOs. As many scholars have noted, these NGOs may be able to engage with government policy-making (Kennedy 2005; Teets 2013; Zhang 2015, 2017) and to contribute to sustainable governance, particularly in areas like environment protection (Mertha 2009; Wu 2009a, b), aid to vulnerable social groups (Spires 2011), and so on. However, certain questions are still understudied and/or under-theorized, such as how and why NGOs contribute to sustainable development. In this chapter, I attempt to answer this question and to theorize it. I contend that NGOs can contribute to sustainable development through two mechanisms – one of which involves constraint and the other collaboration.

My discussion of these two mechanisms is organized in four sections. In the first I elaborate on the two mechanisms, and I construct a theoretical framework for analysing the relationship between NGOs and sustainable development. In the second I discuss two cases, one involving a GONGO and the other a grassroots NGO. In the third section, I use quantitative data to examine the external validity of the argument on the basis of my case studies. In the fourth section, finally, I present my conclusions.

12.1 NGOs and Sustainable Development

I take my definition of urban sustainability governance from D. H. Lund (Chap. 1):

> …..[U]rban sustainability is a highly complex matter. Not only is the city a complex system of material flows, human activities, and governance structures that link the city with is immediate surroundings and with global networks. It is also the case that the conditions for effective governance are becoming more complex, as greater numbers of actors gain influence through both global and highly local initiatives. If we are to move towards sustainable urban development, steering and coordination will be essential. Due to the virtually all-encompassing nature of the efforts required, moreover, all of the different agents who influence the city will have to take part.

Sustainable development involves many issues, but I focus in this chapter on just one: the implications of changing governance structures for sustainable development. I argue that the structure of urban governance in China is becoming more open and pluralistic and that social organizations are trying to influence government policies more and more. These changes may enhance the prospects for sustainable development.

Theoretically, there are two mechanisms through which NGOs can contribute to sustainable development: they can act as a constraint on the government, preventing it from pursuing unsustainable development policies; alternatively, they can assist the government in making and implementing sustainable policies. I refer to these as the "constraining" and "collaborating" mechanisms, respectively.

The constraining mechanism matches the picture painted by pluralist theory and modernization theory. Robert Dahl (1971), for instance, argued that democracy is a system marked by participation and contestation, in which social organizations check the power of government and defend the interests of society. Seymour Martin Lipset declared (1959), in his seminal paper, "Some Social Requisites of Democracy", that economic development leads to the emergence of social organizations which constrain the power of the state and mediate between it and society – thereby making government more responsive. Rapid economic growth and the emergence of NGOs provide the conditions for checking government power. One concept associated with this literature is "civil society", which comprises organizations autonomous of the state which serve to constrain its power.[1]

The collaborating mechanism reflects the theory of mutual empowerment developed by Joel Migdal (1994), as well as Peter Evans's ideas of state-society synergy (1997) and embedded autonomy (1995), well summarized and elaborated in a review article by Xu Wang (1999). However, the mutual empowerment theory differs from the notion of a constraining mechanism, which is based on a "state-versus-society framework that rests on a view of power as a zero-sum conflict between state and society". The mutual empowerment theory, by contrast, portrays states as varying "in effectiveness based on their ties to society. Under certain circumstances

[1] There have been many theoretical and empirical studies of civil society over the last 20 years. For a review of civil society's implications for transitional countries, see Ding (1994). For some edited volumes on the development of civil society in China, see Miller (1992) and Brook and Frolic (1997). I have reviewed some of this literature in Zhang (2015), so I will not discuss the details in this chapter.

states and social forces may be mutually empowering. In other words, some interactions between the state and social forces can have the effect of creating more power for both sides" (Wang 1999, 232). Evans (1995) elaborates a framework known as "embedded autonomy" for specifying the conditions for achieving collaboration or mutual empowerment (the focus is on government-business relationships, but the framework also applies to state-society relationships). Evans argues that, when there is a joint project involving the state on the one hand and economic/social actors on the other, a relationship of embedded autonomy may be generated – if the state is coherent and ties between the two sides are close. Such embedded autonomy forms the institutional foundation for collaboration.

The overall effects of the two mechanisms also find support in the work of Robert Putnam. Putnam finds that social capital, measured by civic activities or people's engagement in civic organizations, can improve government performance (Putnam 1993). We must bear in mind, however, that the two mechanisms will only have their desired effect if the NGOs in question truly intend to promote sustainable development. But in fact some NGOs, especially business associations in which polluting industries are members, may lobby against sustainable policies. To deal with this challenge, I have carried out both qualitative and quantitative studies of the question. Moreover, I have differentiated business associations, environmentally oriented groups, and NGOs focused on the needs of vulnerable groups from each other, as well as from other social organizations.

12.2 NGOs and Sustainable Development in China: Qualitative Evidence

Many China scholars see the role of NGOs in furthering sustainable development as consisting in the provision of social services, such as orphan care, labour rights protection, environmental conservation, or infectious disease control (Schwartz and Shieh 2009). Others see the function of environmental NGOs as going beyond providing public services: i.e. as directly or indirectly advocating and/or organizing collective petitions to push local governments to protect the environment and the well-being of communities (Cooper 2006; Wu 2009a, b; Zhan and Shui 2013). As noted elsewhere in this volume, Green Zhejiang plays a significant role in promoting environmental protection in Zhejiang Province and its capital city, Hangzhou.

As shown in Table 12.1, NGOs in China may be grouped into four categories. Given limitations of space, as well as the availability of studies of the role played by other types of NGO in environmental protection, I will examine only two types in

Table 12.1 Four types of NGOs

Registered under the Bureau of civil affairs		Registered as business	Unregistered
Government-owned/organized NGOs (GONGO)	Self-organized NGOs	Companies that work as NGOs, e.g. Global Village	Grassroots NGOs

this section: GONGOs and grassroots NGOs. In the process, I will illustrate two ways in which NGOs can contribute to sustainable development. Then, on the basis of existing research (include other chapters in this volume), I will make a comparison with the other two types.

12.3 GONGOs and Sustainable Development: The Case of the Association of Energy Conservation and Resource Utilization[2]

Under a system of corporatism, government bodies may organize NGOs to assist them in making and implementing policy. An NGO organized in this way serves as a "bridge and belt" between the government (as its sponsor) and its own members.

The Association of Energy Conservation and Resource Utilization (hereafter the AECRU) is a professional association established in 1988 by city government A. It has group members (*tuanti huiyuan* 团体会员) from relevant enterprises, research institutes, and public service units (*shiye danwei* 事业单位). Since the majority of its members are enterprises, it resembles a trade association in its daily functions. After several rounds of government reform, its supervisory unit is the city government's Committee of Development and Reform (hereafter the CDR), which is a very powerful government institution involved in environmental protection.

The goal of the association is to promote environmental protection through improved energy conservation and recycling. Its operations are described as follows: "Working with small and medium-sized enterprises, the large enterprises run the Association, serve the members and the government, and work as the bridge and belt between city government and enterprise". According to the CDR, the association's mission is to do the following:

1. To implement the city government's energy policies and guidelines, to assist it in improving energy conservation, to build a bridge and belt between government and enterprise, and to convey its members' views to the authorities
2. To do research on energy conservation and integration utilization, to collect relevant information, and to help the city government enact and implement energy-related laws, regulations, and standards
3. To publicize the city government's energy conservation guidelines and policies, to conduct academic communication, and to edit and publish relevant material
4. To promote new technology and products in the area of energy conservation
5. To provide consultative services on energy conservation to its member enterprises and to press the city government for more professional training
6. To promote technological improvement by organizing member enterprises to share information and to improve cooperation
7. To promote international cooperation on energy conservation
8. To take on other tasks delegated by the city government

---

[2] Material for this case comes from an interview with the association's leaders and staff members on 13 September 2015 and from materials provided by the association.

The association's secretary, Mrs. Huang, had worked in the city government's CDR for many years before moving on to the association. When she was in charge of environmental protection in the late 1990s, she noticed that her colleagues regarded environmental protection as a minor issue. Her office was seen as a marginal apparatus of the CDR. "But I was not frustrated. I believed that environment protection was going to be a very important challenge for this city and for China in the near future". She moved to the AECRU from the CDR 3 years before retirement and worked there as secretary and then chairwoman. Using her network within and outside government, as well as her professional knowledge, she led the AECRU competently and did much to improve energy conservation.

The current leadership of the CDR consists of five members: the chairperson, vice-chair, executive chair, secretary, and deputy secretary. Four of these persons are very senior and have "significant" influence within the industry (including the membership and the relevant government apparatus). There are 15 full-time staff members working in 4 departments. Four of the five leaders are staff members (except Mrs. Huang, who is a retired part-time staff member).

In 2013 and 2014, the AECRU did much to carry out the eight tasks listed above. I will focus here on two of these: to take on tasks delegated by the city government and assist it in its work and to act as a representative for its members and an advocate for policies benefitting them.

On account of the AECRU's professional knowledge, the city government, especially the CDR, delegated many tasks to it each year. I will give a few examples here. In 2013 and 2014, the AECRU worked with the city government and the evaluation team composed by experts to collect, check, and evaluate application materials from 203 enterprises. It also provided consulting and instruction to enterprises to help them meet the city government's requirements for obtaining privileges in the area of industrial policy (including tax breaks). Another case concerned data collected by the AECRU on enterprises' use of electronic devices and their demand for new energy-saving electronic machinery. Using these statistics, the association helped the enterprises upgrade their machinery and technology and lobbied the city government to support them. In the third case, the AECRU worked with a district government to monitor enterprises' electrical and other energy usage, with an eye to evaluating whether and to what extent they were following government policies for energy conservation. It also made a suggestion, adopted by the district government, on how to improve energy conservation.

The leadership and staff of the AECRU also know how important it is to serve the members in order to get their support (especially financial support: membership fees furnish about 60 percent of the association's annual income). They work hard, therefore, to represent the interests of members. In 2103, for example, when the city government suddenly started enforcing its strict environmental protection policies (which up to that point had not been seriously enforced), the AECRU asked it to grant a 1-month extension for enterprises to make improvements and to get the necessary certificates. The AECRU also carried out surveys among enterprises and made specific suggestions as to how government regulations and policies could be revised in order to improve environmental protection.

The AECRU thus plays the role of "bridge and belt" between government and enterprise, thereby contributing to environmental protection. The mechanism at work in this case is the collaborating mechanism, with the city government, the trade association, and the enterprises working together to promote environmental protection. In my fieldwork in city A, during 2013–2015, the AECRU was not the only successful case: there were also a few other cases where government-sponsored organizations helped implement policies to protect the environment, saving energy, reducing pollution, protecting rare and endangered animals and plants, etc. While some of these GONGOs mobilize social support to a degree, some of them suffer several weaknesses. First, top-down policy-making or government-led agenda-setting may exclude many important issues. There is a need for some bottom-up agenda-setting that reflects citizens' interests and concerns. Second, some GONGOs are not staffed or financed very well. In most cases, they are staffed by two to three retired government officials and receive financing from a government "project fund".

12.4 Grassroots Environmental NGOs

In a study of dam building in southwestern China, Andrew Mertha argues that, as China has become increasingly market-driven, decentralized, and politically heterogeneous, the control and management of water has been transformed from an unquestioned economic imperative into a lightning rod for bureaucratic infighting, societal opposition, and open protest. In an associated change, moreover, the policy-making process has moved from "fragmented authoritarianism" to "fragmented authoritarianism 2.0": "[A]though bargaining has always been present in Chinese politics, more recently the media, nongovernmental organizations, and other activists – actors hitherto denied a seat at the table – have emerged as serious players in the policy-making process" (Mertha 2009).

One approach that grassroots NGOs can take is to work with local governments – especially their environmental protection bureaus – to protect the environment and achieve sustainable development. Unlike GONGOs, however, grassroots NGOs have greater autonomy, so they can work against local governments disposed to develop the economy at the cost of sustainability. In other words, both collaborating and constraining mechanisms work for grassroots NGOs, although their success in using them depends on many factors besides their own efforts – especially support from high-level government and from the media.[3] An official report describes the relationship between environmental NGOs and the government as cooperative:

[3] This fits the state-in-society approach well: neither the state nor society are unitary actors – they are divided into different segments. Different state actors may have very different interests and ideas, as may different social actors. To achieve their goals, certain state actors ally with certain social actors, sometimes against alliances between other state actors and social actors. See Migdal (2001) for a theoretical discussion.

"(NGOs should) To help but not make trouble, to participate but not intervene, to monitor but not replace (the role of the government), and to obey the law" (Bureau of Environment Protection 2007). There are many studies on this topic (e.g. Mertha 2009; Steindhart and Wu 2016), so I will review just two cases briefly.

"Green Home Volunteers" (GHV), an initially unregistered environmental NGO, was established in 1996 by Mrs. Wang Yongchen. The volunteers worked together to plant trees along the Yellow River, to plant and protect grass in the desert, and to protect the white-flag dolphin. In 2000, they realized that journalists are valuable for protecting the environment, because of the influence exerted by the media, the coverage provided by them, and the professional skills and knowledge of journalists. The GHV decided, therefore, to organize a journalists' salon. In 2006 it started a project called the "10-year tour of rivers", in an effort to call attention to questions of environmental protection, especially the need to protect the natural environment from being destroyed by more and more dams built by local governments and energy companies.[4] Later on the GHV became a registered NGO (type two in Table 12.1: self-organized NGO), but it still enjoys some autonomy.

Another case is "Green River", a registered NGO in Sichuan Province. Evolving from a grassroots NGO, it was founded by Mr. Yang Xin, the first man to raft through the Yangzi River. Green River participated in many environmental protection projects, one recent case being an effort to stop a dam project on the Yangzi River.

The Yangzhanggua Gorge, a beautiful area along the Yangtze River in Sichuan Province, has wonderful scenery and great biodiversity. But the then Ministry of Water Resources had drawn up plans for a dam, known as the Yage Dam, to be built there in 2006. When this came to Yang Xin's attention, he decided to devote himself to stopping the project. He thought the economic value of the dam is too low to justify sacrificing the gorge's scenery and biodiversity.

Yang Xin realized the best strategy for mobilizing public concern about biodiversity in the gorge would be to collect hard evidence. He also realized the snow leopard would be the best "logo" for biodiversity. It would attract intensive media attention, thus possibly stopping the dam project. He assembled two teams, comprising zoologists, botanists, electronic engineers, anthropologists, photographers, doctors, and explorers. They built two camps and placed many cameras and infrared cameras in the area, in order to find leopards and other animals. They found 9–14 leopards in the gorge! There were many other rare animals besides. The staff of Green River wrote up a report on the subject in November 2014. While it was not a very professional report, it attracted intensive coverage in the media, and the dam building project was halted.[5]

However, while some environmental NGOs have organized successful protests, thereby curbing misbehaviour on the part of the government, they suffer from many problems. These include an inability to raise funds due to their legal status and an

[4] Wang Yongchen, "Do Not Make a River as Many Pools", 28 November 2013, http://www.21ccom.net/articles/zgyj/gmht/article_2013112896083.html

[5] "Green River: The Strategy of Survival of a Grassroots Environment Organization." http://xiamag.com/29444.html

inability to attract capable staff because they can only pay low salaries. Most environmental NGOs are therefore unprofessional.[6]

12.5 A Quantitative Study of NGOs' Policy Lobbying in China[7]

In this section, I use quantitative methods to test the external validity of the thesis that NGOs contribute to sustainable development on the basis of the two mechanisms identified earlier. The data used here originates from a survey of the registered NGOs[8] of two provinces and a national-level city in China, namely, Zhejiang (a province in eastern China), Heilongjiang (a province in northeastern China), and Beijing (the capital region and a national-level city).[9] The survey was conducted by the Center for Civil Society Studies (CCSS) of Peking University. The CCSS research team chose the aforementioned three areas for the following reasons: Beijing, being China's capital, should have more control over NGOs and Zhejiang Province, being the richest area of China with the most active private economy, has a vibrant civil society. Surveys in Beijing and Zhejiang were conducted in 2010 and completed in 2011. In 2011 Heilongjiang Province was added to the data, because its economy – which formerly was planned – is considerably less developed than that of Beijing or Zhejiang. In terms of per capita GDP, Beijing ranked second in 2010, Zhejiang fourth, and Heilongjiang eighteenth in China. The poorest areas of China were thus excluded from the population sample, mainly due to funding limitations (see Table 12.2).

12.5.1 Dependent Variables

The effect of advocacy is measured in the survey by two questions on whether a given NGO has successfully initiated, changed, or stopped a policy pursued by the local government. Answers are classified in terms of a dummy variable, where "1" denotes effective advocacy and "0" stands for ineffective advocacy. Missing values are coded as "0" (no effect), on the assumption that NGOs which had been successful in their advocacy would have responded (Table 12.3).

---

[6] Report on China's nonofficial environment NGOs 2008 and 2010, cited in Ren, Chong and Yaya Jiang, "The Status of Grassroots Environment NGOs in China", *Jiefang Daily*, 13 December 2014

[7] This is a simplified model; for detailed analysis and the rationale for using this quantitative study, see Zhang (2017).

[8] Many social organizations in China are not legally registered, due to China's restrictive registration laws on NGOs. See White (1993) for a discussion of four types of social organizations.

[9] National-level cities in China have the same administrative level as provinces.

Table 12.2 Distribution of cases and response rates by regions and administrative levels

Region	Administrative level	Sample size	Number of NGOs that responded	Response rate (%)
Beijing	Total	500	302	60.40
	City level	155	96	61.94
	District/county	345	206	59.71
Zhejiang	Total	764	417	54.58
	Province	105	46	43.81
	Prefecture	369	196	53.12
	County	290	175	60.34
Heilongjiang	Total	715	331	46.29
	Province	129	54	41.86
	Prefecture	243	122	50.21
	County	343	155	45.19
Total		1979	1050	53.06

Note: District/county level in Beijing equates to prefecture level in the provinces

Table 12.3 Descriptive statistics for dependent variable

		Frequency	%
Policy advocacy effectiveness	No	955	90.8
	Yes	97	9.2
	Total	1052	100.0

This method of measurement involves a risk of oversimplification, inasmuch as I have given the same value ("1") to each of the following three things: persuading the local government to adopt a new policy, persuading it to cancel a policy, and persuading it to change a policy. This elides the differences between these forms of influence, two of which involve constraint (persuading the local government to cancel or to change a policy) and one of which involves collaboration (persuading it to adopt a policy).

12.5.2 Independent Variables

On the basis of the society-centred approach,[10] I propose certain independent variables. The focus of this approach is on the societal attributes of social actors (including interest groups), e.g. their socio-economic status and resources (Dahl 1982; Verba and Nie 1997). Tsujinaka (1989) lists the following elements: organizational

[10] Scholars present a variety of arguments on whether social resources are equally distributed among social groups. While pluralist theorists like Dahl (1982) hold that resources are distributed in a dispersed way, Lowi (1969) finds that the resources are distributed in an accumulated manner in society – meaning that the advantaged few hold almost everything and the disadvantaged hold little.

size (number of members), economic and financial resources, characteristics of staff members (number, education, and skill), social influence, and leadership. This perspective assumes that power involves a process of exchange between different parties, in which people exchange their own resources for things they want which are owned by others. In the particular case at hand, the NGO exchanges its resources for support from the government, such as financial aid, favourable regulations and policies, and so on. The government, in turn, may seek the NGO's support. Such support may take the form of votes, or assistance in policy implementation, or information which is useful in the policy-making process. Some government actors also seek rents. Jessica Teets (2013) finds that autonomous NGOs can use their resources to provide consulting and expertise for the government, which lacks the necessary resources to implement many policies. This can lead to various outcomes in terms of the state-society relationship: for example, if the state is much stronger than society, there will be a predatory state. If there is a balance between the state and society, there will be a tendency towards mutual empowerment and reciprocal supervision between the state and society. Finally, if society is stronger than the state (or if some parts of it are), we get a captured state that serves particularistic interests and leads to economic stagnation (Olson 1982).

Some scholars, however, prefer a state-centric perspective. Greenstone (1975) argues that the activities, demands, and even preferences of a given interest group will be greatly influenced by the character of the polity. Similarly, Margarita Estevez-Abe (2003) sees the needs and resources of state officials as the key to explaining group behaviour.

Based on the above theoretical discussion, I use the following independent variables:

1. Number of members. As member-based organizations, NGOs are supposed to represent the interests of their members. On the one hand, more members means more social influence – and thus a greater ability to influence government policy. On the other hand, as Mancur Olson (1965) kindly warns, a larger group tends to be less likely to act collectively than a smaller group.
2. Number of staff members (logged). The more staff members an NGO has, the greater will be its capacity and the greater its value for government policy-making. It is dropped from the model because it is not statistically significant, while its large missing values significantly reduce the number of valid cases.
3. The NGO's financial resources, as measured by its annual budget.
4. Government funding ratio: the percentage of an NGO's funding that comes from government.
5. Government consultation with the NGO in question. This measures the government's intention to consult the NGO on policy issues, as indicated by whether the government elicited policy suggestions from it. This is a dummy variable: "1" means yes, there was government consultation.
6. The NGO's intention, at the time it was established, to influence policy-making and to help the local government implement policy. Some NGOs – even very large ones – have no interest in policy advocacy, and they do not mobilize their

resources for political purposes. Some NGOs are only interested in helping the local government implement policy. I code these cases with a dummy variable.
7. The manner in which the NGO was organized. Was it organized by the government (n = 1414), or was it organized on a voluntary basis (n = 912)? The answer is assigned a dummy variable, where "1" denotes voluntary organization. I assume that voluntarily organized NGOs will tend to be more active.
8. Field of operation. This measures the policy areas in which NGOs try to lobby. I recode these areas into six categories: 1) science, education, sports, and culture (N = 384); 2) social services (N = 97); 3) business (N = 196); 4) rural issues (N = 141); 5) legal, environmental, religious, and foreign-related (N = 55); and 6) others. Categories one through five are coded as dummies. Different policy areas have different implications for sustainable development: the fifth type (legal, environmental, religious, and foreign-related) is somewhat sensitive and is the most relevant; the second and fourth are also very relevant. Business associations, finally, may have a varying impact, depending on the nature of the industry in question.

12.5.3 Findings

I carried out a logistical regression using all of the independent variables as well as other variables. However, the regression model shown in Table 12.4 is a simplified model, in which only the significant variables are used.

Table 12.4 Logistical regression model of NGOs' consulting and policy-lobbying effectiveness

Independent\dependent variables		Policy-lobbying effectiveness
Models		Simplified model
	Member number log	−0.382** (0.143)
	Staff member log	
	Annual budget log	0.527** (0.203)
	Government-organized	
	Government consulting	2.402*** (0.432)
	Government funding ratio	−0.911* (0.534)
	Policy influence intention	0.721′ (0.439)
Policy fields	Education, culture, sports, and health	1.097** (0.529)
	Environmental, legal, religious, and foreign-related	1.613* (0.825)
	Social services	1.219* (0.693)
	Business	0.972* (0.576)
	Rural	
	Constant	−4.082*** (0.561)
	−2 log likelihood	209.209[a]
	Nagelkerke R squared	0.347
	N	436

Having more members, we find, leads to less influence. This suggests that, in the absence of elections, the mobilizing power of NGOs is almost useless. The logic of the theory of collective action works well here: having more members leads to less collective action.

NGOs with greater financial resources tend to be more influential. However, having a higher government funding ratio tends to reduce NGOs' policy influence. This suggests that, if an NGO raises more of its funds independently of the government, it will have greater policy influence. Staff size is statistically insignificant. The policy influence intention of an NGO is very important, because NGOs that seek to influence policy tend to mobilize resources for policy lobbying.

Providing policy consulting to the government is a very important factor. It reflects the fact that, under an authoritarian regime, the government is open to external influence only by certain NGOs and only on certain issues. This would seem to mean that the effectiveness of NGOs' efforts to promote sustainable development is still substantially conditional on the government's intentions.

Whether an NGO was organized by the government or on a voluntary basis has no statistically significant impact on the effectiveness of its policy advocacy.

Different policy areas have different implications for the success of efforts to promote sustainable development. NGOs that focus on promoting sustainable development tend to have more influence on policy than rural NGOs and other NGOs. Business associations have a varying impact, depending on the nature of the industry in question.

12.6 Conclusion

In this chapter, I have taken a mixed-method approach to the study of NGOs' policy lobbying in China. I find that NGOs are lobbying the government more frequently today, as a result of the associational revolution taking place in China. This increasing resort to policy advocacy may in turn contribute to sustainable development, because greater involvement by citizens and organizations in policy-making may help to make policy-making more oriented to sustainable development. This suggests that the structure of governance in China is changing from a state-led model into a polycentric governance structure:

"Polycentric" connotes many centers of decision making that are formally independent of each other. Whether they actually function independently, or instead constitute an interdependent system of relations, is an empirical question in particular cases. To the extent that they take each other into account in competitive relationships, enter into various contractual and cooperative undertakings or have recourse to central mechanisms to resolve conflicts, the various political jurisdictions in a metropolitan area may function in a coherent manner with consistent and predictable patterns of interacting behavior. To the extent that this is so, they may be said to function as a "system" (Ostrom et al. 1961, 831–32).

However, the transition to polycentric governance is still ongoing. There are also some negative signs: First, the government's intention and selective openness are still very important in determining the policy-lobbying effectiveness of NGOs. Second, resource-rich NGOs tend to be more influential than resource-poor ones, and most environmental NGOs are resource-poor. This recalls Lowi's critique of elite-dominant pluralism in the United States – that interest groups consolidate the existing distribution of power rather than challenging it (Lowi 1969). To promote sustainable development, therefore, the government must empower relevant NGOs and become more open to social demands.

I have also identified two mechanisms through which NGOs contribute to sustainable development: a constraining mechanism and a collaborating one. In my qualitative study, I have reviewed these mechanisms in detail, and in my quantitative study, I have confirmed their external validity. We can achieve more theoretical leverage by putting my theoretical analysis under a comparative framework. My two mechanisms differ from Najam and Young's typologies of government-NGO relationships. Najam (2000) provides a typology of government-NGO relationships: confrontation, complementarity, cooperation, and co-optation.[11] Young (2000) provides a different typology: supplementary, complementary, and adversary.[12] While the GONGO case is close to a complementary model defined by Young, the grassroots NGO case is a combination of supplementary (with environment bureaus) and confrontation (with the energy company and local government).

References

Bureau of Environment Protection. (2007). A report of China's nonofficial environment NGOs. *Xuehui, 3*, 24.

Brook, T., & Frolic, M. (eds). (1997). *Civil society in China* M.E Sharpe. New York: Armonk.

Cooper, C. (2006). "This is Our Way In": The civil society of environmental NGOs in South-West China. *Government and Opposition, 41*(1), 109–136.

Dahl, R. (1971). *Polyarchy: Participation and opposition*. New Haven: Yale University Press.

Dahl, R. (1982). *Dilemmas of pluralist democracy: Autonomy vs control*. New Haven: Yale University Press.

Ding, X. (1994). Institutional amphibiousness and the transition from communism. *British Journal of Political Science, 24*(3), 293–318.

Estevez-Abe, M. (2003). State-society partnerships in the Japanese Welfare State. In Schwartz & Pharr (Eds.), *The state of civil society in Japan*. Cambridge: Cambridge University Press.

[11] The two criteria Najam uses are the match of the government and NGO's respective (1) goals and preferences and (2) the means.

[12] Supplementary model refers to "nonprofits are seen as fulfilling the demand for public goods left unsatisfied by government". Complementary model refers to "nonprofits are seen as partners to government, helping to carry out the delivery of public goods largely financed by government". And adversarial model refers to "nonprofits prod government to make changes in public policy and to maintain accountability to the public. Reciprocally, government attempts to influence the behaviour of nonprofit organizations by regulating its services and responding to its advocacy initiatives as well (Young 2000: 150–151)".

Evans, P. (1995). *Embedded autonomy: States and industrial transformation*. Princeton: Princeton University Press.

Evans, P. (1997). *State-society synergy: government and social capital in development*. California: International and Area Studies, University of California Press.

Greenstone, F. (1975). Interest Groups. In F. Greenstein & N. Polsby (Eds.), *Handbook of political science* (Vol. 4). Reading: Addison-Wesley Series in Political Science. Addison-Wesley Pub. Co.

Kennedy, S. (2005). *The business of lobbying in China*. Cambridge, MA: Harvard University Press.

Lipset, S. (1959). Some social requisites of democracy. *American Political Science Review, 53*(1), 245–259.

Lowi, T. (1969). *The end of liberalism; ideology, policy, and the crisis of public authority*. New York: Norton.

Mertha, A. (2009). *Fragmented authoritarianism 2.0: Political pluralization of the Chinese Policy Process. The China Quarterly, 200*, 1–18.

Migdal, J. (1994). Introduction. In J. Migdal, A. Kohli, & V. Shue (Eds.), *State power and social forces: Domination and transformation in the Third World*. New York: Cambridge University Press.

Migdal, J. S. (2001). *State in society: Studying how states and societies transform and constitute one another*. Cambridge/New York: Cambridge University Press.

Miller, R. (Ed.). (1992). *The development of civil society in communist systems*. North Sydney: Allen & Unwin.

Najam, A. (2000). The four-c's of third sector-government relations: Cooperation, confrontation, complementarity, and co-optation. *Nonprofit Management and Leadership, 10*(4), 375–396.

Olson, M. (1965). *The logic of collective action; public goods and the theory of groups*. Cambridge, MA: Harvard University Press.

Olson, M. (1982). *The rise and decline of nations: Economic growth, stagflation, and social rigidities*. New Haven: Yale University Press.

Ostrom, V., Tiebout, C., & Warren, R. (1961). The organization of government in metropolitan areas: A theoretical inquiry. *American Political Science Review, 55*(4), 831–842.

Putnam, R. (with R. Leonardi and R. Nanetti). (1993). *Making democracy work: Civic traditions in modern Italy*. Princeton: Princeton University Press.

Schwartz, J., & Shieh, S. (Eds.). (2009). *State and society responses to social welfare needs in China: Serving the people*. London: Routledge.

Spires, A. (2011). Contingent symbiosis and civil society in an authoritarian state: Understanding the survival of China's grassroots NGOs. *American Journal of Sociology, 117*(1), 1–45.

Steindhart, C., & Wu, F. (2016). In the name of the public: Environmental protest and the changing landscape of popular contention in China. *The China Journal, 75*, 61–82.

Teets, J. (2013). Let many civil societies bloom: The rise of consultative authoritarianism in China. *The China Quarterly, 213*, 19 38.

Tsujinaka, Y. (1989). *Liyijituan* [Interest Group]. Beijing: Jingji ribao chubanshe.

Verba, S., & Nie, N. (1997). *Political participation in America: Political democracy and social equality*. Chicago: University of Chicago Press.

Wang, X. (1999). Mutual empowerment of state and society: Its nature, conditions, mechanism, and limits. *Comparative Politics, 31*(2).

White, G. (1993). *Riding the tiger: The politics of economic reform in post-Mao China*. Stanford: Stanford University Press.

Wu, F (2009a) *Environmental activism and civil society development in China*. Harvard-Yenching Institute Working Paper Series.

Wu, F. (2009b). Environmental politics in China: An issue area in review. *Journal of Chinese Political Science, 14*(4), 383–406.

Young, D. (2000). Alternative models of government-nonprofit sector relations: Theoretical and international perspectives. *Nonprofit and Voluntary Sector Quarterly, 29*(1), 149–172.

Zhan, X., & Shui, Y. T. (2013). Political opportunities, resource constraints and policy advocacy of environmental NGOs in China. *Public Administration, 91*(2), 381–399.
Zhang, C. (2015). Non-governmental organization's policy advocacy in China: Resources, government intention, and network. *China: The International Journal, 13*(1), 181–199.
Zhang, C. (2017). Nongovernmental organizations' policy advocacy and government responsiveness in China. *Nonprofit and Voluntary Sector Quarterly*. https://doi.org/10.1177/0899764017705735 Online first.

Epilogue: New Perspectives on China's Emerging Green Urban Governance

China's rapid industrialization and urbanization have caused unprecedented environmental devastation across the Chinese territory. This has had wide-reaching implications for China's urban and rural citizens alike, as well as for the rest of the planet. Since the early 1980s, China's socialist market economy has fast-tracked this development. Until recently, the leadership of the Chinese party-state privileged this development model politically by rewarding party-state leaders at all levels of administration on the basis of their ability to generate economic growth in their locality. Now, China's central leadership has decided that it must tackle the environmental crisis. The leadership has acted as "radicals at the centre" (Delman 2011) and become proponents of a green transformation of China's development model at all levels of operation. Even more, there is a growing recognition that the public needs to be better informed about the nature and severity of environmental risks and how to deal with them, and that the public must be productively engaged as the critical eye of local governments and as a partner in co-governance. These points were re-emphasized by Xi Jinping in his authoritative report to the 19th Communist Party of China Congress in October 2017 (Xi 2017).

This volume shows that numerous policies, experiments, and reforms have been put in place in China to stimulate the greening of China's cities, and that they entail a rich portfolio of new urban governance instruments, tools, and practices. It also shows that these changes are not coming about by themselves solely because of an enlightened leadership in Beijing. At the local level, there has been a rapid diversification of urban stakeholder interests and some of these have not shied away from challenging local party-state practices associated with authoritarian politics, state planning, and the socialist market development model. They are influenced—in many different ways—by China's phenomenal urbanization process, and stimulated by multiple types of policy mobilization and policy transfer that have lead to rapid diffusion of knowledge and learning that have again resulted in new ways of doing things (cf. Peck and Theodore 2010).

In Chap. 1, Lund asked what type of governance serves to best promote sustainable urban development. She sought to answer that by pinpointing three governance

J. Delman et al. (eds.), *Greening China's Urban Governance*, ARI - Springer Asia Series 7, https://doi.org/10.1007/978-981-13-0740-9

paradigms as main elements in an urban transition to sustainability: traditional public administration, new public management, and new public governance/network governance. She concluded that the prospects for sustainable urban governance depend on the ability of key stakeholders in cities to put together the appropriate hybrid governance paradigms with different types of instruments and tools that entail different normative foundations. These theoretical insights and the associated debates about different approaches to green urban governance have inspired the contributors in this volume to open up new empirical fields of inquiry in urban China. Many of the authors have turned toward particularism, deconstruction, and case-study singularity (cf. Peck 2015). Yet, from a theoretical perspective, the case studies contribute to explain how a multiplicity of national and urban stakeholders in China perceive, make sense of, and address their deteriorating urban environments; how they respond to the specific challenges of China's modernization, urbanization, and urban environmental disruption; and how they enact new forms of green governance to make urban development, structures, space, and life "greener" and more sustainable.

We find that the Chinese party-state continues to be the dominant player in greening China's urban governance (He et al. Chap. 3; Luova Chap. 4; Liu et al., Chap. 5; Schlæger and Zhou, Chap. 7; Delman, Chap. 8; Wang et al., Chap. 11). However, while the key green policies adopted at the national level are clearly important for China's urban leaderships, due to the hierarchical nature of the party-state system, the individual studies also show that China's emerging urban green governance brings together—mostly on a pragmatic basis, but also as a result of both strategic policy design, civil participation, and protest—diverse programmatic building blocks and instruments from China and elsewhere in search of new urban solutions. The governance approaches examined use instruments and tools from the toolboxes of traditional public administration, new public management, and new public governance/network governance respectively, on their own or in different mixes as discussed by Lund in Chap. 1 (cf. Liu et al., Chap. 5; Delman, Chap. 8). The specific combination applied in any particular case varies considerably across sectors (Liu et al., Chap. 5) and projects (J. Liu, Chap. 9; Wang et al., Chap. 11).

The approaches taken at the city level are often quite distinct, because they reflect local contexts and priorities. Delman (Chap. 8), Burell and Almén (Chap. 10), and Wang et al. (Chap. 11) deal with cases in Hangzhou, the capital of Zhejiang Province. These cases are special in some ways, but they also reflect comparable tendencies observable in other cities. The three chapters were written independently of each other, and their research questions and theoretical foci are distinct. Together, however, they demonstrate how different state and non-state actors contribute to the transformation of urban environmental governance in Hangzhou. In Chap. 11, Wang et al. used the 2014 protest against the construction of a waste-incineration plant in Hangzhou to study a new environmental consciousness which has emerged in the absence of green justice. The cause of the conflict, they point out, lay in a closed decision-making process and an omission to provide information to the public. Applying a normative perspective, they argued that instituting a more open and accountable governance process may make it possible to avoid such conflicts in the future.

The case examined by Burell and Almén in Chap. 10 is the same as that examined by Wang et al. (Chap. 11). They focus on the outcomes of local environmental governance rather than on the causes of change. In the absence of open and effective channels through which citizens can seek redress, they found that protest became a means by which the concerned citizens could exert a tangible influence on environmental governance. Further, J. Liu (Chap. 9) shows, with a number of cases from around China, how social media can be used by ordinary citizens to engage in policy learning, as well as to achieve palpable policy results in different cities in the face of similar projects.

The effect of citizen action, especially of protest, is to force urban governments to rethink their approach to environmental governance. The city government is not a passive actor in this respect, however. In Chap. 8, Delman examined the party-state's performance-review system in Hangzhou in detail. With its strong focus on social review and social participation, this system can be seen as the result of an effort on the part of the city leadership to address some of the challenges posed by an increasingly vigilant, active, and even confrontational citizenry. It also addresses some of the problems raised by Wang et al. in their exploration of green justice in Chap. 11. Even more, Delman (Chap. 8) pointed out that the social dimensions of the assessment system could lead to more inclusive governance through participation, greater transparency, and more public accountability.

The performance review and personnel management system of the party-state is in fact becoming a major incentive structure for local leaders. Delman (Chap. 8) notes that this combination of a traditional public administration approach with *new public management* instruments puts an increasingly heavy stress on the record of local leaders in handling environmental issues, exactly through the use of social participation and public accountability mechanisms. While Burell and Almén (Chap. 10) could find no direct connection between protests and the performance-grading system of Hangzhou's Environmental Protection Bureau, they still contend, like Delman, that the mere existence of this system puts pressure on local officials to avoid provoking dissatisfaction among Hangzhou residents. In Chap. 7, Schlæger and Zhou discuss how such practices can be identified in different forms across China. They find that authorities in China seek to open up platforms for public deliberation and consultation, with an eye to mitigate environmental disruption and encourage environmental vigilance.

Due to the well-recognized functional weakness of formal channels of social participation under the Chinese party-state, engaged citizens thus use protests, digital media, and NGOs in order to pressure the government to adjust its policies and governance approaches. Protests can be a useful tool for influencing local authorities, not least because the authorities are willing at times to offer concessions to citizen demands since they are concerned about social stability. J. Liu's study (Chap. 9) shows how the use of digital media strengthened the hand of environmental protesters, by enhancing their capacity for quick diffusion and learning across widely spread urban areas. Local governments are thus faced with increasingly well-informed and strategic citizens who contribute to the development of green urban governance. Zhang (Chap. 12) even found that NGOs, environmental ones included, can influence government policy-making through lobbying, irrespective of their degree of autonomy from the state.

Altogether, the chapters in this volume account convincingly for attempts by the authoritarian party-state to engage the citizenship and vice versa. They tell the story of an engaged society that presents new challenges to local governments, and which in different, but not always obvious ways is becoming an integral part of green urban governance in China. Evidently, China's environmental crisis has challenged urban governments to change their approach to environmental management in their cities. Zhang (Chap. 12) even hints that the structure of urban governance in China could profitably change from a state-led model into a polycentric governance structure with many centers of decision-making, formally independent of each other. Whether this is actually happening already and whether they could function independently or alternatively constitute an interdependent co-governance system is an empirical question in the particular case. This is not so because of decisions made at the center, but primarily because the approaches to development of green urban governance regimes in China's cities will continue to vary considerably within the national policy framework based on the local context.

There are also differences across sectors. In their comparative study of climate resilience and water management in Beijing and Copenhagen, Li et al. (Chap. 5) explained that, in a fairly conservative and traditional public-management sector such as water, the government in Beijing mainly applies the mechanisms of traditional public administration. Unlike in Copenhagen, the public in Beijing has not yet been much involved. The choice of different governance paradigms in the two cities reflects, in the view of the authors, the larger difference between the two national contexts in regard to political systems, planning traditions, and land-ownership patterns. The authors also found, however, that there is a good deal of room in both capital city cases to introduce more bottom-up practices for the sustainable improvement of water supply and wastewater management, and for Beijing to engage with policy learning from Copenhagen with regard to citizen participation.

The unique insights afforded here into very diverse processes and practices of emerging green urban governance help comprehend how China's urban governance regimes cope with new environmental and social challenges, and with popular demands for environmental well-being, justice, and fairness. The traditional top-down administrative approach has come under severe pressure. The focus on rational resource allocation, on command-and-control approaches, on guided policy experimentation, and on the execution of increasingly complex planning methods (see He et al., Chap. 3) has proven unpractical for satisfying the demands of increasingly critical citizens and for the emerging middle classes (cf. Ren and Liu, Chap. 6) who strive for better living conditions and less damage to their urban living environment. Many of the studies in this volume point up the need for stronger public environmental monitoring to make urban green governance more evidence based and more timely (cf. Schlæger and Zhou, Chap. 7). While it is apparent that urban governments are held gradually more accountable for their own environmental performance as well as that of projects implemented in collaboration with businesses and civil society, there is still a need for more experiments to find new solutions and new models of governance that minimize environmental risks and protect citizens' environmental rights and benefits. The transformation from a state-driven planned economy to a more liberal socialist market economy and the need for continuous adjustments in the distribution of costs and benefits among different social groups are the drivers behind such new models, and they clearly reflect the advantages of different approaches suited to local circumstances.

We would argue that several forces have contributed to shaping China's emerging green urban governance regimes. The first force is the combined effects of disastrous environmental degradation and rising popular demands for better life quality. During China's rapid urbanization process, the country has faced the challenges of increasing pollution, deteriorating quality of air and water, environment-induced health challenges, higher mortality rates, and so on. The smog that is choking China's urban areas is an unprecedented environmental catastrophe in the Chinese context. Therefore, when the Chinese leadership developed programs and set the targets for national strategic programs in recent years such as: "Comprehensive Harmonious Society" (*quanmian xiaokang shehui* 全面小康社会), "New Urbanization" (*xinxiang chengzhenghua* 新型城镇化), and "Construction of Ecological Civilization" (*shengtai wenming jianshe* 生态文明建设), the consideration was not only to restore China's environmental order but evidently also to mitigate popular dissatisfaction with the state of the environment.

Secondly, there is a strong path dependency in relation to the historical institutional legacy when it comes to governance approaches in urban China (see Introduction to this volume). The hierarchical and centralized imperial public culture of the past, the hierarchical top-down socialist institutions of the party-state since 1949, and the power of planning over economy and society still exist and exert their strong influence on urban governance. The current approach is a government-led pattern of governance which could be seen—from the perspective of the party-state—as a strategy of efficiency to achieve the realization of development targets through top-down command-and-control mechanisms. But it could also be seen as an effective approach to implement green policies through strong and relatively unified political action. The new approaches to governance discussed in the chapters of this volume show that that this model has been modified considerably over time to include governmental mobilization of diverse policy instruments, i.e. new types of planning, programs with city labeling, performance assessments, social reviews, government-guided social participation, and so on. Meanwhile, the government-led governance model is increasingly suffering from obvious shortcomings in relation to inefficient distribution of resources and justice, and sometimes it leads to decision-making that contravenes the specific needs of the environment as well as of different urban citizen groups.

The global dissemination of knowledge and the embedding of China into global green governance frameworks and regimes combine as the third force that influences the development of China's green urban governance. Many of the new ideas and approaches pursued in the local contexts have been inspired by international debates, ideas, concepts, and models, collaboration with external partners, and so on. China has also become actively involved in global green governance. We would even argue that we are already witnessing a transformation that could make China and its cities part of a global vanguard or alliance that promotes green urban governance. During this process, China not only benefits from, but it also contributes to intensified knowledge exchange and global collaboration to strengthen the country's technical and institutional capacity for green governance.

Finally, new front-end technological developments play a significant role for the development of environmental governance. Comprehensive development of ICT, social media platforms, mobile devices, data storage and processing techniques, as well as the potential offered by big data will undoubtedly lead to new leaps in environmental governance. The study by Schlæger and Zhou in Chap. 7 demonstrates that governments, other stakeholders, and citizens alike could use digital technology even more effectively based on their own learning to improve environmental monitoring and political and administrative interventions. Undoubtedly, China and the world are entering a new epoch characterized by a deep information society. Governance mediated by data-based and data-driven technology is already in the offing and will play an increasingly important role in the future. What the studies in this volume tell us is that there is a great demand for social participation and justice in relation to the provision of environmental services and the exploitation of eco-resources in China. If the new technologies are not used to satisfy and leverage these demands as they are in some of the case studies in this volume (Schlæger and Zhou, Chap. 7; Delman, Chap. 8; J. Liu, Chap. 9), then civil contention or protests may be difficult to contain.

China's urban agglomerations offer many opportunities for feasible improvements in environmental infrastructure and services. Inasmuch as the urban setting has become the central place for innovation and knowledge accumulation, it provides an ideal platform for exploring new practices of urban green governance through policy mobilization and policy transfer (cf. Peck and Theodore 2010) as well as for policy experimentation at the local level (Heilmann 2008). New technologies and innovative new social systems for sustainability usually arise in urban areas as documented in this volume. They can spread from there outward to dependent areas through interaction and integration. In this respect, environmental education can be helpful for developing new ideas and environmental ethics, and for instilling them in society (cf. Luova, Chap. 4). There may also be changes in lifestyle and consumption patterns as a consequence of better livelihoods and higher environmental awareness (cf. Ren and Liu, Chap. 6), and they may in turn lead to changes in the system of production. As this volume demonstrates, these changes are already on the way in China. However, as argued by Brehm and Svensson (Chap. 2), if new urban green governance practices do not become widespread, pollution may just move to other areas than those that benefit from these practices. This calls for more regional and national coordination of green policies and governance regimes.

Furthermore, to move forward, there is a requirement for clear targets for sustainability and environmental welfare. In this context, operational environmental and sustainability indices could be useful (van Dijk and Zhang 2005). On the basis of such indices, stakeholders in urban green governance regimes could measure how urban governments, their organizations, their partners, and their constituencies are managing the increasing environmental risks—often referred to as "urban pathologies" (*chengshibing* 城市病)—that arise in the course of urbanization: i.e., pollution, traffic jams, carbon emissions, land degradation, excessive amounts of garbage, the destruction of river systems, and so on. It is also noteworthy that an increasing number of social organizations are working on environmental issues, holding promise for greater societal involvement (Zhang, Chap. 12). These organizations may contribute significantly to the green turn in urban governance—not only by helping to achieve better public consensus but by becoming both public partners and public movements as well.

There is, we would contend, a broad recognition in China that the green turn in urban governance is necessary and welcome. We have shown how the battle lines have been drawn and how the battle is shaping up through development of new governance regimes, instruments, and tools needed for the transition. While the Chinese experience with developing urban green governance happens under the continued dominance of an authoritarian party-state, it is noteworthy, as convincingly shown in several of the chapters, that the party-state is often receptive to popular demands and pressure at the local level and that in some case it has actually accepted the need for more transparency and accountability in execution of government, and this is augmented by increasing public participation.

Theoretically, then, this volume contributes to explaining the development of green urban governance in China against the backdrop of a global, a national, and a local environmental crisis and of contingent policy processes at all of these levels that aim to shape new green policies, instruments, experiments, and models in urban areas. The various case studies presented here contribute fresh insights into how new forms of green urban governance develop, how they play out on the ground, and what are their challenges and impacts. Even more, China's emerging green urban governance regimes continue to be based not only on local experience but also on international knowledge and practices. The outcome so far has by no means always been neat, but it appears to be driven by productive dynamics.

Finally, Chinese experiences with green urban governance are of acute interest because we need to understand what is being done in China at different scales within its urban territories to deal with environmental and social challenges that often possess regional or global implications. Even more, China's urban experience and urban governance are increasingly writing themselves unto a new agenda of research and practice that will provoke the rest of the world to rethink the governance of urban environmental processes at multiple scales and in multiple nodes within sprawling, densely populated, and dramatically growing urban agglomerates.

References

Delman, J. (2011). "Radicalism at the Centre": Regime Legitimation Through Climate Politics and Climate Governance. *Journal of Chinese Political Science, 16*(2), 183–205.

Heilmann, S. (2008). From local experiments to national policy: The origins of China's distinctive policy process. *The China Journal, 59*(January), 1–30.

Peck, J. (2015). Cities beyond compare? *Regional Studies, 49*(1), 160–182.

Peck, J., & Theodore, N. (2010). Mobilizing policy: Models, methods, and mutations. *Geoforum, 41*, 169–174.

van Dijk, M. P., & Mingshun, Z. (2005). Sustainability indices as a tool for urban managers, evidence from four medium-sized Chinese cities. *Environmental Impact Assessment Review, 25*, 667–688.

Xi, J. (2017). Xi Jinping Report to 19th Congress (full text) (习近平十九大报告(全文)). *Xinhua News* (新华网), 18 October, http://finance.sina.com.cn/money/bank/bank_hydt/2017-10-18/doc-ifymviyp2268296.shtml. Accessed 18 Oct 2017.

Index

J. Delman et al. (eds.), *Greening China's Urban Governance*, ARI - Springer Asia Series 7, https://doi.org/10.1007/978-981-13-0740-9

MIX
Papier aus verantwortungsvollen Quellen
Paper from responsible sources
FSC® C105338

If you have any concerns about our products,
you can contact us on
ProductSafety@springernature.com

In case Publisher is established outside the EU,
the EU authorized representative is:
Springer Nature Customer Service Center GmbH
Europaplatz 3, 69115 Heidelberg, Germany

Printed by Libri Plureos GmbH
in Hamburg, Germany